AF389494

Advances in Industrial Robotics and Intelligent Systems

Advances in Industrial Robotics and Intelligent Systems

Editors

António Paulo Moreira
Félix Vidal
Pedro Neto

MDPI • Basel • Beijing • Wuhan • Barcelona • Belgrade • Manchester • Tokyo • Cluj • Tianjin

Editors

António Paulo Moreira
University of Porto
INESCTEC
Porto
Portugal

Félix Vidal
Asociación de Investigación
Metalúrgica del Noroeste
Porriño
Spain

Pedro Neto
University of Coimbra
Coimbra
Portugal

Editorial Office
MDPI
St. Alban-Anlage 66
4052 Basel, Switzerland

This is a reprint of articles from the Special Issue published online in the open access journal *Robotics* (ISSN 2218-6581) (available at: https://www.mdpi.com/journal/robotics/special_issues/Industrial_Robot).

For citation purposes, cite each article independently as indicated on the article page online and as indicated below:

LastName, A.A.; LastName, B.B.; LastName, C.C. Article Title. *Journal Name* **Year**, *Volume Number*, Page Range.

ISBN 978-3-0365-7540-7 (Hbk)
ISBN 978-3-0365-7541-4 (PDF)

Cover image courtesy of Pedro Neto

Contents

About the Editors

António Paulo Moreira

António Paulo Moreira graduated with a degree in Electrical Engineering from the University of Porto in 1986, an M.Sc. degree in Electrical Engineering Systems in 1991, and a Ph.D. degree in Electrical Engineering in 1998. António is an Associated Professor with tenure at the Electrical Engineering Department of the University of Porto and is a coordinator of the Centre for Robotic and Intelligent Systems Centre of INESC TEC—Institute for Systems and Computer Engineering, Technology and Science. António has participated in 25 scientific projects, being the coordinator in 7 projects. The knowledge developed in these research projects has generated 40 development contracts and technology transfer to companies, and António has been the coordinator of 18 of these projects. To date, António has contributed to 308 scientific publications, 202 of which are in international conferences, books, and theses, and 78 are in international journals with referees.

Félix Vidal

Félix Vidal has an M.Sc. in Industrial Engineering (automation and electronics) and is an International Welding Engineer (IWE). Currently, he is working in AIMEN as the Head of Smart Systems and Smart Manufacturing, being the project coordinator of several European R&D projects related to digital transformation. He has more than 20 years of experience as a researcher in European and national projects dealing with the digitization of industry in the EU. In recent years, he has authored papers in major conferences and journals related to smart manufacturing. He is a member of EFFRA and euRobotics and a reviewer for different papers in international manufacturing and mechatronics conferences and journals.

Pedro Neto

Pedro Neto, Ph.D., is an Associate Professor at the Mechanical Engineering Department and coordinator of the Collaborative Robotics Laboratory at the University of Coimbra. He is a two-time recipient of the Impact and International Publications Award at the Faculty of Science and Technology, has received more than a dozen awards, and is in the World's Top 2% Scientists according to Elsevier. Pedro Neto served as vice president of the Portuguese Robotics Society, is a member of the IEEE Committee on Factory Automation, is an associate editor in refereed journals, and is a scientific committee member of flagship conferences. His current research interests include human–robot interaction and collaboration, machine learning, soft robots, and advanced robot applications in the manufacturing domain. He has co-authored more than 100 papers and collaborated with more than 100 companies in projects and consulting, some of them having been granted by the European Commission in the prestigious HORIZON framework.

Editorial

Special Issue on Advances in Industrial Robotics and Intelligent Systems

António Paulo Moreira [1], Pedro Neto [2,*] and Félix Vidal [3]

1   INESC TEC-INESC Technology and Science, Faculty of Engineering, University of Porto, 4099-002 Porto, Portugal; amoreira@fe.up.pt

2   CEMMPRE, Department of Mechanical Engineering, University of Coimbra, POLO II, 3030-788 Coimbra, Portugal

3   Asociación de Investigación Metalúrgica del Noroeste, 36410 Porriño, Spain; fvidal@aimen.es

*   Correspondence: pedro.neto@dem.uc.pt; Tel.: +351-239790767

Citation: Moreira, A.P.; Neto, P.; Vidal, F. Special Issue on Advances in Industrial Robotics and Intelligent Systems. *Robotics* 2023, 12, 45. https://doi.org/10.3390/robotics12020045

Received: 15 March 2023
Accepted: 16 March 2023
Published: 20 March 2023

Robotics and intelligent systems are intricately connected, each exploring their respective capabilities and moving towards a common goal. In industry, it is common to see robotic systems aided by machine learning and vice versa in applications including robot navigation, grasping, human–robot interaction/collaboration, safety and team management, among others. Achievements targeting the industrial domain can be directly applied in robotic systems operating in other domains. While significant advances have been made in the last few years, industrial robotics and intelligent systems face several scientific and technological challenges related to their integration with other systems, interaction with humans, safety, flexibility, reconfigurability and autonomy. These challenges are especially relevant for robots operating in unstructured industrial environments and sharing a workspace with human coworkers and other robots.

This Special Issue presents recent research and technological achievements in the field of advanced intelligent robotic systems. The contributions included cover the coordination of multiple robots navigating a factory floor, path planning strategies, human–robot interaction, robot redundancy and kinematics, system integration, grasping and manipulation.

Mobile multi-robot systems able to operate on a factory floor have recently emerged. This scenario brings several challenges, such as the coordination of the robots, path planning and the robots' behavior in reacting to communication faults [1]. An interesting study presents a time-based algorithm able to dynamically control a fleet of Autonomously Guided Vehicles (AGVs) in an automatic warehouse, integrating a routing algorithm based on the A* heuristic search to generate collision-free paths and a scheduling module to improve the routing results [2]. Since robots share the working space with humans, the authors explored and evaluated humans' perception of different autonomous mobile robots' courtesy behaviors at industrial facilities, particularly at crossing areas [3]. Localization and state estimation are key in developing autonomous mobile robots. In [4], a slip-aware localization framework for mobile robots experiencing wheel slip is proposed, which fuses infrastructure-aided visual tracking data and proprioceptive sensory data from a skid-steer mobile robot to enhance accuracy. In another study, a team of mobile manipulators within a compact planar workspace with obstacles is proposed to achieve autonomous object transportation [5].

Recent advances in dynamic programming redundancy resolution, applied to generic kinematic structures, are reported in [6]. In this study, a novel Robot Operating System (ROS) architecture is proposed and demonstrated on a 7-DOF robot. In [7], a method of calculating forward kinematics using a recursive algorithm that builds a 3D computational model from the configuration of a human-inspired mobile manipulator is presented. Path planning is studied in [8], addressing a complex trajectory evaluation of robotic arm trajectories containing only robot states defined in the joint space without any time parametrization (velocities or accelerations). In [9], a bin-picking solution that uses simulation to create bin-picking environments in which a procedural generation method builds entangled tubes is proposed. A new hyperloop transportation system design is proposed in [10]. The study elaborates on the design and integration of propulsion components for a linear motion system, providing high-speed transportation means for passengers and freights by utilizing linear synchronous motors.

Acknowledgments: Thanks to all the authors and peer reviewers for their valuable contributions to the Special Issue 'Advances in Industrial Robotics and Intelligent Systems'.

Conflicts of Interest: The authors declare no conflict of interest.

References

1. Matos, D.; Costa, P.; Lima, J.; Costa, P. Multi AGV Coordination Tolerant to Communication Failures. *Robotics* **2021**, *10*, 55 [CrossRef]
2. Santos, J.; Rebelo, P.M.; Rocha, L.F.; Costa, P.; Veiga, G. A* Based Routing and Scheduling Modules for Multiple AGVs in an Industrial Scenario. *Robotics* **2021**, *10*, 72. [CrossRef]
3. Alves, C.; Cardoso, A.; Colim, A.; Bicho, E.; Braga, A.C.; Cunha, J.; Faria, C.; Rocha, L.A. Human–Robot Interaction in Industrial Settings: Perception of Multiple Participants at a Crossroad Intersection Scenario with Different Courtesy Cues. *Robotics* **2022** *11*, 59. [CrossRef]
4. Flögel, D.; Bhatt, N.P.; Hashemi, E. Infrastructure-Aided Localization and State Estimation for Autonomous Mobile Robots *Robotics* **2022**, *11*, 82. [CrossRef]
5. Vlantis, P.; Bechlioulis, C.P.; Kyriakopoulos, K.J. Mutli-Robot Cooperative Object Transportation with Guaranteed Safety and Convergence in Planar Obstacle Cluttered Workspaces via Configuration Space Decomposition. *Robotics* **2022**, *11*, 148. [CrossRef]
6. Ferrentino, E.; Salvioli, F.; Chiacchio, P. Globally Optimal Redundancy Resolution with Dynamic Programming for Robot Planning: A ROS Implementation. *Robotics* **2021**, *10*, 42. [CrossRef]
7. Gonçalves, F.; Ribeiro, T.; Ribeiro, A.F.; Lopes, G.; Flores, P. A Recursive Algorithm for the Forward Kinematic Analysis of Robotic Systems Using Euler Angles. *Robotics* **2022**, *11*, 15. [CrossRef]
8. Dobiš, M.; Dekan, M.; Beňo, P.; Duchoň, F.; Babinec, A. Evaluation Criteria for Trajectories of Robotic Arms. *Robotics* **2022**, *11*, 29 [CrossRef]
9. Leão, G.; Costa, C.M.; Sousa, A.; Reis, L.P.; Veiga, G. Using Simulation to Evaluate a Tube Perception Algorithm for Bin Picking *Robotics* **2022**, *11*, 46. [CrossRef]
10. Bhuiya, M.; Aziz, M.M.; Mursheda, F.; Lum, R.; Brar, N.; Youssef, M. A New Hyperloop Transportation System: Design and Practical Integration. *Robotics* **2022**, *11*, 23. [CrossRef]

Article

Multi AGV Coordination Tolerant to Communication Failures

Diogo Matos [1,*], Pedro Costa [1,2], José Lima [1,3] and Paulo Costa [1,2]

1   Centre for Robotics in Industry and Intelligent Systems (CRIIS)—INESC TEC, 4200-465 Porto, Portugal; pedrogc@fe.up.pt (P.C.); jllima@ipb.pt (J.L.); paco@fe.up.pt (P.C.)
2   Faculty of Engineering, University of Porto, 4200-465 Porto, Portugal
3   Research Centre in Digitalization and Intelligent Robotics (CeDRI), Instituto Politécnico de Bragança, 5300-252 Bragança, Portugal
*   Correspondence: diogo.m.matos@inesctec.pt

Abstract: Most path planning algorithms used presently in multi-robot systems are based on of-fline planning. The Timed Enhanced A* (TEA*) algorithm gives the possibility of planning in real time, rather than planning in advance, by using a temporal estimation of the robot's positions at any given time. In this article, the implementation of a control system for multi-robot applications that operate in environments where communication faults can occur and where entire sections of the environment may not have any connection to the communication network will be presented. This system uses the TEA* to plan multiple robot paths and a supervision system to control com-munications. The supervision system supervises the communication with the robots and checks whether the robot's movements are synchronized. The implemented system allowed the creation and execution of paths for the robots that were both safe and kept the temporal efficiency of the TEA* algorithm. Using the Simtwo2020 simulation software, capable of simulating movement dynamics and the Lazarus development environment, it was possible to simulate the execution of several different missions by the implemented system and analyze their results.

Keywords: multi-AGV control; path planning; Timed Enhanced A*; tolerance to communication faults

Citation: Matos, D.; Costa, P.; Lima, J.; Costa, P. Multi AGV Coordination Tolerant to Communication Failures. *Robotics* 2021, 10, 55. https://doi.org/10.3390/robotics10020055

Academic Editor: Xinjun Liu

Received: 7 February 2021
Accepted: 23 March 2021
Published: 27 March 2021

Publisher's Note: MDPI stays neutral with regard to jurisdictional claims in published maps and institutional affil-iations.

1. Introduction

The constant technological development felt at present creates a need for a constant adaptation on the part of the industrial sector in order to fulfil the demands of the corporate market. To remain competitive, industries must generate a demand for new innovative solutions in an attempt to create value. These solutions do not always reflect a direct valorization of the final product. In most industries, the production costs have a significant impact in their market competitiveness, which leads to a significant evolution of the automated systems. These systems grant the possibility of reducing the labour cost and simultaneously optimizing production time. As a product of this evolution, Automated Guided Vehicles Systems (AGV) were created and saw their first use in an industrial environment in 1954. Since then, the use of this type of system has seen a steady increase, and is a common sight in industry at present [1].

Currently, the AGV are predominantly used in the moving of products, being mainly used as a mean of transporting material between the production lines and the storage sectors. However, their use is not restricted to the industrial sector; AGV, for example, can also be used in hospitals and distribution centres. The mass use of AGV creates questions about their efficiency and productivity in scenarios where multiple robots are operating in restricted environments and exposed to communications faults.

The coordination of a fleet of autonomous vehicles is a very complex task, most multi-robot systems rely currently on static and pre-configured interactions between the robots [2]. When the unpredictability associated with communication flaws is added, this task becomes even more difficult. Due to this fact, the study of trajectory planning algorithms allied to methods of detection and mitigation of communication faults, has increased in the later years. These algorithms use the robots' localization information, either predicted or measured, to control the traffic of the robot fleet, and attempt to plan safe and optimized routes in an industrial environment.

This work provides an implementation of a traffic control system, in a multi AGV environment, based on the TEA* algorithm [3] and robust communication faults. The main focus consists of obtaining a cooperative movement between all robots that are part of the system, avoiding, simultaneously, any situation that could lead to a mutual block and subsequently lead to the not finishing all the assigned tasks. This system should also be able to operate efficiently when subjected to communication faults of one or more robots. This system should also keep the efficiency and time optimizations offered by the TEA* in [4], expanding the use of this algorithm to environment where communication flaws are common, by using binary semaphores [5] to control the access to areas affected by communication faults.

Until this point, the TEA* algorithm has never been implemented in an industrial environment where both delays in robots movements as well as communication faults are common occurrence. As such, the paper's main contribution is the implementation of a traffic control system based on this algorithm capable of operating under these circumstances. This system will be composed of two parts: the TEA* planning algorithm and the Supervision module.

The rest of this paper is organized as follows: Section 2 presents a brief overview of other works related to the topic. Section 3 presents the overall system architecture as well as the lower level control implemented in each robot. Section 4 details the modifications implemented to the TEA* algorithm. The supervision module will be presented in Section 5. Section 6 will validate the performance of the implemented system. Finally, a few concluding remarks are made in Section 7.

2. State of the Art

The coordination of a fleet of robots falls under the *Multi-Agent Path Finding* (MAPF) category of problems. This problems comprise of finding a set of paths for the agents that are encompassed by the system. These paths have the objective of moving the agents from their current vertices to their targeted vertices while avoiding conflicts and reducing as much as possible the cost for the movement of the agents [6]. MAPF has practical applications in video games, traffic control, and robotics [7]. Solving MAPF problem in an optimal way is NP-Hard (non-deterministic polynomial-time hardness) in terms of complexity [8,9].

Presently, several different methods are commonly used to plan the movement and actions of fleet of mobile robots. This methods can range from the use of potential fields [10] to the use of Reinforcement Learning [11,12].

From all these methods, the ones based on Graph Search algorithms are of special importance not just because they can discover the possible paths but also because they allow the discovery of the most efficient possible paths for the robots [13].

As the name itself suggests, these methods consist of two stages: the initial decomposition of the workspace into a graph, and the application of a graph search algorithm in order to discover efficient paths for the robots. The initial decomposition of the workspace, in most cases, can be easily achieved by either using an approach based on Roadmaps [14,15] or Cell Decomposition [16].

After obtaining the graph, necessary the implementation of a method of searching it is necessary. Of particular importance in this field is the A* algorithm and its derivatives [17]. The standard A* algorithm, introduced by [17], uses both the cost already incurred from the initial node and the cost until the finish node. This characteristic allows this algorithm to obtain optimal and complete solutions.

Later, new variants of the A* algorithm were developed with the intention of being used in Multi-AGV systems. Such as the Dynamic A* algorithm proposed by [18], this algorithm stands out as it is capable of operating in situations where the environment is unknown or just partially known.

Another notable variant is the Lifelong Planning A* algorithm, also called LPA*, proposed by [19]. This algorithm allows the use of previous information in order to reduce the processing time. More recently, in 2011, the 3D A* algorithm was introduced by [20], which adds the temporal dimension, allowing for a coordinate representation of (x,y,t). This detail allows for the representation of stationary movement, where the x and y coordinates maintain the same through iterations and the t coordinate is increasing.

Another variant of the A* algorithm is the M* [21], this variant was created since planning trajectories for large numbers of robots is very computationally expensive. Therefore, the M* algorithm allows for the creation of a relatively cheap path for the robots by sacrificing a small portion of the overall efficiency of the system. Although this method does not generate the most efficient solution possible for the robot routing problem, it generates good enough solutions that avoid conflicts between the robots.

Recently, a new graph search algorithm was introduced for the implementation in Multi-AGV systems. The Time Enhanced A* algorithm, also known as TEA*, introduced by [3] and later modified by [4], is a graph search algorithm created due to the need for an algorithm suitable for Multi-AGV systems, which avoids collisions, deadlocks and guarantees the efficient execution of a set of tasks [3]. The TEA* algorithm was created by introducing the concept of temporal layers to the A* algorithm, which gives the possibility of the algorithm of executing the path planning in an online mode [3] using the estimation of the position of the robots for any given time instance. This makes the TEA* algorithm capable of taking into account possible changes that can occur in the environment, unlike its predecessor. The TEA* algorithm is also capable of dynamically shifting the priorities of each robot in order to resolve possible conflict situations [4], unlike the static priorities used by [22].

All the methods based on graph search algorithms referred to above rely on a centralized architecture [2,23], which expands on the idea of a decentralized multi-robot system by using a service-oriented architecture. However, this idea is still in its infancy.

In the field of tolerance to communication faults, most advances are focused either on regeneration of communication faults [24] or in applying estimation methods to deal with sudden sporadic communication faults [25]. However, there are few studies on multi robot systems that are capable of dealing both with sporadic faults, as well as, dealing with entire zones of the environment where there is no communication with the robots.

3. Overall System Architecture and Lower Level Control

To better comprehend the function and implementation of both the TEA* algorithm and the supervision module, the system's overall architecture is explained, as well as its overall composition.

The implemented system is based on a centralized control architecture [23]. This configuration was chosen, since the TEA* Graph Search Algorithm needs prior knowledge of all the current robots positions in order to correctly generate a set of paths for all the robots controlled by the systems [4].

The implemented system is responsible for controlling the movement of a team of robots. This team of robots would consist of four robots similar to the ones used in the Factory lite 2019 edition [26]. These robots locomotion system is made up of two differential traction wheels and are simulated using the Simtwo2020 simulation software (CRIIS, Porto, Portugal), created by Paulo Costa. This software is capable of simulating differential traction robots [27], as well as already having implemented the simulation model for the robots from the Factory Lite competition [26], these robots were developed and built by Centre for Robotics in Industry and Intelligent Systems (CRIIS).

In Figure 1, a top level representation of the implemented system is shown. As seen in this figure, the implemented control system is comprised by three core modules. The TEA* adapted library and supervisor modules located in an application that represents the central control unit and the trajectory control module located in each robot control unit. Information is transmitted between the four robots and the central controlling application via the UDP/IP communication protocol.

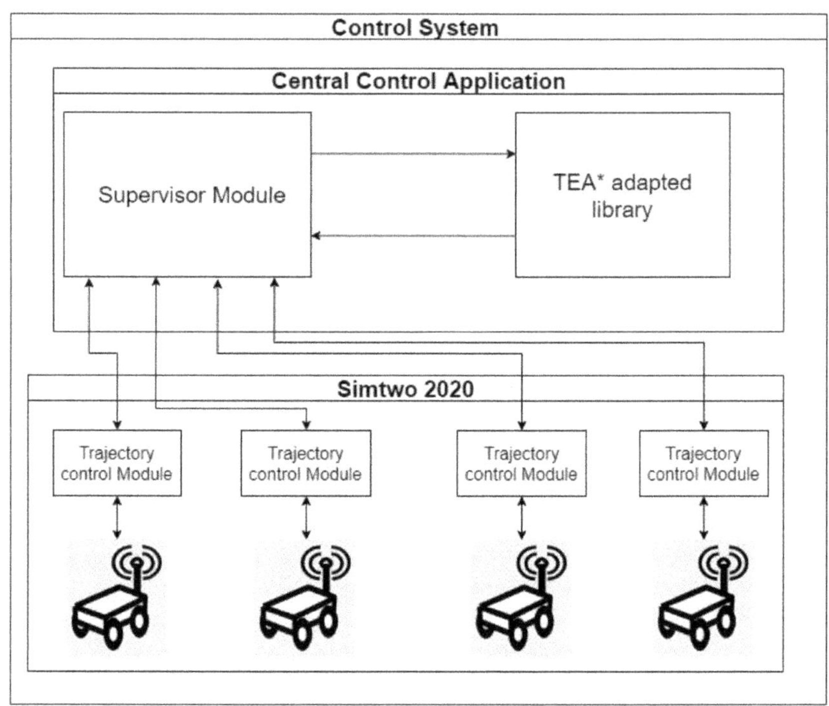

Figure 1. Top level view of the Control system (TEA*—Time Enhanced A*).

The central control unit uses the interaction between the supervision module and the modified TEA* algorithm, to control each robot planned path, this path is then transmitted to the robot. The robot control algorithm is only responsible for controlling the velocities of each of the robots wheels, so that it can accurately execute its assigned path.

The simulation of communication faults in this simulated environment was executed by a special module that was implemented between the simulation platform and the centralised control platform. In the case of static communication faults, this module will analyze the position of the robot and checks if it is within the area defined by the user. If this condition is verified, the module will not transmit any data regarding that robot to the control system. For sporadic faults, this module uses a fixed probability to decide whether to transmit or not the data to the central control unit. It is also capable of simulating sporadic faults within a defined area or in the entire environment. All the data used by this module is isolated from the rest of the system to maintain the validity of the experiment results.

4. Modified TEA* Algorithm

As described before, several modifications to the algorithm presented by [4] were implemented during the design and creation of the central control unit. Three major modifications were implemented into this algorithm, these modifications are the following:

- Implementation of a Task Scheduling function;

- Modification of the method that associates the current robot position with a node from the graph;
- Implementation of an algorithm that creates a binary semaphore when a communication fault is detected.

These modifications had the objective of allowing the application of the algorithm into industrial scenarios, and also allowing the system to safely plan paths for the robots during communication faults.

4.1. Task Scheduling

The main difference between the method used by [4] and the implemented Task Scheduling function, was that the later would move the robot to a charging post when all of the tasks assigned to the robot were concluded. This would allow for the robot to recharge while no new tasks are assigned to it.

To make the robots move to their charging stations when no tasks are assigned to them, both the function that obtains the next task, as well as the function that adds new tasks to the robot need to be modified.

When a robot reaches its target point, this new function checks if there are anymore tasks assigned to the robot and in the case that there are no more missions, it orders the robot to move to the closest recharging station; this attribution is represented via a flowchart in Figure 2.

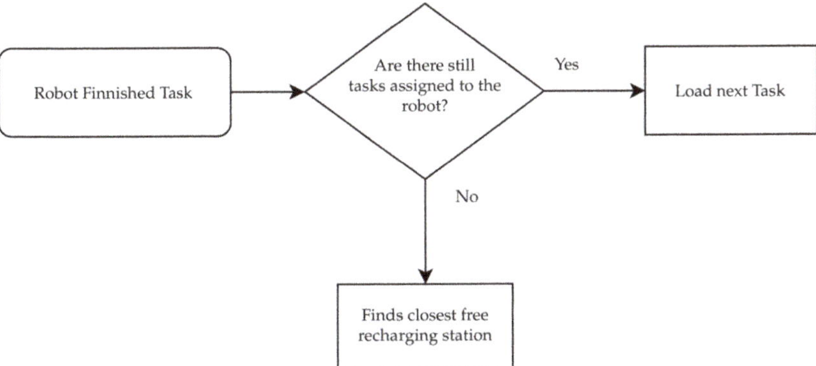

Figure 2. Flowchart representation of the attribution of charging stations to the robots

When a new task is added, the system checks whether the robot is either moving to or stopped at a charging station or if the robot is executing another task. If it ends up being the latter, it will add the new task to the subtask array as the last element and will also increment the number of tasks. Otherwise, it will change the target point to the new task node and add that task to the subtask array. It is also necessary to increment both the current task point as well as the number of tasks.

At the beginning of each planning cycle the nodes corresponding to the positions of all robots that are currently charging, will be considered to be "Obstacle Inaccessible Robot". This new node status represents a node that is currently occupied by a robot which is currently recharging or that has suffered a communication fault, therefore meaning that this robot cannot be moved to give way to another robot. This allows the planning of path around the position of these robots and avoiding the movement of these robots.

For this reason, the recharging positions should be placed in locations where they will not block traffic, such as at the extremities of dead end pathways. After a robot has completed all of its tasks, the robot will move to the nearest unoccupied recharging station and will remain there until another task is assigned to it.

4.2. Planning during Communication Faults

It is also necessary to modify the TEA* library so that it is able to plan safe and efficient paths even when one or more robots are experiencing communication faults. The library created by [4] does not take into account this possibility, therefore, when one or more robots are subjected to communication faults, it is not able to guarantee a safe and efficient planning, having a high probability of occurring collisions and deadlocks.

The method implemented in this project, in order to create a system robust to communication faults and to areas where there is no communication, consists in the interaction between both modules of the central control unit.

Each communication fault will have a set of nodes associated with it, these nodes are generated by the supervision module and their generation is explained in Section 5. Then the modified TEA* library places the nodes associated with the current active faults as immovable obstacles, for all robots and during all steps of the planning, until the affected robot exits the communication fault area and communication can be re-establish. This prevents any other robot from entering the communication fault zone. Therefore, creating a binary semaphore [28] that regulates the traffic of robots entering and leaving the zone. The new library will also not plan any new paths for the robots experiencing communication faults, this serves to reduce the risk of the TEA* algorithm inadequately assuming that these robots are executing a different path than the one they are actually taking, which could lead to a loss of efficiency when planning the paths of the remaining robots.

This additional control is important since when a robot is experiencing a communication fault situation there are no guarantees that any alterations to the planned path will reach and be executed by the robot, therefore it is necessary to limit the presence of several robots inside the same communication fault zone. However, in some cases this is impossible as it is shown in Section 6.

5. Supervision Module

This module is responsible for controlling when the robots paths need to be replanned and for detecting, measuring and handling communications faults. This module is composed by two sub-modules hierarchically related with each other. Each of these sub-modules is responsible for handling one of the tasks mentioned earlier. These sub-modules will be referred as the Planning Supervision Sub-Module (PSSM) and the Communication Supervision Sub-Module (CSSM), respectively.

The Figure 3 represents the hierarchical relation between all the parts that constitute the implemented control system. Represented in blue in this figure are both the sub-modules that constitute the System Supervision Module. In this figure, it is also possible to observe that the CSSM is located above the PSSM, showing that the CSSM commands have a higher authority than the PSSM ones.

The relation between the two sub-modules is based on a loop and interrupt-based control. On environments where no communication faults occur the CSSM does not intervene and the replanning of paths is handled by the PSSM. In this case the CSSM's only job is to detect that a communication fault is occurring. However if a communication fault is detected, the PSSM is overruled and the replanning of the robots paths is then controlled by the CSSM. When the fault has ceased, the control is handed back to the PSSM.

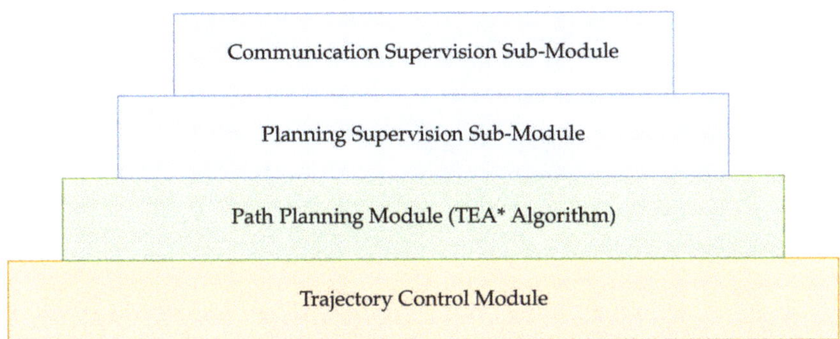

Figure 3. Representation of the hierarchical relation between all the parts of the system.

5.1. Planning Supervision Sub-Module (PSSM)

This sub-module is responsible for controlling when the robots need to have their path replanned in situations where no communication fault is detected. In order to accomplish this task, it needs not only to detect when one of the robots is delayed or ahead of time but also detect when a robot has completed the current step of his path.

Firstly, it is necessary to define when a robot is considered delayed or ahead of time. When using the TEA* algorithm, it is necessary that all robots are synchronized, meaning that all robots must be in the same step of their path, in order to avoid collisions and deadlocks. Therefore, delayed and ahead of time robots are robots that are one or more steps behind and one or more steps forward than the remaining robots.

However, in practice, this is not so linear since it would be impossible to synchronise all the robots in the system, thereby originating a very ineffective system where the paths are constantly being replanned. For this reason, a detection system must be created that allows small delays when those delays do not constitute a problem for the overall safety of the robot. To accomplish this goal, the PSSM executes three different verifications every time the robots position is updated. These verifications are useful to decide if it is necessary or not to replan the paths for all robots .

- Check whether any robots are too distant from their supposed position;
- Check whether the maximum difference between steps is 1;
- Check whether any robot is moving to a position currently occupied by another robot.

However, before any of these verifications can be executed it is necessary to know the type of movement the robot is executing. This movement could be classified into one of three categories:

- Robot is stopped;
- Robot is rotating;
- Robot is currently moving along a link.

After obtaining the type of movement of the robot, the sub-module checks all of the robots current steps in order to obtain the lowest step value possible. It then compares the current robot coordinates to the coordinates that the robot would have if it was currently executing the same step as the robot with the lowest step value. If the distance between this coordinates is greater than the stipulated threshold, the robots paths are replanned.

However, this function only checks for robots that are currently moving between nodes, not taking into consideration any delays during the rotation or even during communication with the robot. Another problem of this function is that it assumes that all the links have a dimension superior to the stipulated threshold.

Therefore, another function is also required to guarantee that the robots that are rotating or stopped are also synchronized with the other robots, as well as, guarantee that the synchronization and safe planning of paths for the robots can also be executed, even if the link dimension is shorter than threshold distance. This function checks whether the maximum step different between all the robots in the system is superior to one. If this condition is verified it means that the robots are not synchronized therefore their paths must be replanned.

Even in this example, the desynchronization was artificially induced; in an industrial environment it can be caused by a lot of factors, from small differences in the wheels diameter to differences in the grip coefficient.

In some cases, even a small desynchronization can cause a collision, as illustrated in Figure 4. In these cases, the most viable solution in order to maintain the safety of the system is to constantly replan the robots' paths as soon as the position of one of the robots is updated. Therefore the sub-module must detect these cases and act accordingly.

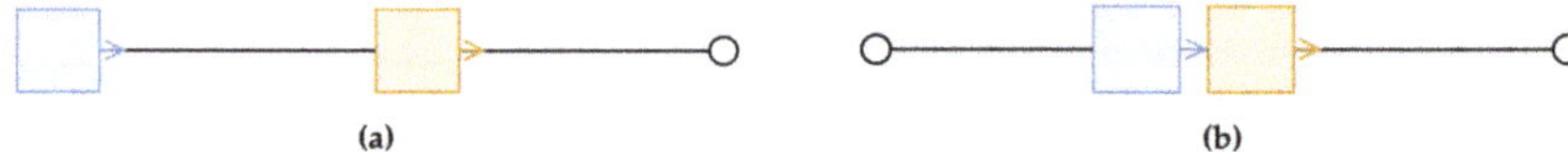

(a) (b)

Figure 4. Example of a small desynchronization leading to a collision. The subfigures a,b show the movement of the robot in this example, these subfigures use square to represent the location and orientation of the different robots. In this figure the circles represent the different nodes associated with that part of the map. (**a**) Initial position of the robots; (**b**) Collision originated by a delay of the movement of the orange robot.

To detect these specific cases, a new function was implemented; this function checks whether the node that corresponds to the current step of the robot path is equal to the current position of any of the remaining robots. In the case of a robot being currently located in a link between two nodes, the function checks the coordinates of both nodes that are part of the link, in order to guarantee that no other robot is moving to those nodes. If any of the described conditions happen, the robots paths are immediately replanned so that collisions are avoided.

When no replanning of the robots path is being executed, this sub-module needs to detect when a robot has completed its current path step and increment that step. However, these points can be a problem in the case where some robots are waiting for other robots to finish their movement. Therefore, it is necessary to stipulate when the steps of these robots can be increment in a safe way.

If the robot movement is a rotation or if the robot is travelling between two nodes determining when the robot has finished its current step of the planned path can be done by comparing the current robot position and orientation with the desired robot position and orientation at the end of that step.

In the case where the robot has stopped and is waiting for other robots to finish their movement, the robot step counter only increments when all moving robots have finished that step, preventing the movement of the robot before the path is cleared. This method assumes the worst case scenario where the movement of the stopped robot depends on the movement of all the moving robots of the system. It would be relatively costly in computational terms, to check all of the systems robots path and determine which robots needs to move in order to allow the movement of the stopped robot.

5.2. Communication Supervision Sub-Module (CSSM)

In order to create a system robust to communication faults the Communication Supervision Sub-Module (CSSM) was implemented. It is responsible for detecting communication faults, calculating their size and forcing the replanning of the robots paths to take into consideration said faults. This sub-module will predominantly be faced with two different fault situations:

- Situations where one area in the factory floor map consistently has no communication with the central control unit;
- Situations where temporary loss of connection happens with the central control unit

In either case, the system will not have any prior knowledge of the existence and location of these faults. For this reason it will have to constantly map the factory floor map, since faults can be dynamically created and destroyed. This allows the modified TEA* algorithm to then plan a path accordingly, as explained in Section 4.2. Figure 5 shows a flowchart of the normal functioning of this sub-module.

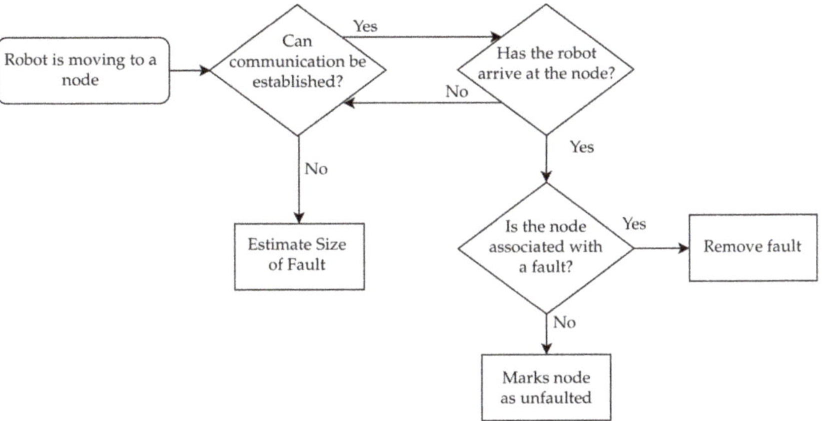

Figure 5. Flowchart representation of the Communication Supervision Sub-Module.

On the detection of a communication fault in one or more of the robots, the CSSM generates an interrupt, where the normal functioning of the central control unit is halted and the size of the fault is calculated. Afterwards, it forces a new replanning event, in order to take into account the estimated fault size; and return the system back to its normal functioning. During this replanning event, the nodes associated with the active faults are placed as occupied, therefore preventing other robots' entry into the zone affected by an active fault.

As shown in Figure 5, while the robot is unaffected by communication faults the current position of the robot is recorded has locations where it is possible to establish communication with the robot. These nodes will be referred to as unfaulted nodes and will be used in the estimation stage of the method to delimit the area subjected to faults. However, this designation does not mean that it is always possible to establish communication with a robot located in them. These nodes can still suffer from sporadic faults, as well as the appearance of new communication faults in areas where previously communication could be established. Therefore, their status as unfaulted nodes may not be permanent. Figure 6 shows an example of the mapping of unfaulted.

The estimation stage of this method happens when a robot enters a fault state. When entering this stage there are three possible situations that can happen: either the robot is entering a fault that is still not mapped, it is entering a fault that has already been mapped previously, or it is entering a unmapped extension of a previously mapped fault. Each of these situations is treated differently, as shown in Figure 7.

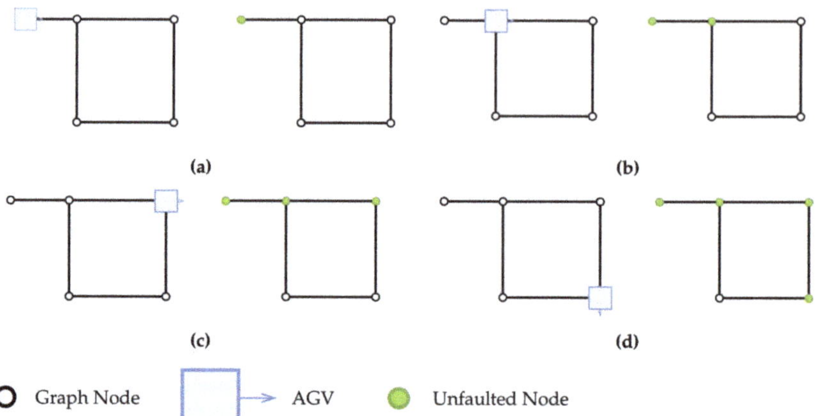

Figure 6. Example the mapping of unfaulted nodes. The progression of this mapping is shown via the subfigures a,b,c,d showing the Autonomous ground vehicle (AGV) movement on the left image and the status of the mapped nodes on the right one. (**a**) Initial state of the system; (**b**)System state at the end of the first step; (**c**)System state after the second step; (**d**) System state after the third step.

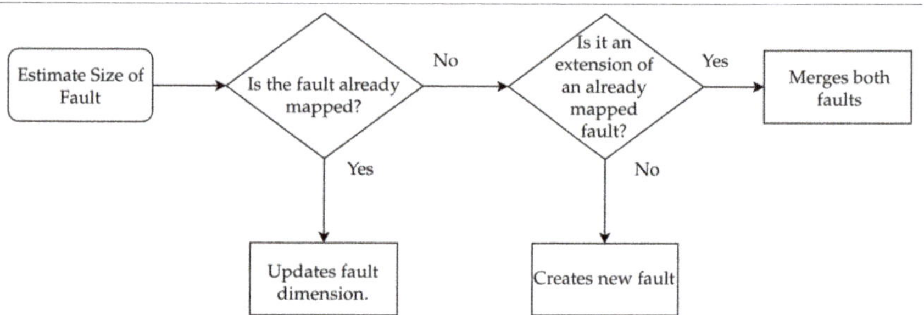

Figure 7. Flowchart representation of the different types of faults that the estimation method can encounter.

To estimate the dimension of a fault, the planned path for the robot is analyzed, so that it can determine the most likely exit node, node where communication can re-establish. To achieve this, each of the nodes corresponding to the next step of the planned path is sequentially compared with the records of the unfaulted nodes. This comparison goes on until either one of the path nodes is found to be an unfaulted node or the planned path ends. Afterwards, this method records all of the nodes that comprised the robot path between the current robot position and the exit node as a set of faulted nodes. Each set of faulted nodes therefore represents an area where no communication can be established with the robots, and have associated with it an array of possible entry/exit nodes. After the structure of nodes that represent the fault is created, the id value of the robot that detected it is associated with the fault.

In this method, when a fault is detected in the middle of a link, the entry/exit points of a fault zone are chosen in a worse case scenario; therefore these points are the last known nodes with good communication with the robot and the first point where communication could be re-established, respectively.

This method also accounts for the fact that a robot can exit sooner than the estimated exit node, especially in a situation where only a small percentage of the graph node has already been mapped. In these cases, an interrupt is generated when communication is re-establish with the robot. During this interrupt, both the faulted node set, as well as the entry/exit node set associated with that fault are corrected. This correction is executed by removing from the faulted node set all the robots path nodes that are between the actual exit point and the estimated exit node. The status of the removed nodes are changed to *not mapped*. Finally, the node where communication was re-established is placed into the entry/exit set. On exiting the fault, the robot will no longer be associated with that fault.

Figure 8 shows an example of the detection and mapping of a new fault; in this example the nodes represented in green are nodes considered to be unfaulted nodes, the nodes represented in orange are part of the entry/exit set associated with this fault and the nodes represented in red are faulted nodes associated with this fault. In this figure, the button graph representation serves to illustrate the current status attributed to each node.

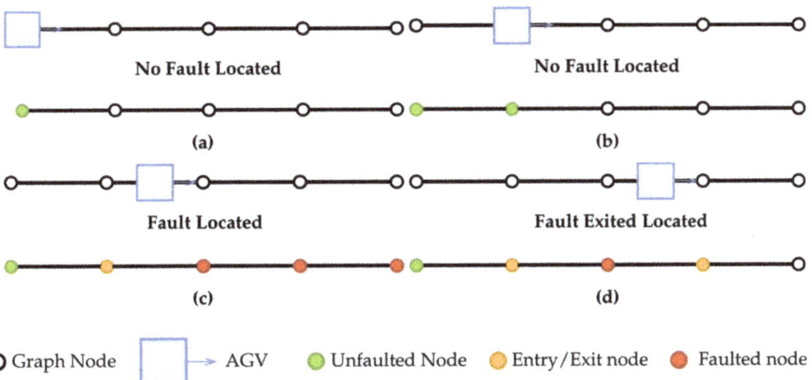

Figure 8. Example the mapping of a new fault. The progression of this mapping is shown via the subfigures showing the robot movement on the top image and the status of the mapped nodes on the bottom one. (**a**) Initial state of the system; (**b**) Mapping status and robot location at the end of the first step; (**c**) Mapping status and robot location when the communication fault is detected, initial estimation of the fault size; (**d**) Mapping status and robot location when communication is reestablished, readjustment of the fault size.

After the estimation of the fault dimension, the group of nodes associated with that fault is analyzed to see if they belong to an already existing fault or if they can be associated with an already existing fault as an extension of it, as shown in Figure 7.

First, the method starts by checking whether at least two of the nodes marked as entry/exit nodes are also entry/exit nodes of an already mapped fault. If this is not the case, the method will analyze each of the nodes that comprise the affected node set, individually and see if they belong to other already mapped faults. If any of these conditions are true both node sets are merged. In the rare cases where the estimated fault nodes sets are associated with more than one already mapped fault, all the faults associated with the estimated fault are merged into only one fault.

The mapping methodology presented is a methodology where the efficiency is directly proportional to the time spent, by moving the robots. An increased movement of the robots to different nodes would lead to more nodes being mapped, therefore leading to a more efficient estimation of the affected area upon the detection of a fault. To deal with sporadic faults, this idea was of an increase in efficiency with the passing of time was improved upon.

For starters, the CSSM would only assume that a robot is in a communication fault status, if there is no new message received after a time equivalent to the execution to four of the robot control cycles has passed since the last message from that robot was received. In the implemented system, this means that a gap of approximately 400 ms elapsed, taking into account the central unit control cycle period, before any action is taken. In terms of the robots' movement, this gap is too small, when used in conjunction with the PSSM, to cause any collisions or any significantly loss of efficiency, as shown in Section 6. Assuming a linear nominal velocity of 10 cm per second the gap would correlate with a movement of 4 cm if the robot was going full speed.

During this gap, the robot currently experiencing the fault will attempt to resend its data at least three times, therefore reducing any affect that random packet loss could induce a communication fault situation.

Applying the idea of increasing the efficiency of the algorithm with the passing of time a method for dealing with sporadic faults was implemented. This method is comprised of two parts, the first one is executed every time a communication is established with the robot in a node that was previously flagged as suffering communication faults and the second one is executed when a communication fault is detected in an unfaulted node.

The first part of this function verifies whether the node where the robot is communicating from is considered a faulted node. If so, the fault associated with that node is removed and all the faulted nodes from that are placed as not mapped, except for the node from where the communication originated which is placed as unfaulted. All the entry/exit nodes associated with that fault are still kept as unfaulted nodes. Figure 9 shows an example of the removal of a fault; in this figure both the faulted nodes, represented in red, originated from to a sporadic fault that is currently inactive. Therefore, when a robot enters this fault it does not experience any communication loss, this leads to all the nodes being considered to be unfaulted nodes, represented in green.

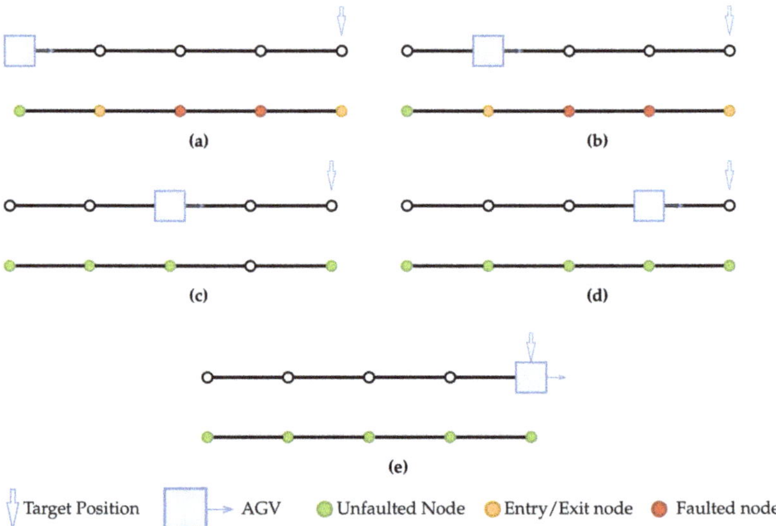

Figure 9. Example of the removal of fault.(**a**) Initial state of the system; (**b**) Mapping status and robot location at the end of the first step; (**c**) Mapping status and robot location after the robot enters the area previously affected by a communication fault, the nodes associated with that fault are placed as not mapped; (**d**) The robot proceeds to remap the nodes; (**e**) System state when the robot reaches its target point with all of the nodes fully mapped.

This first part serves to remove any sporadic faults that might occur. This removal is especially important for the efficient planning of the paths since it avoids the unnecessary creation of binary semaphores.

In situations when a sporadic fault occurs on an entry/exit node of another already mapped fault, the sporadic fault is associated with that fault. When the sporadic fault is removed, the nodes of the already mapped fault are placed as not mapped. This allows for the storing of the dimension associated with that fault, permitting an efficient remapping of the fault if a robot enters it.

The Figure 10e shows an example of the removal of a sporadic fault that has been merged with a fault that corresponds to an area where there is no communication with the robot. As with Figure 9, the robot starts by placing both the faulted nodes as unmapped nodes. However, when the robot enters the second unmapped node it experiences a communication fault for being in an area inaccessible by the communication network Figure 10d, therefore the fault dimension estimation algorithm is called leading to the remapping of that fault.

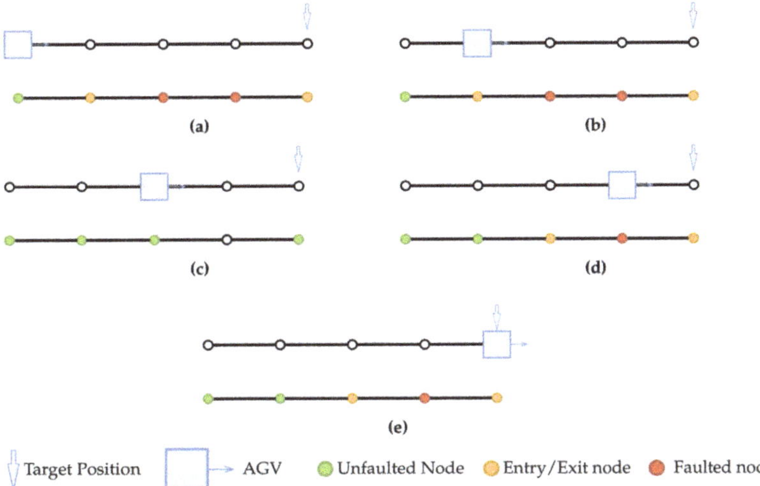

Figure 10. Example of the removal of a sporadic fault in an entry/exit node. (**a**) Initial state of the system; (**b**) Mapping status and robot location at the end of the first step; (**c**)The robot enters an area previously affected by a communication fault, since no fault is detected all nodes associated with the previously mapped fault are placed has not mapped; (**d**) The robot proceeds to map the remaining nodes that are still affected by communication faults; (**e**) Final system state, the new size of the fault is now fully mapped.

The second part of this function runs when a fault is detected in an unfaulted node. When this happens, this fault is assumed as a sporadic fault, so its dimension is limited to the node where it happened. This node is removed from the unfaulted set and the mapping methodology is applied. Similar to the static faults when communication is reestablished with the robot, the size of the fault is updated.

6. Tests and Validation

In this section, all the implemented modules as well as the interaction between them will be tested and validated. It will also present the situations where this system is incapable of guaranteeing a safe execution of the mission.

These tests will be divided into two categories, those with static communication faults and those with sporadic and static communication faults. The results of these tests, as well as other tests executed to the system, are represented in the appendixes of this document. For the representation of the results of these tests, a methodology where the robots are represented by a full square, their destination point is represented by an empty square of the same colour as the robot and their executed path between images is represented in the same colour as the robot. For ease of comprehension, numbers with the same colour as the robot have been added to the figures, in order to demonstrate the task order associated with each robot.

For these tests, four robots will be used. Each robot will be assigned a set of tasks that it needs to complete. When all tasks assigned to all of the robots that comprise the system are completed and all robots have reached their charging post the test is deemed as finished.

All tests were executed using the same factory floor map. This layout was inspired by a real-life factory.

6.1. Planning Supervision and Overall System Performance when Subjected to Static Communication Faults

The first category will analyze the behaviour of the system when only exposed to areas where there is no communication. The second set of tests will expose the system not only to areas where there is no communication but also to packet loss and sporadic communication faults.

For the first set of tests, the initial conditions represented in Figure 11 will be used.

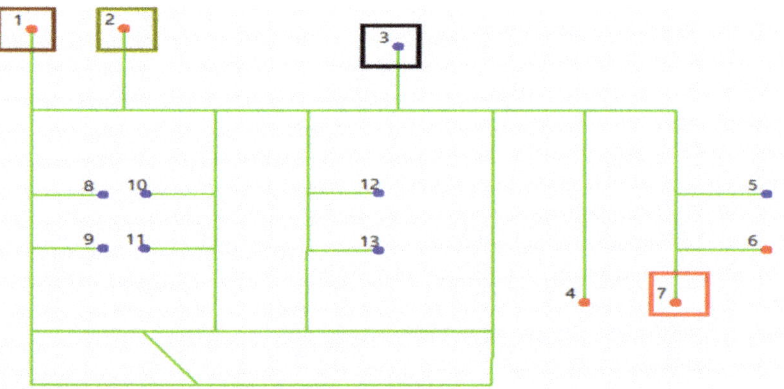

Figure 11. Representation of the workstation distribution used in this set of tests. The coloured squares indicate the initial position of the robots.

In this set of tests the no communication section chosen encompasses both workstations 10 and 11, this area is represented in Figure 12 by the colour red.

To test the system response, the creation of a new mission is also necessary. As the intent of these tests is to analyze the implemented system response to areas where there is no communication, a mission comprised of only one task to each robot is enough. Table 1 shows each robot assigned task as well as its initial priority.

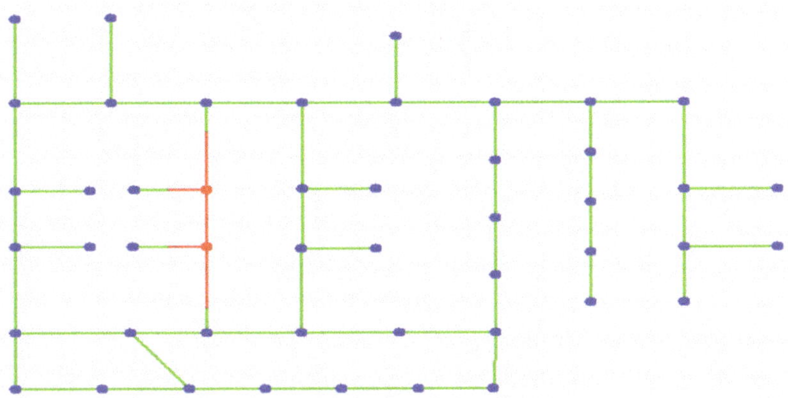

Figure 12. Representation of the chosen area where no communication can be established with the robots.

Table 1. Tasks and initial priority of each robot. The numbers under the task column represent the workstation, as identified in Figure 11, to which the robot is ordered to move to.

Id Robot	Colour	Initial Priority	Task
1	Brown	1	9
2	Dark Green	2	10
3	Black	3	11
4	Red	4	5

Figure A4 shows the paths travelled by the robots during the execution of this mission, as well as showing the mapping of the factory floor in order to determine which nodes are experiencing communication faults. In this mapping, the nodes represented in green are considered unfaulted nodes, the nodes represented in yellow are considered entry/exit nodes of the faults that have been mapped, the nodes represented in purple are considered possible entry/exit nodes of the faults that have yet to be mapped, and finally the red nodes are the faulted nodes.

By analyzing Figure A4, it is possible to observe that the system managed to successfully complete the assigned mission avoiding any collisions or deadlocks, even when two of the tasks required the robots to travel through the area that wasn't covered by the communication network. The system was also capable of correctly mapping the fault by using the data from the two robots that travelled through it.

The system first detected the fault in Figure A4e; initially the critical zone was considered to be part of the fault as seen in Figure A4f. On detecting this fault a replanning event is triggered and this leads the black robot to stop and wait for the dark green robot to exit the fault.

When the dark green robot exits the fault, Figure A4i, the fault dimensions are adjusted and the status of the node representing workstation 10 is changed from faulted node to an entry/exit node, as seen in Figure A4j. The black robot can now advance in its course, since the fault is no longer occupied, therefore a new replanning event is executed given the current position of the dark green robot, the black robot is now forced to plan a path that accesses the area via the bottom side instead of using the top side as initially planned.

When the dark green robot re-enterw the fault, in Figure A4k, the fault is once again placed as occupied. However, this time since the black robot is accessing the fault via the bottom side there is no need for it to stop immediately. An important thing to note is that in this moment only the top node of the fault has been mapped as faulted node since it was the only one used by the robots until this point, as seen in Figure A4l. This leads to the black robot only stopping when it reaches the bottom entrance of the zone as seen in Figure A4m.

After the dark green robot once again exits the fault, in Figure A4o, the black robot enters the fault from the other side and therefore maps the other faulted node, as seen in Figure A4p. Since one of the new mapped faulted nodes is the same as one of the possible entry/exit nodes of the previously mapped fault, the system merges both faults.

A link to a video showing the execution of this test can be found in Appendix E.

This test was repeated 10 times in order to show that the presented sample was not a sporadic success. Analyzing the data from these tests, it is possible to conclude that the system is capable of successfully dealing with areas that are not covered by the communication network, since in all samples the system executed the mission successfully. It also allows for the stipulation of a baseline execution time for this mission of 1 min and 47 s. When this mission is executed without any communication faults, its mean execution time is of 1 min and 34 s. As expected, the overall efficiency of the system decreases when the system is exposed to communication faults.

6.2. Planning Supervision and Overall System Performance when Subjected to Static and Sporadic Communication Faults

The second set of tests has the objective of verifying that the implemented system is capable of completing an assigned mission, in a safe and efficient way, when exposed to both sporadic faults, as well as, zones where there is no communication with the robot.

To accomplish this, the CSSM must first not only detect the area where there is no communication but also detected the occasional loss of communication in a specific node. After the detection, this sub-module must also check if any of the detected faults have ceased to exist, as explained in Section 4.2.

In order to compare results of both sets of tests, the same mission and initial conditions used in the first set will be used, as shown in Figure 13 and Table 2.

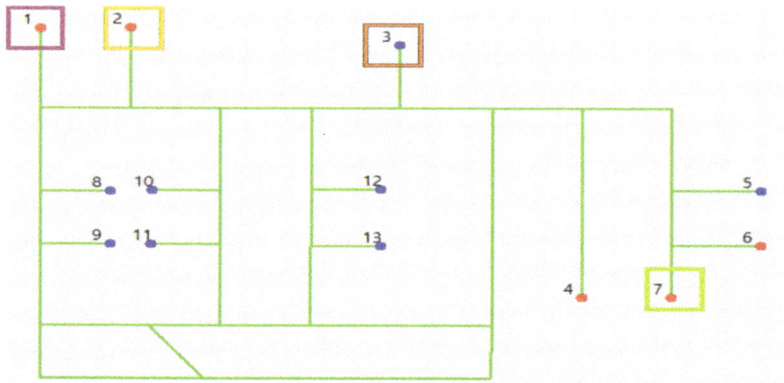

Figure 13. Representation of the workstation distribution as well as of the initial position of the robots used in this set of tests. The coloured squares indicate the initial position of the robots.

Table 2. Tasks and initial priority of each robot. The numbers under the task column represent the workstation, as identified in Figure 13, to which the robot is ordered to move to.

Id Robot	Colour	Initial Priority	Task
1	Purple	1	9
2	Yellow	2	10
3	Orange	3	11
4	Green	4	5

The sector of the map that is not covered by the communication network is also the same as the one used in the previous set of tasks, as shown in Figure 14.

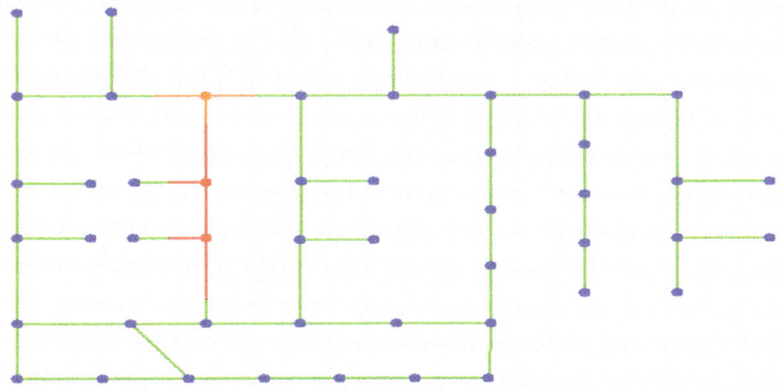

Figure 14. Representation of the chosen area where no communication can be established with the robots.

These sets of tests will test the occurrence of a sporadic fault in an entry/exit node of the fault that represents the area where no communication can be established with the robot, this area is represented in orange in Figure 14.

The representation of the results of the first part of this set of tests is represented in Figure A6. As with the previous tests, the nodes represented in green are considered unfaulted nodes, the nodes represented in yellow are considered entry/exit nodes of the faults that have been mapped, the nodes represented in purple are considered possible entry/exit nodes of the faults that have yet to be mapped and finally the red nodes are the faulted nodes.

By analyzing Figure A6, it is possible to observe that when the yellow robot moves to the top entry node of the fault corresponding to the section of the factory that is not covered by the communication network, referred as the static fault for brevity, a sporadic fault occurs in this node. Since this fault occurs in an entry/exit node this node is associated with the static fault, as seen in Figure A6a–f.

Similar to the previous tests, the orange robot advances until the edge of the fault and then stops and waits for the yellow robot to exit the fault, and therefore places the binary semaphore then corresponds to that fault in the free state before proceeding, seen in Figure A6e.

On exiting the fault a second time, as shown in Figure A6g, the yellow robot sees that the sporadic fault has ceased since the system can now communicate with the robot in a zone where it previously couldn't. This leads to the system reclassifying the node where communication was re-established with the robot as an unfaulted node. The system also reclassifies the node belonging to the static fault as an unmapped node, due to the impossibility of the system knowing if it belongs to the static fault or to the sporadic fault.

Similar to the previous tests after the yellow robot exits the fault, the orange robot is cleared to proceed with the execution of its task. This leads to the orange robot entering the static fault, as shown in Figure A6i. The executing of this robot tasks leads to the remapping of the static fault and this time the dimension is correct since no sporadic fault occurs when the orange robot exits the fault for a second time. This process is shown in Figure A6i–p.

A link to a video showing the execution of this test can be found in Appendix E.

The execution of this test was repeated 10 times. In all samples, the system was able to successfully operate in environments where sporadic communication faults as well as static communication faults can occur, and manage to safely execute its assigned tasks.

6.3. Cases Where the Implemented System Can't Guarantee a Safe Execution of the Mission

Even due the overall results of the implemented system tests were very successful, there are some situations where the implemented system is incapable of guarantee a safe execution of its assigned mission.

These situations normally occur due to either external interference or due to a communication fault occurring in an unfortunate time or having a large dimension. These situations are the following:

- Two or more robots entering the same fault at the same time;
- Impossibility of a robot to move away from the exit node of a fault.

In the first one, if two or more robots enter the same unmapped static fault at the same time, a collision can occur as the system has no way to control the synchronization between robots as it is incapable of communicating with the robots. This normally occurs when a communication fault has a large dimension.

The second situation where this may occur is if the entry/exit of a fault is located in a link between two nodes. There is a chance of a situation where one robot is currently traversing that link and another robot enters the fault through another entry point. The first robot will be in a deadlock situation since both of its possible movements choices are now locked until the second robot exits the fault.

7. Conclusions

The article presents the implementation of a control system capable of controlling the traffic of a fleet of robots, i.e., plan and control the movement of a fleet of robots that have multiple tasks assigned. The implemented system must plan safe and efficient paths, avoid deadlocks and be immune to network failures. It uses the TEA* algorithm proposed in [4], as a base for the path planning and has a high tolerance to communication faults. The implemented system is comprised of several modules. This system not only planned the robots paths but also supervised their execution and was on guard against any communication failures.

Several modifications of the TEA* algorithm, proposed by [4], were also implemented in order to make it compatible with the new environment conditions that the algorithm would have to face. This modifications also intended to move the TEA* algorithm closer to a real life industrial application.

As shown in Section 6.2, the implemented system is capable of safely executing a set of tasks in environments where both static and sporadic communication faults occur. On previous implementations of the TEA* algorithm [4], when the robots were subjected to communication faults, there would be a high probability of the occurrence of deadlocks since the TEA* algorithm requires the synchronization of the movements and positions of all the robots of the system in order to be effective.

Although the system has a good tolerance to communication faults, it still lacks total effectiveness when dealing with communication faults, as demonstrated by in Section 6.3.

Author Contributions: The contributions of the authors of this work are pointed as follows: Conceptualization: D.M., P.C. (Pedro Costa) and J.L.; Methodology: D.M., P.C. (Pedro Costa) and J.L.; Software D.M., P.C. (Pedro Costa), J.L. and P.C. (Paulo Costa); Validation: D.M., P.C. (Pedro Costa) and J.L.; Writing—Review and Editing: D.M., P.C. (Pedro Costa) and J.L.; Supervision: P.C. and J.L. All authors have read and agreed to the published version of the manuscript.

Funding: This work is financed by National Funds through the Portuguese funding agency, FCT—*Fundação para a Ciência e a Tecnologia* within project UIDB/50014/2020. This work has been supported by the European Regional Development Fund (FEDER) through a grant of the Operational Programme for Competitivity and Internationalization of Portugal 2020 Partnership Agreement (PRODUTECH4S&C, POCI-01-0247-FEDER-046102).

Institutional Review Board Statement: Not applicable.

Informed Consent Statement: Not applicable.

Data Availability Statement: This study didn't report any new data.

Acknowledgments: The authors of this work would like to thank the members of INESC TEC iiLab for all the support rendered to this project.

Conflicts of Interest: The authors declare no conflict of interest.

Abbreviations

The following abbreviations are used in this manuscript:

TEA*	Time Enhanced A*
AGV	Autonomous ground vehicle
MAPF	Multi-Agent Pathfinding
LPA*	Lifelong Planning A*
Mutex	Multiple exclusion object
NP	Non-deterministic Polynomial-Time
PSSM	Planning Supervision Sub-Module
CSSM	Communication Supervision Sub-Module

Appendix A. Execution of the First Mission with no Communication Faults

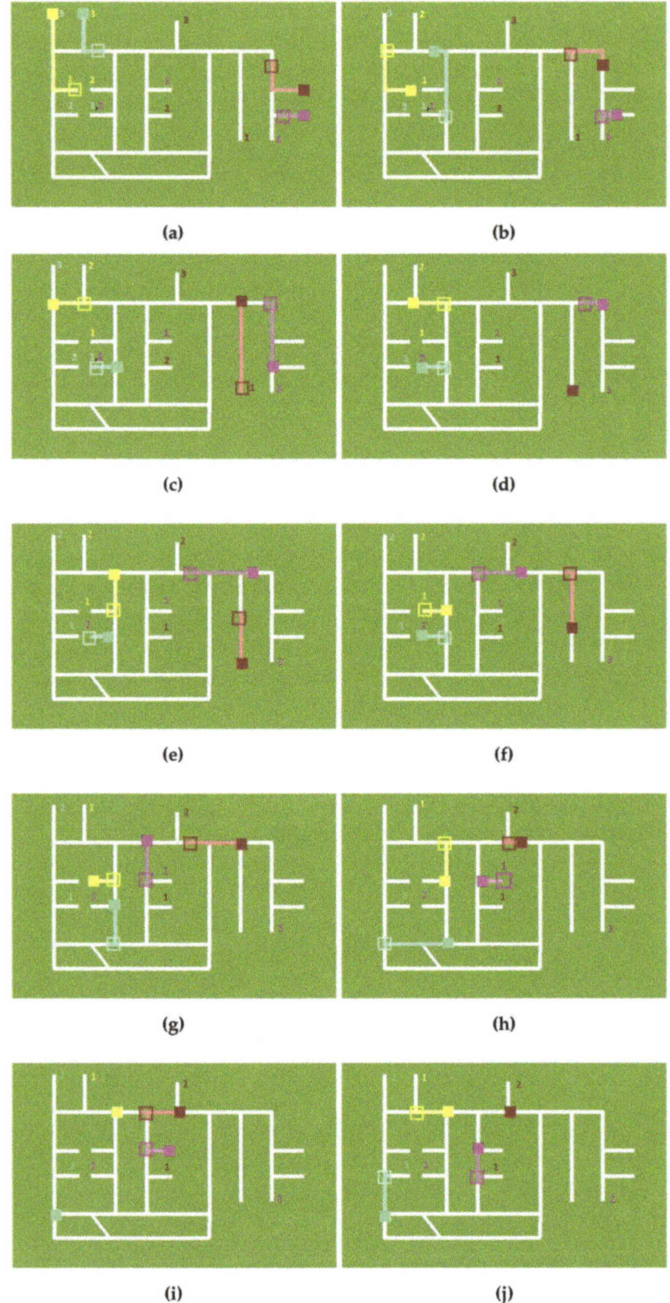

Figure A1. *Cont.*

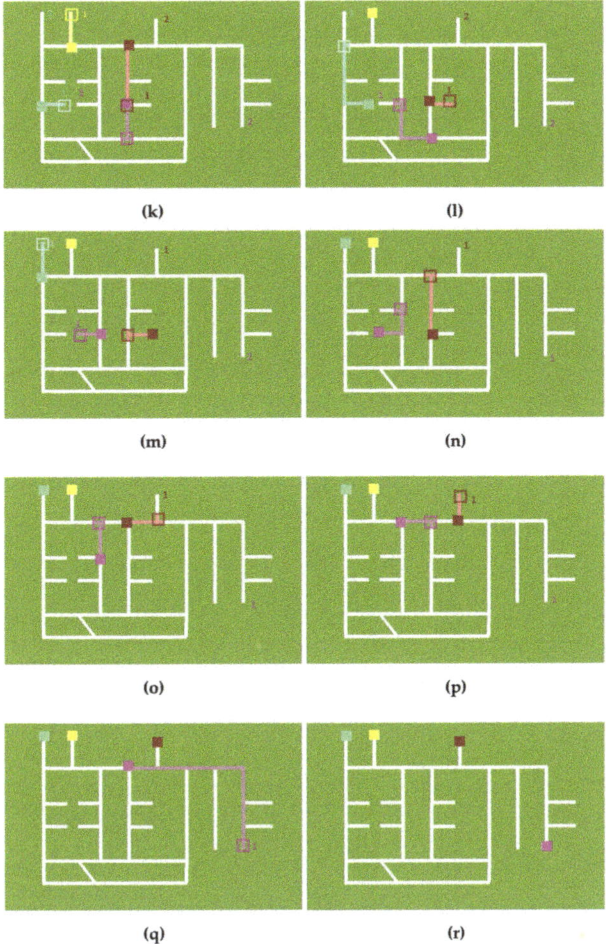

Figure A1. Movement of the robots during the execution of the first mission, each robot as two tasks assigned to it. In subfigures (a-r), the robot is represent by the full coloured square, while its movement between each of the subfigures is represented by the outlined square of the same colour. The numbers presented in the subfigures represent the task order assigned to each robot. (a) Initial positions of the robots; (b) Positions of the robots after one task is completed by the yellow robot; (c) Position and movement of the robots during the execution of the first task assigned to both the red and blue robots; (d) The purple robot must wait until the red robot starts its movement before it can advance; (e) A delay in the movement of the light blue robot originates a replanning event; (f) The robots keep executing their planned paths without any need of replanning; (g) Position and movement of the robots during the execution of the second tasks assigned to the yellow robot; (h) The robots keep executing their planned paths without any need of replanning; (i) Position and movement of the robots during the execution of the first tasks assigned to the purple; (j) The red robot waits until the purple robot moves from its current stop to initiate its movement; (k) The red robot shadows the movement of the purple robot until it reaches the entry node of its target point; (l) Movement of the yellow robot into its charging post and execution of the second task assigned by to the blue robot; (m) Movement of the blue robot into its charging post and execution of the second task assigned by to the red robot; (n) Positions of the robots when the purple robot finishes its second task; (o) All tasks are completed and the robots move to their charging posts; (p) Both robots move towards their charging stations; (**q**) Movement of the red and purple robots towards their charging posts; (**r**) Final positions of the robots after all tasks are completed.

Appendix B. Execution of the Second Mission with no Communication Faults

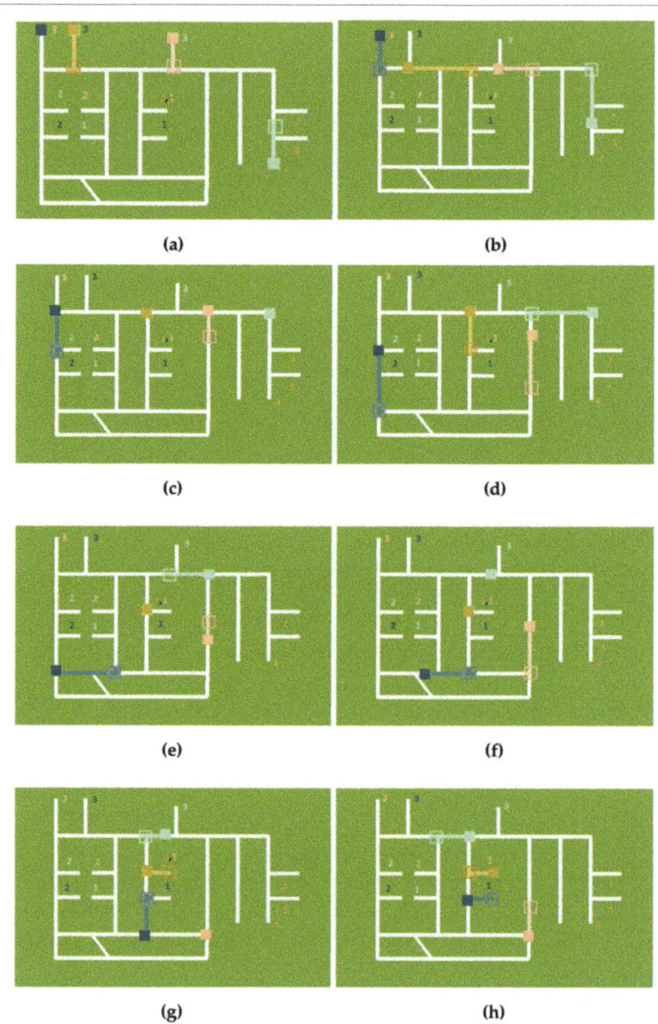

Figure A2. Movement of the robots during the execution of the second mission, first figure. In subfigures a–p, the robot is represent by the full coloured square, while its movement between each of the subfigures is represented by the outlined square of the same colour. The numbers presented in the subfigures represent the task order assigned to each robot. (**a**) Initial positions of the robots; (**b**) All robots keep following their planned paths; (**c**) Since the brown robot has a higher priority than the pink one, this last one is force to take the longer path in order to avoid any conflicts; (**d**) The light blue robot had to stop and wait for the pink robot to move from the intersection before resuming its movement; (**e**) Delays in the movement of the dark blue robot force the system to replan the path of the pink robot to avoid pathing conflicts; (**f**) Delays in the rotation of the brown robot force further replanning of the robots paths; (**g**) No replanning is necessary since the robots are still synchronised; (**h**) Both the brown and dark blue robot finish their fist task.

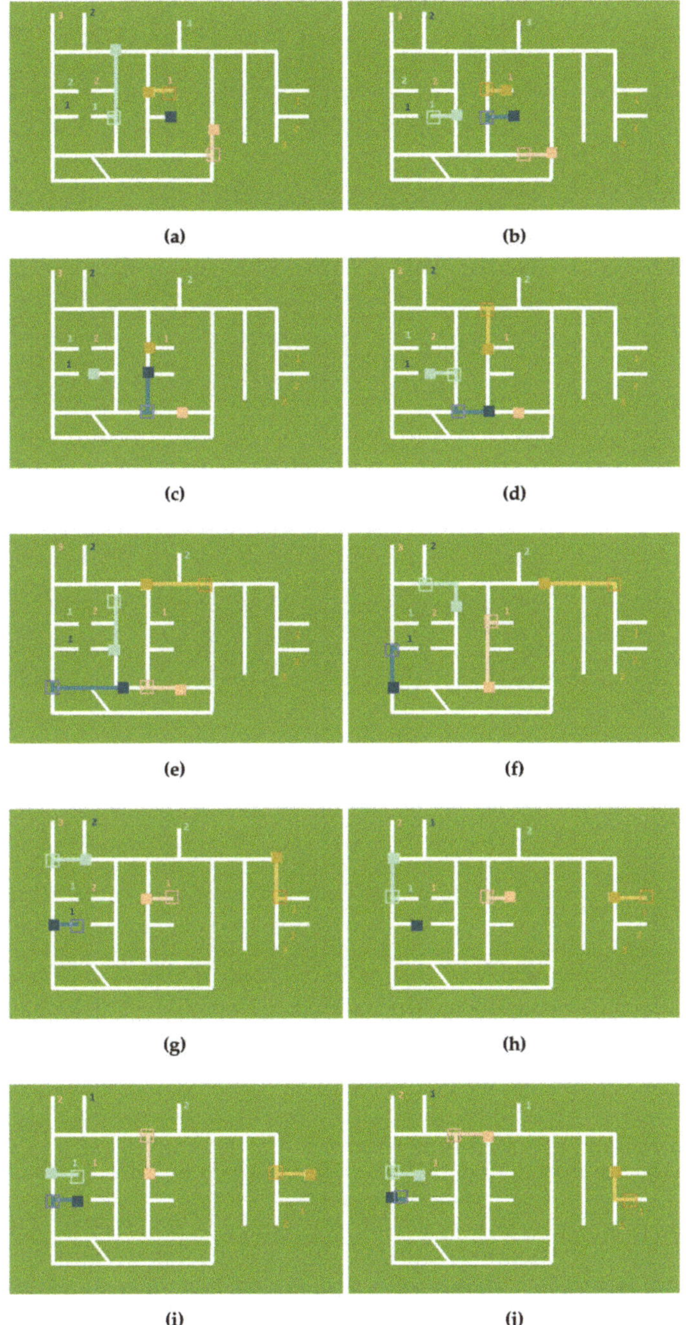

Figure A3. *Cont.*

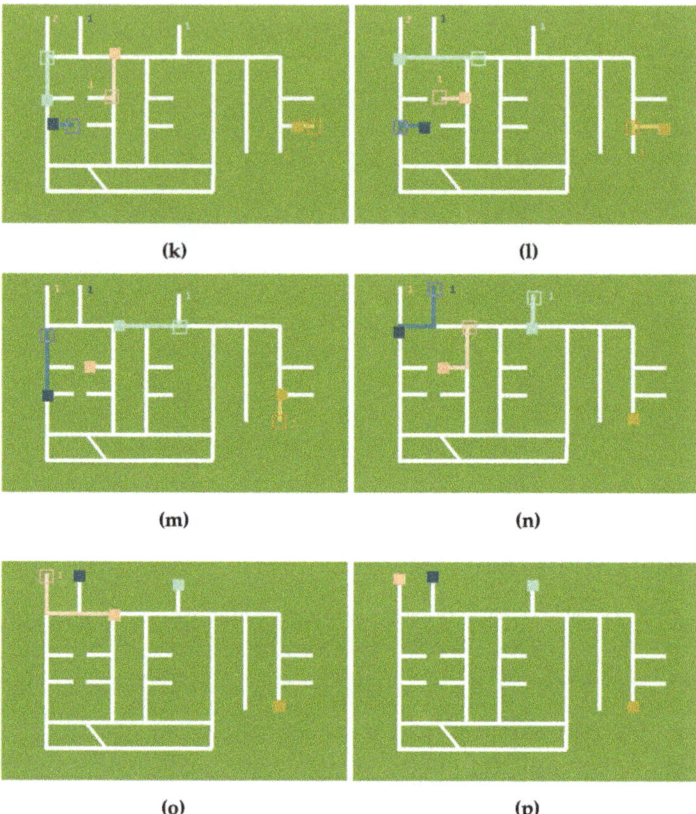

Figure A3. Movement of the robots during the execution of the second mission, second figure. In subfigures a–p, the robot is represent by the full coloured square, while its movement between each of the subfigures is represented by the outlined square of the same colour. The numbers presented in the subfigures represent the task order assigned to each robot. (**a**) Futher delays in rotation of the dark blue robot originate another replanning event; (**b**) The dark blue robot finishes its second task; (**c**) The pink robot has to wait until the dark blue robot exits from that branch of the map; (**d**) The light blue robot finishes its first task; (**e**) The robots keep executing their planned paths without any need of replanning; (**f**) Given the position of the robots the risk of a delay causing a pathing conflict is very low; (**g**) The robots keep moving to their target positions; (**h**) The dark blue robot and the pink robot finish their second and first task, respectively; (**i**) The pink and dark blue robots start moving to their charging positions; (**j**) The light blue robot and the brown robot finishes their second task; (**k**) New replanning event occurs due to a delay in the light blue robot movement; (**l**) The dark blue robot moves back in order to avoid a deadlock situation with the light blue robot; (**m**) The brown and pink robots finish their third and second tasks, respectively; (**n**) All three robots move towards their charging stations; (**o**) All robots move to their charging stations; (**p**) Final positions of all robots after all tasks are completed.

Appendix C. Execution of a Mission with Static Communication Faults

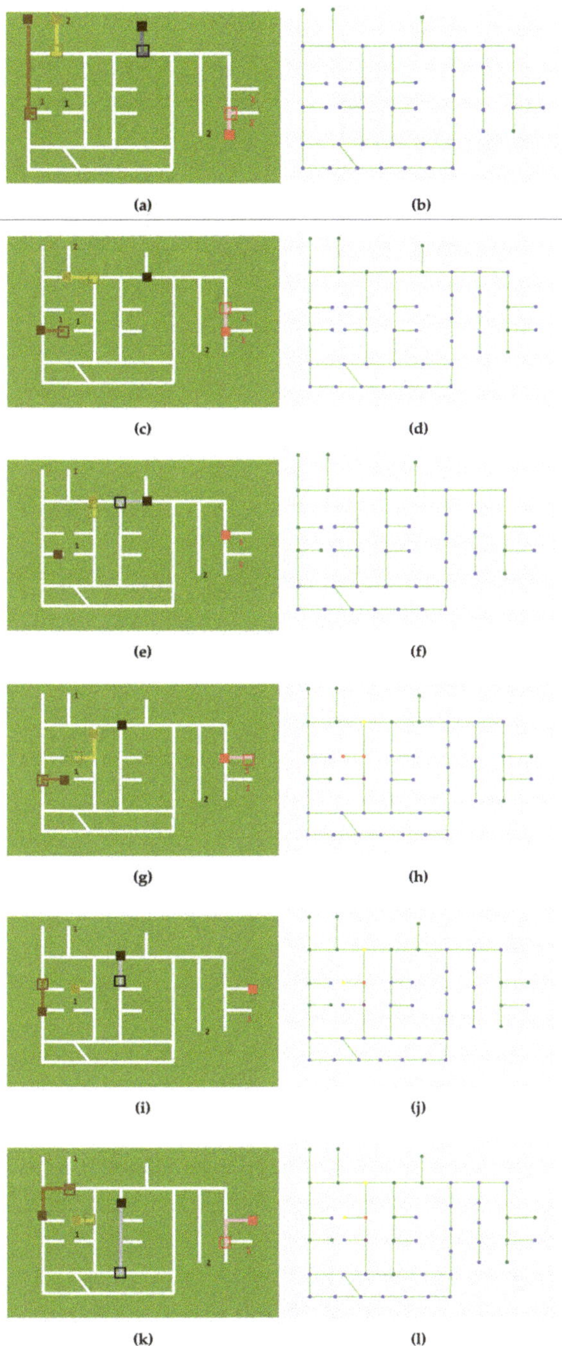

Figure A4. *Cont.*

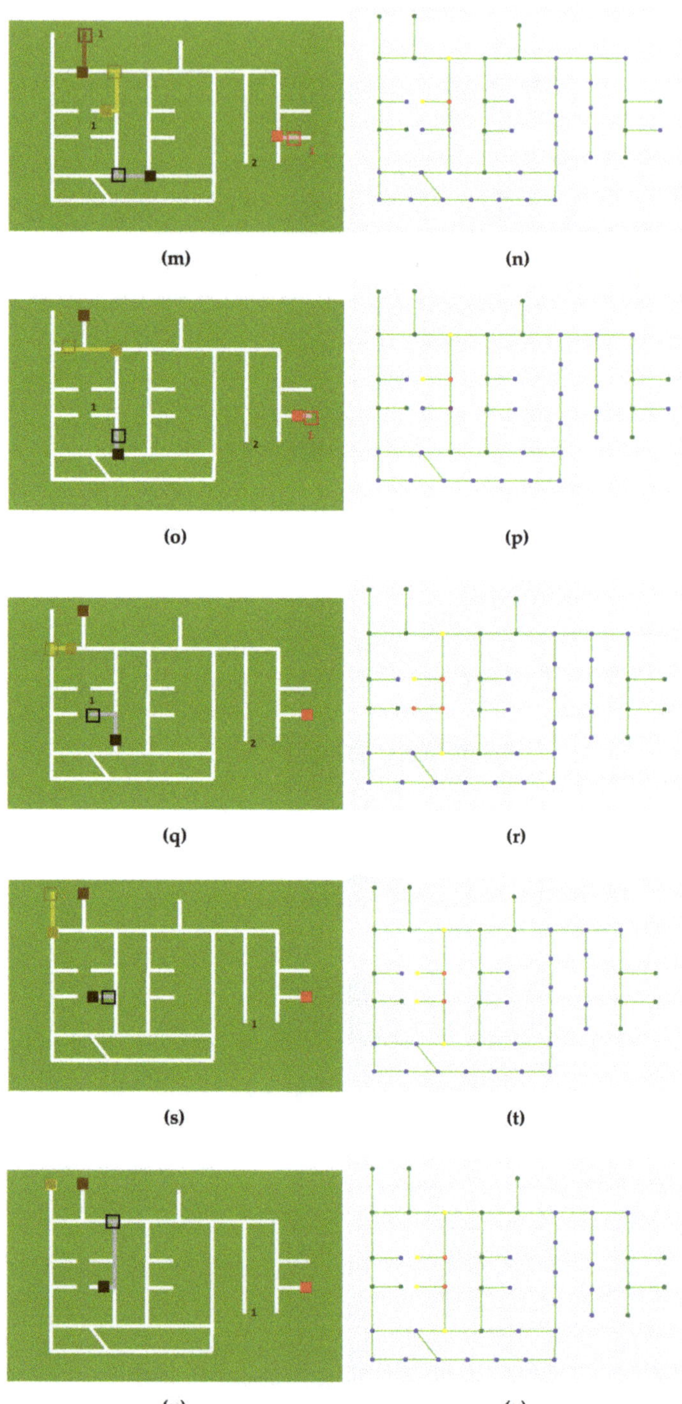

Figure A4. *Cont.*

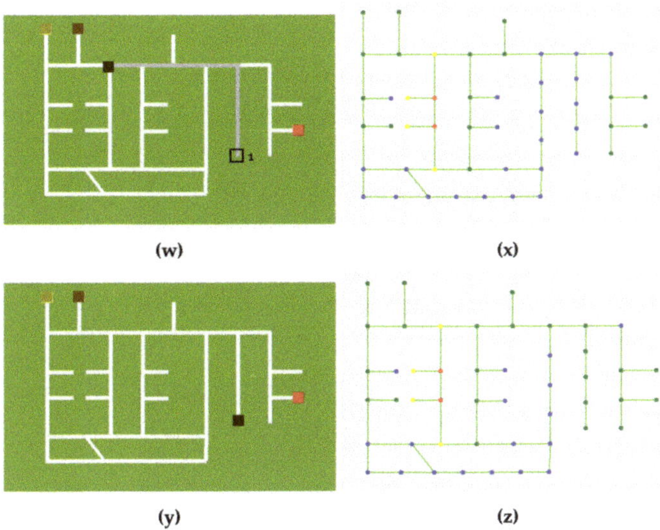

Figure A4. Behaviour of the system when subjected to an area without communication. In right subfigures, the robot is represent by the full coloured square, while its movement between each of the subfigures is represented by the outlined square of the same colour. The numbers presented in the subfigures represent the task order assigned to each robot. In the left subfigures the status of each node is displayed, blue equals unmapped, red equals faulted, yellow represents a entry/exit node and green represents an unfaulted node. (**a**) Initial positions of the robots; (**b**) Initial status of all the nodes of the system; (**c**) Position and movement of the robots during execution of the first task by the brown robot; (**d**) Status of the nodes during the movement of the robots; (**e**) The brown robot finishes its first task; (**f**) Status of the nodes during the after the brown robot finishes its first task, several nodes are now mapped as unfaulted nodes; (**g**) Entry of the light brown robot into the area affected by the communication fault; (**h**) Initial estimation of the size of the area affected by the communication fault; (**i**) The light brown robot exits the fault, the red robot also finishes its first task; (**j**) The size of the fault is now adjusted using the most recent data from the light brown robot; (**k**) The light brown robot reenters the area affected by the fault; (**l**) Status of the nodes when the light brown robot reenters the area affected by the fault; (**m**) The light brown robot moves through the area affected by the fault; (**n**) The new data is analysed however no modifications to the fault size is necessary; (**o**) The black robot is force to wait until the light brown robot exits the area affected by the communication fault; (**p**) Status of the nodes during this wait; (**q**) The black robot now enters the area affected by the fault via a previously unmapped entry; (**r**) The fault size is adjusted based on the last know location of the black robot and its planned path; (**s**) Communication is reestablished with the black robot; (**t**) The fault size is readjusted to take into account the new data; (**u**) The black robot reenters the fault; (**v**) no adjustments are necessary since now the fault is fully mapped, all of its entry/exit nodes have been mapped; (**w**) he black robot now moves to its charging station; (**x**) No more faults are detected during this movement; (**y**) Final position of the robots in the system; (**z**) Final status of all the nodes in the system.

Appendix D. Execution of a Mission with Static and Sporadic Communication Faults

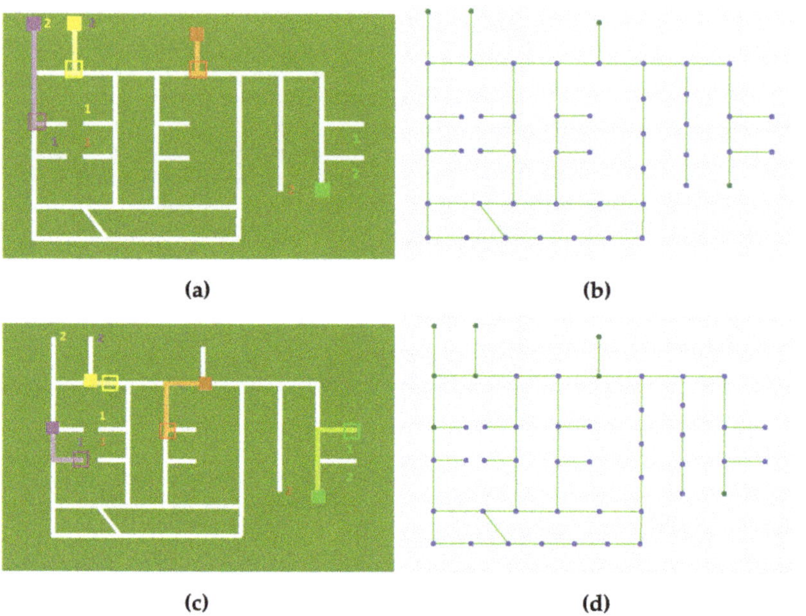

Figure A5. Behaviour of the system when subjected to a sporadic fault in an entry/exit node, first image. In right subfigures, the robot is represent by the full coloured square, while its movement between each of the subfigures is represented by the outlined square of the same colour. The numbers presented in the subfigures represent the task order assigned to each robot. In the left subfigures the status of each node is displayed, blue equals unmapped, red equals faulted, yellow represents a entry/exit node and green represents an unfaulted node. (a) Initial positions of the robots; (b) Initial status of all the nodes of the system; (c) The yellow robot enters the area affected by the communication faults; (d) Status of the nodes before the yellow robot enters the fault.

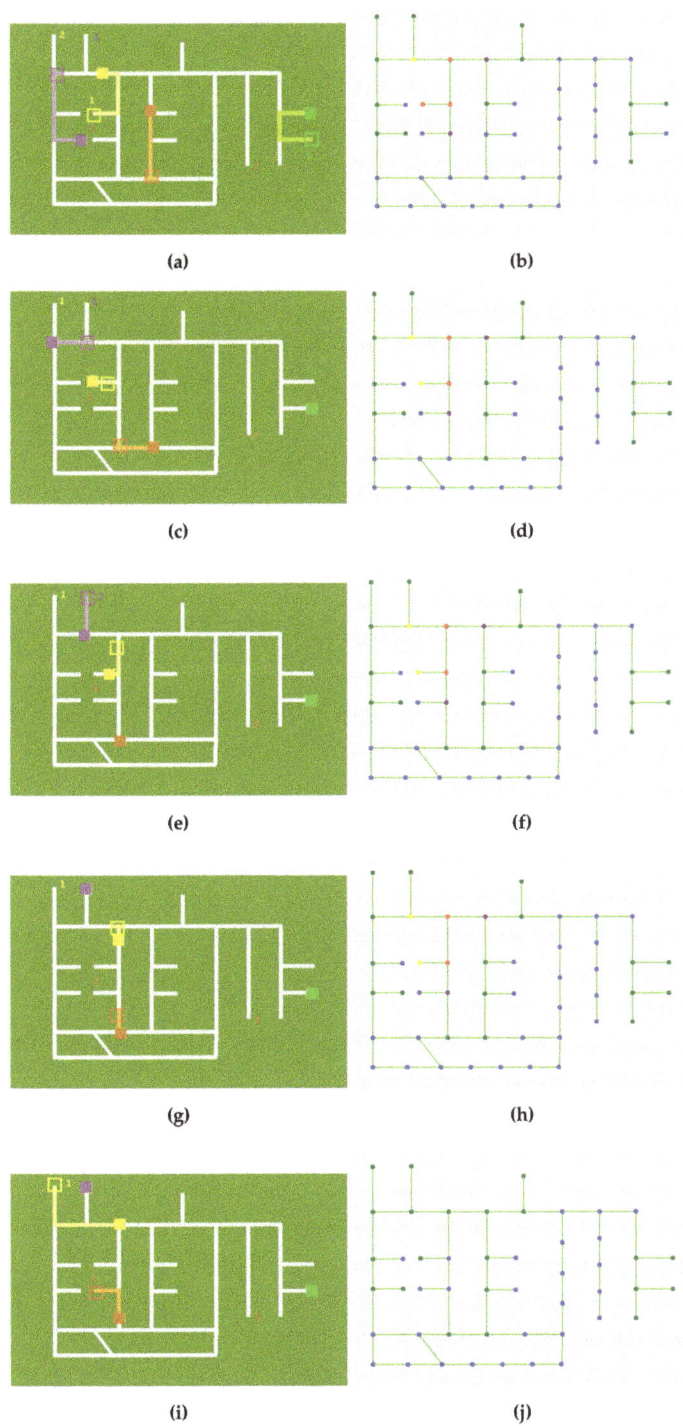

Figure A6. *Cont.*

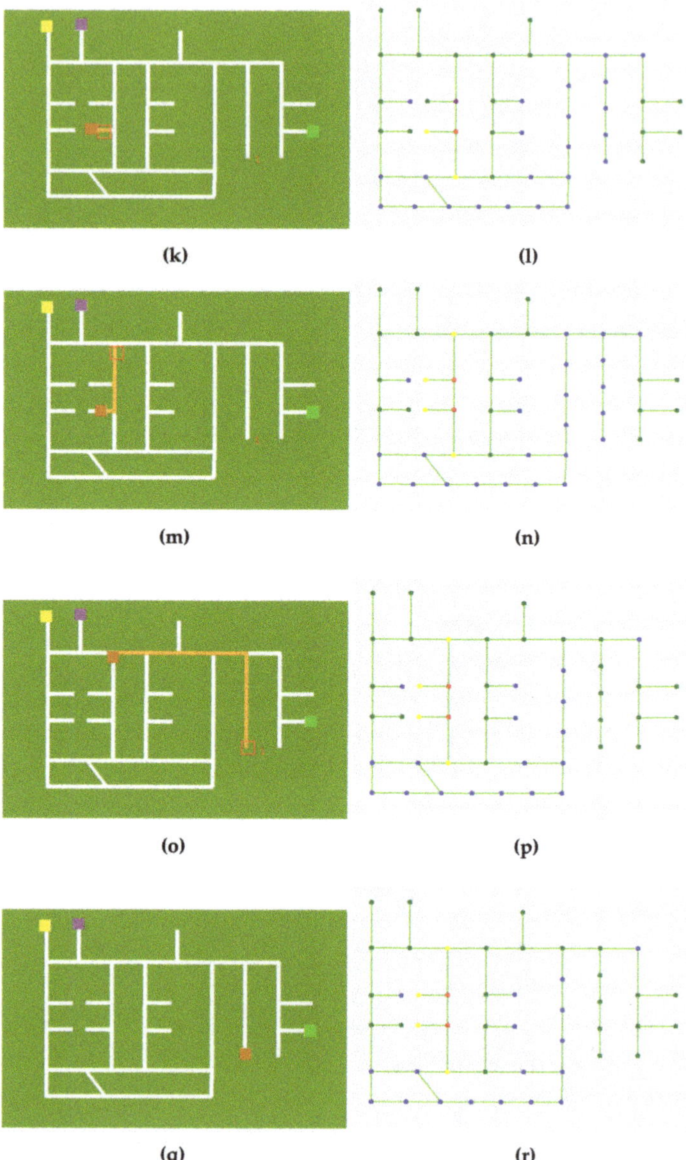

Figure A6. Behaviour of the system when subjected to a sporadic fault in an entry/exit node, second image. In right subfigures, the robot is represent by the full coloured square, while its movement between each of the subfigures is represented by the outlined square of the same colour. The numbers presented in the subfigures represent the task order assigned to each robot. In the left subfigures, the status of each node is displayed, blue equals unmapped, red equals faulted, yellow represents an entry/exit node and green represents an unfaulted node. (**a**) The yellow robot moves through the faulted area; (**b**) Initial estimation of the fault size based on the path planned for the robot and the already mapped nodes; (**c**) Communication is reestablished with the yellow robot; (**d**) The size of the fault is adjusted; (**e**) The yellow robot reenters the fault, which forces the orange robot to wait until communication can be reestablish; (**f**) Status of the nodes before communication is reestablished; (**g**) The yellow robot moves through the faulted area; (**h**) No changes are executed to the size of the mapped fault; (**i**) Communication is reestablish with the yellow robot in a node that was previously affected by a communication fault; (**j**) Since communication was establish in a previously faulted node the nodes associated with that communication fault are placed as unmapped; (**k**) The orange robot reenters the fault and starts remapping it; (**l**) Status of the nodes before the orange robot enters the fault; (**m**) The orange robot moves through the fault remapping it; (**n**) Half of the fault size is remapped; (**o**) The orange robot exits the fault and moves towards its charging station; (**p**) The fault size is fully mapped; (**q**) Final position of the robots in the system; (**r**) Final status of all the nodes in the system.

Appendix E. Video Links

Link to the video of the execution of the first mission by the implemented system, in an environment where no communication faults occur: https://youtu.be/7MnuJ7nOKug (accessed at 23 March 2021).

Link to the video of the execution of the second mission by the implemented system, in an environment where no communication faults occur: https://youtu.be/oxt4x80ydPw (accessed at 23 March 2021).

Link to the video of the execution of the a mission by the implemented system, in an environment where only static communication faults occur: https://youtu.be/rlSbgYZXpAg (accessed at 23 March 2021).

Link to the video of the execution of the a mission by the implemented system, in an environment where both static and sporadic communication faults occur: https://youtu.be/dl5ZbOh_vUg (accessed at 23 March 2021).

References

1. Ullrich, G. The History of Automated Guided Vehicle Systems. *Autom. Guid. Veh. Syst. Prim. Pract. Appl.* **2015**. [CrossRef]
2. Siefke, L.; Sommer, V.; Wudka, B.; Thomas, C. Robotic systems of systems based on a decentralized service-oriented architecture. *Robotics* **2020**, *9*, 78. [CrossRef]
3. Santos, J.; Costa, P.; Rocha, L.F.; Moreira, A.P.; Veiga, G. Time enhanced A*: Towards the development of a new approach for Multi-Robot Coordination. In Proceedings of the IEEE International Conference on Industrial Technology (ICIT), Seville, Spain, 17–19 March 2015; pp. 3314–3319. [CrossRef]
4. Moura, P.; Costa, P.; Lima, J.; Costa, P. A temporal optimization applied to time enhanced A*. *AIP Conf. Proc.* **2019**, *2116*, 1–4. [CrossRef]
5. Downey, A.B. *The Little Book of Semaphores*, 2nd ed.; v2.2.1; Green Tea Press: Needham, MA, USA, 2016.
6. Atzmon, D.; Stern, R.; Felner, A.; Wagner, G.; Barták, R.; Zhou, N.F. Robust multi-agent path finding and executing. *J. Artif. Intell. Res.* **2020**, *67*, 549–579. [CrossRef]
7. Felner, A.; Stern, R.; Shimony, S.E.; Boyarski, E.; Goldenberg, M.; Sharon, G.; Sturtevant, N.; Wagner, G.; Surynek, P. Search-based optimal solvers for the multi-agent pathfinding problem: Summary and challenges. In Proceedings of the 10th Annual Symposium on Combinatorial Search, SoCS 2017, Pittsburgh, PA, USA, 16–17 June 2017; pp. 29–37.
8. Surynek, P. An optimization variant of multi-robot path planning is intractable. In Proceedings of the National Conference on Artificial Intelligence, Atlanta, GA, USA, 11–15 July 2010; Volume 2, pp. 1261–1263.
9. Yu, J. Intractability of optimal multirobot path planning on planar graphs. *IEEE Robot. Automat. Lett.* **2016**, *1*, 33–40. [CrossRef]
10. Falcó, A.; Hilario, L.; Montés, N.; Mora, M.C.; Nadal, E. A Path Planning Algorithm for a Dynamic Environment Based on Proper Generalized Decomposition. *Mathematics* **2020**, *8*, 2245. [CrossRef]
11. Bae, H.; Kim, G.; Kim, J.; Qian, D.; Lee, S. Multi-robot path planning method using reinforcement learning. *Appl. Sci.* **2019**, *9*, 3057. [CrossRef]
12. Yu, J.; Su, Y.; Liao, Y. The Path Planning of Mobile Robot by Neural Networks and Hierarchical Reinforcement Learning. *Front. Neurorobot.* **2020**, *14*, 1–12. [CrossRef] [PubMed]
13. Da Costa, P.L.C.G. Planeamento Cooperativo de Tarefas e Trajectórias em Múltiplos Robôs. Ph.D. Thesis, Faculdade de Engenharia da Universidade do Porto, Porto, Portugal, 2011; p. 231.
14. Latombe, J.C. Introduction and Overview. Robot Motion Planning. In *Robot Motion Plan*; Springer: Boston, MA, USA, 1991; pp. 1–57. [CrossRef]

15. Lozano-Pérez, T.; Wesley, M.A. An Algorithm for Planning Collision-Free Paths Among Polyhedral Obstacles. *Commun. ACM* **1979**, *22*, 560–570. [CrossRef]
16. LaValle, S.M. *Planning Algorithms*; Cambridge University Press: Cambridge, UK, 2006; 826p. [CrossRef]
17. Hart, P.E.; Nilsson, N.J.; Raphael, B. The Heuristic Determination. *IEEE Trans. Syst. Sci. Cybern.* **1968**, 100–107. [CrossRef]
18. Stentz, A. Optimal and Efficient Path Planning/or Partially Known Environments. In *Intelligent Unmanned Ground Vehicles*; Springer: Boston, MA, USA, 1997.
19. Koenig, S.; Likhachev, M. Incremental A*. In *Advances in Neural Information Processing Systems 14 (NIPS 2001)*; Dietterich T., Becker S., Ghahramani Z., Eds.; MIT Press: Cambridge, MA, USA, 2002.
20. Wang, W.; Goh, W.B. Multi-robot path planning with the spatio-temporal A* algorithm and its variants. In *Advanced Agent Technology, Proceedings of the AAMAS 2011, Taipei, Taiwan, 2–6 May 2011*; Dechesne F., Hattori H., Mors A., Such J.M., Weyns D., Eds.; Springer Publishing: New York, NY, USA, 2011; pp. 313–329. [CrossRef]
21. Wagner, G.; Choset, H. Subdimensional expansion for multirobot path planning. *Artif. Intell.* **2015**, *219*, 1–24. [CrossRef]
22. Jansen, R.; Tg, C.; Sturtevant, N. A new approach to cooperative pathfinding. In Proceedings of the 7th International Joint Conference on Autonomous Agents and Multiagent Systems, Taipei, Taiwan, 2–6 May 2008; Volume 3, pp. 1401–1404.
23. Cao, Y.U.; Fukunaga, A.S.; Kahng, A.B. Cooperative Mobile Robotics: Antecedents and Directions. *Auton. Robot.* **1997**, *4*, 7–27. [CrossRef]
24. Raynal, M. Fault-tolerant Agreement in Synchronous Message-passing Systems. In *Synthesis Lectures on Distributed Computing Theory*; Morgan & Claypool Publishers: Williston, VT, USA, 2010; Volume 1, pp. 1–189. [CrossRef]
25. Kandath, H.; Senthilnath, J.; Sundaram, S. Mutli-agent consensus under communication failure using Actor-Critic Reinforcement Learning. In Proceedings of the 2018 IEEE Symposium Series on Computational Intelligence, SSCI 2018, Bangalore, India, 18–21 November 2019; pp. 1461–1465. [CrossRef]
26. Braun, J.; Fernandes, L.A.; Moya, T.; Oliveira, V.; Brito, T.; Lima, J.; Costa, P. Robot@Factory Lite: An Educational Approach for the Competition with Simulated and Real Environment. In Proceedings of the Robot 2019: Fourth Iberian Robotics Conference, Porto, Portugal, 20–22 November 2019; pp. 478–489. [CrossRef]
27. Costa, P.; Gonçalves, J.; Lima, J.; Malheiros, P. Simtwo realistic simulator: A tool for the development and validation of robot software. *Theory Appl. Math. Comput. Sci.* **2011**, *1*, 17–33.
28. Silver, D. Cooperative pathfinding. In Proceedings of the Artificial Intelligence and Interactive Digital Entertainment Conference (AIIDE) 2005, Marina del Rey, CA, USA, 1–3 June 2005; pp. 117–122.

Article

A* Based Routing and Scheduling Modules for Multiple AGVs in an Industrial Scenario

Joana Santos [1], Paulo M. Rebelo [2,*], Luis F. Rocha [2], Pedro Costa [1,2] and Germano Veiga [1,2]

[1] Faculty of Engineering, University of Porto (FEUP), 4200-465 Porto, Portugal; jooanaraquel@gmail.com (J.S.); pedrogc@fe.up.pt (P.C.); germanoveiga@fe.up.pt (G.V.)
[2] Institute for Systems and Computer Engineering, Technology and Science (INESC TEC), 4200-465 Porto, Portugal; luis.f.rocha@inesctec.pt
* Correspondence: paulo.m.rebelo@inesctec.pt

Abstract: A multi-AGV based logistic system is typically associated with two fundamental problems, critical for its overall performance: the AGV's route planning for collision and deadlock avoidance; and the task scheduling to determine which vehicle should transport which load. Several heuristic functions can be used according to the application. This paper proposes a time-based algorithm to dynamically control a fleet of Autonomous Guided Vehicles (AGVs) in an automatic warehouse scenario. Our approach includes a routing algorithm based on the A* heuristic search (TEA*—Time Enhanced A*) to generate free-collisions paths and a scheduling module to improve the results of the routing algorithm. These modules work cooperatively to provide an efficient task execution time considering as basis the routing algorithm information. Simulation experiments are presented using a typical industrial layout for 10 and 20 AGVs. Moreover, a comparison with an alternative approach from the state-of-the-art is also presented.

Keywords: multi-robot coordination; automated guided vehicles; routing; scheduling; motion planning; simulation; robotics

Citation: Santos, J.; Rebelo, P.M.; Rocha, L.F. ; Costa, P.; Veiga, G. A* Based Routing and Scheduling Modules for Multiple AGVs in an Industrial Scenario. *Robotics* **2021**, *10*, 72. https://doi.org/10.3390/robotics10020072

Academic Editor: Xinjun Liu

Received: 12 April 2021
Accepted: 13 May 2021
Published: 19 May 2021

Publisher's Note: MDPI stays neutral with regard to jurisdictional claims in published maps and institutional affiliations.

1. Introduction

In recent years, and due to the mandatory need to continuously adapt the production flow, industrial companies are increasingly adopting fully automated internal logistic systems, namely based on AGVs, instead of manual or inflexible mechanical solutions (e.g., forklifts, conveyors, and others).

Considering the Industry 4.0 initiative, both AGVs, as well as mobile manipulators, are seen as strategic tools in the Factories of the Future. In a very competitive industrial environment, these can contribute to increase productivity and reduce the costs associated with the internal logistic system, ensuring an efficient material flow. Likewise, their introduction also allows human operators to be reallocated to more complex and ergonomic tasks, with increasing value to the final product. These set of characteristic makes AGV's appealing for a wide range of industrial applications, such as goods transportation, end-of-line automation chain, warehouse and distribution. However, and despite their versatility, there is the need to deploy advanced multi-robot coordination algorithms in order to ensure the AGV's continuous operation, guaranteeing the minimum tasks execution time and the smoothness of the vehicle's movements.

The AGV's fleet coordination problem has received wide attention from both the research and industrial fields. Typically, two main systems comprise any multi-AGV application: free-collision routing system [1,2] and scheduling system, that encompass both task scheduling and dispatching [3]. The vehicle routing system is responsible for computing trajectories that minimize the total distance traveled by AGVs considering different constraints such as each vehicle's carrying capacity and the plant layout where vehicles can circulate, while ensuring free-collision routes. Some approaches are based on time windows as proposed by the authors in [4–8]. Here, the AGV route is constructed considering that one point can only be visited one time for only one vehicle at a given time interval. The feasibility of each route is evaluated, checking windows overlapping. In its turn, the scheduling system is associated with the task scheduling and their attribution to individual AGVs. This scheduling and dispatching should take into account some decision criteria, namely the task deadlines and traffic status, among others.

Bearing these ideas in mind, this paper proposes an integrated approach for both AGV route planning and task scheduling and dispatching. More in detail, the proposed routing algorithm called Time Enhanced A* (TEA*) is an extension of our previous work addressed in [9,10], that is, in this paper, further integrated with a scheduling module in order to minimize the tasks execution time. The main feature of TEA* is the addition of a temporal component to the known A* algorithm that generates routes efficiently, considering that each robot knows other robots' positions during the time. Furthermore, TEA* is an on-line approach allowing its integration in dynamic environments.

The major contributions of our paper are to propose a Multi-Robot Coordination System which includes a time-based algorithm to generate free-collision routes based on the A* algorithm and a scheduling module that use the routing algorithm information to minimize a cost function, dependent on the following parameters:

1. Average execution time of all tasks;
2. Number of stoppages;
3. Execution time for the last vehicle;

This paper is organized as follows. In Section 2 the state-of-the-art of path planning algorithms and multi-robot systems are presented. Section 3 describes the proposed multi-robot routing algorithm. Section 4 describes the industrial case scenario, followed by the comparison between the proposed approach with an alternative state-of-the-art [11] Sections 5 and 6 present the results achieved using Tabu-Search Method. Finally, some conclusions and the contribution of this paper are presented in Section 7.

2. Related Work

In the last decade, the multi-robot coordination problem has been a target of many scientific studies. To solve it, several authors have proposed the use of meta-heuristic approaches to address both the problem of AGV routing and task scheduling. The authors in [12–16], propose using a Genetic Algorithm to find the optimal or sub-optimal solution, which satisfies the routing system goals, including the minimization of the tasks completion time, minimal distance, among others. Likewise, ref. [17] proposes an integrated solution that comprises a routing module based on genetic algorithms and a scheduling system based on bid auctions. Despite the potential of the proposed methods, normally these approaches treat each robot as an individual agent (without physical constraints) [17,18], simplifying the problem at hand. Furthermore, the results do not include industrial case scenarios. Alternatively, particle swarm strategies can also be applied to multi-AGV systems as in [19,20]. Ref. [21] proposes a Particle Swarm Optimization (PSO) based algorithm, called Fractional Order Robotic Darwinian Particle Swarm Optimization (FORDPSO), integrated with a fuzzy system to optimize the driving of multi-robots in unknown environments. However, these methodologies are not yet sufficiently tested in real industrial environments.

Considering solely the use of a bid auction strategy, the authors in [3,22] propose a solution where each robot constructs 'bids' for each task and a central module receives the bids and assigns tasks considering the fitness function's maximization. Similarly ref. [23] proposed different combinatorial bidding strategies, comparing its performance with single-item auctions.

Recognize the use of analytical methods, the authors in [24] define a robot mission through Linear Temporal Logic formulas (LTL). An LTL approach considers that the truth of a declaration can be changed during the time. This work focus on minimizing the cost function, which is the maximum time between candidate solutions of an optimizing proposition. In [25] the authors use a model based on Integer Linear Programming (ILP) to find paths that minimize the time until the last robot reaches its goal or minimizes the total traveled distance.

Generally, two different approaches define the architecture of any multi-robot system: a Centralized [26] or Distributed [27] methodology. The authors in [28,29] use a centralized architecture where one of the robots is the leader, and the others are the followers. Here, the major challenges are related to ensuring communication robustness and the algorithm flexibility to change the leadership. Other works use distributed architecture like in [27,30], where each robot calculates its path independently using, for example, a D* Algorithm [31] and then, the path is broadcast for all robots, making that every robot knows all path information.

Regarding only task scheduling system, it typically aims for minimizing an objective function which includes system characteristics, such as the number of vehicles, the order in which the vehicles execute their missions, etc. In [32], Kelen et al. compares two scheduling methods that determine the ideal number of vehicles for a given industrial scenario: the Shortest Job First and Tabu Search. Additionally, a routing method based on an enhanced Dijkstra algorithm, was used to manage the AGV's path. The number of stoppages and the time that each vehicle waits for the mission assignment was not measured.

In [20], Yu Zhang et al. proposes simple heuristics at the high-level layer, referred to as the 'coalition' level, that creates an abstraction layer relatively to specific details in robots' specifications. Simple heuristics can be 'MinProcTime', that gives priority to the missions with shorter processing times, and another one can be the 'MinStepSum', similar to the 'MinProcTime', but determines the best solution incrementally when the ordering of the assignment are not pre-determined. However, in the simulation experiments presented were not considered an industrial scenario with real-world scheduling problems.

In the past half-decade, new approaches based on Artificial Intelligence (AI) are emerging. AI addressed in [33,34], is the science that seeks to study and understand the phenomenon of intelligence and, at the same time, a branch of engineering, as it seeks to build instruments to support human intelligence. In practice, an AI system besides storing and manipulate data can also acquire, represent, and manipulate knowledge. This manipulation concerns the ability to deduce or infer new knowledge from existing knowledge and use representation and manipulation methods to solve complex problems. This area of engineering is vast and has been the subject of huge investment from both business and research institutes. More specifically in Robotics, there is already a recent line of work on Multi-Agent Path-Finding (MAPF) [35–38]. On AGVs, the MAPF problem is to find the best paths, for a fixed number of agents, from their current locations to the final task position, where all agents have a free-collision path, as it is described in [39–41]. Another use of AI in Robotics, is the combination of non-AI algorithms (time-window based or greedy) for the AGV coordination with AI for the prediction of future tasks [42].

Despite the many scientific studies carried out, task scheduling solutions often resort to small heuristics (First-in-First-Out, Shortest-Distance), often decoupled from the trajectory planning system. In turn, trajectories are predefined in offline mode and designed to prevent, as far as possible, the occurrence of deadlocks. This type of solution often leads to the logistics system being oversized concerning the operation's real needs.

3. Proposed Routing Algorithm

To overcome the challenges related with the multi-AGV coordination problem, presented in the previous sections, in this section a new methodology for AGV's route planning is proposed. This novel methodology is called TEA* Algorithm [9,10], in which the paths are recalculated continuously, making it an online method. According to the vehicles' movements and the environment changes, TEA* updates the paths of each AGV in order to avoid collisions and to guarantee the continuous operation of the logistic system. In fact, as in the majority of the industrial logistic systems, it is expectable that changes on initial task information and/or unpredictable events, such as delays in the transportation system as a result of obstacles presence [21], can occur. In these scenarios, the adoption of online path planning algorithms, capable of dealing with such events, becomes mandatory.

As referred before, TEA* is based on the traditional A*, where a third dimension was added, the time. The input map has three dimensions: vertex's coordinates (x and y) and a representation of the time, as shown in Figure 1. The time is represented with temporal layers given by $k \in [0, k_{Max}]$ (k_{max} denotes the maximum number of layers). Each temporal graph is a set of free and occupied/obstacles vertexes.

Figure 1. Representation of the input map focusing the vertexes with the same position of the AGV (over the time) denoted with different colors.

As far as computational complexity is concerned, many factors can influence the system (RAM, CPU, others), however, through [43] where the test conditions are the same for all algorithms, it is possible to validate that A* is the most suitable algorithm for robot fleet management in different environments. Therefore, taking into account the case study carried out in [43], and the results obtained in [9], it can be concluded that TEA* is a practical, versatile and quite optimized algorithm in terms of path planning algorithms.

3.1. The TEA* Method

In multi-AGV systems, time is a crucial component for a better prediction of the vehicles' positions. Besides the constantly recalculated paths, TEA* determines the route for each AGV during the temporal layers. This grants the identification of upcoming collisions, allowing them to be avoided with considerable anticipation. The path information for each AGV is converted to a busy vertex on the following robot's map, allowing collision avoidance.

Consider a graph G with a set of vertexes $V = \{0, 1, ..., NUM\_VERTEXES\}$ and edges $E = \{0, 1, ...NUM\_EDGES\}$ (links between the vertexes), with a representation of the time $[\,0, 1, ..., k_{max}\,]$ (as can be seen in Figure 1). Each AGV can only starts and stops in vertexes and each vertex can only be occupied by only one vehicle at the time.

During the path search for a single AGV the neighboring vertexes are evaluated using a similar approach as A* algorithm [44]. Moreover, each edge in each temporal layer, has a cost function value, denoted as $f(j, k)$, given by the sum of two terms (see Equation (1)).

$$f(j,k) = \alpha g(j,k) + \beta h(j,k),$$

$$k \in [0, k_{max}], j \in [0, NUM\_VERTEXES] \quad (1)$$

Considering the path between j_0 and j_f, the first term $\alpha g(j,k)$, represents the distance between the current vertex j to the initial vertex j_0, in the k temporal layer. The second term, denoted as $\beta h(j,k)$, is a heuristic value that calculates the distance to the final vertex j_f. The terms α and β assign different weights to the distance and the heuristic function.

Each vertex, in each temporal layer, has different values of $g(j,k)$ and $h(j,k)$, according with the Equation (2). Here, $g(j,k)$ is given by the sum of the distance between the current vertex j and the initial vertex j_0, being the edge distance between j and its adjacent vertex $j+1$ denoted as $dis(j, j+1, k)$.

$$g(j,k) = dis(j,j_0,k) + dis(j,j+1,k)$$
$$h(j,k) = dis(j,j_f,k)$$

(2)

The main differences between TEA* and the known A* algorithm are mainly concerned with the addition of the time component and can be defined as follows:

Definition 1. *The neighbor vertexes belong to the next temporal layer. The neighbor vertexes of a vertex j (v^j_{adj}) are given by the set of all adjacent vertexes in the next time component ($k+1$). The number of temporal layers depends on the required iterations to achieve the final point of the mission and the map dimensions. Note that the larger the map, more time layers are required.*

Definition 2. *The neighbor vertexes include the vertex containing the AGV's current position. The set of neighbor vertexes includes not only the adjacent vertexes but also the vertex corresponding to the position in analysis. This property allows a vehicle to maintain its position between consecutive time instants if any neighbor vertex is free. In this case, $dis(j,j,k+1)$ assumes a constant value that corresponds to the cost of keeping its position.*

The Algorithm 1 describes the TEA* approach for a single AGV with the following parameters:

- val^k_j: Value of vertex j in the time layer k (Free—0 or Occupied/Obstacle—1).
- $pos^k_{l,j}$: AGV l occupies the vertex j in the k time layer.
- $O = \{o^k_j, ..\}$: Open list contains the vertex j in the k time instant. Each item contains the respective cost value, $o^k_j.cost$.
- j_0: initial vertex.
- j_f: final vertex.
- v^j_{adj}: adjacent vertex of vertex j.
- $p_{j,k} = (i,\tau)$: The vertex i in the time layer τ is the parent vertex of vertex j in the instant k.
- h^k_j: Heuristic Value for vertex j in k temporal layer.
- g^k_j: Distance Value for vertex j in k temporal layer.
- $dis(j_1,j_2)$: Distance value of the edge (j_1,j_2).

3.2. Smoothing Trajectories

For TEA* to be applied, the shop floor layout requires to be modeled as a set of vertexes and edges (links between the vertexes). Each AGV travels on these graph paths, from one node to another, through a pre-defined set of edges. Each edge is represented as a cubic Bézier curve (as proposed in article [45]), given by Equation (3).

$$x(\lambda) = a_x\lambda^3 + b_x\lambda^2 + c_x\lambda + x_0$$
$$y(\lambda) = a_y\lambda^3 + b_y\lambda^2 + c_y\lambda + y_0$$

(3)

Here, λ denotes an integer value between 0 and 1 according with the AGV's position in the curve, (x_0, y_0) is the initial point of the curve, and a_x, b_x, c_x, defines the spline's curvature. Figure 2 represents a portion of the map which contains Bézier curves and straight lines.

Algorithm 1: TEA* Algorithm

---

1   $O \leftarrow o_{vi}^{0}$;

2   **while** $OpenList.size() \neq 0$ **do**

3     $j = min_O\{o_j^k.cost\}$;

4   **if** $j == j_f$ **then**

5   **return**

6   **for** v_{adj}^j *adjacent vertexes of j* **do**

7   **if** $val_{v_{adj}^j}^{k+1} == 0$ **then**

8   Only the non-visited vertexes have heuristic zero;

9   **if** $h_{v_{adj}^j}^{k+1} == 0$ **then**

10           $CalculateHeuristic(h_{v_{adj}^j}^{k+1})$;

11           $p_{j,k} = (v_{adj}^j, k+1)$;

12           $CalculateCost(o_{v_{adj}^j}^{k+1})$;

13           $O \leftarrow o_{v_{adj}^j}^{k+1}$;

14   **else if** $g_{v_{adj}^j}^{k+1} > g_{v_{adj}^j}^{k} + dis(j, v_{adj}^j)$ **then**

15           $UpdateCost(o_{v_{adj}^j}^{k+1})$;

16           $O \leftarrow o_{v_{adj}^j}^{k+1}$

17   **return** 0;

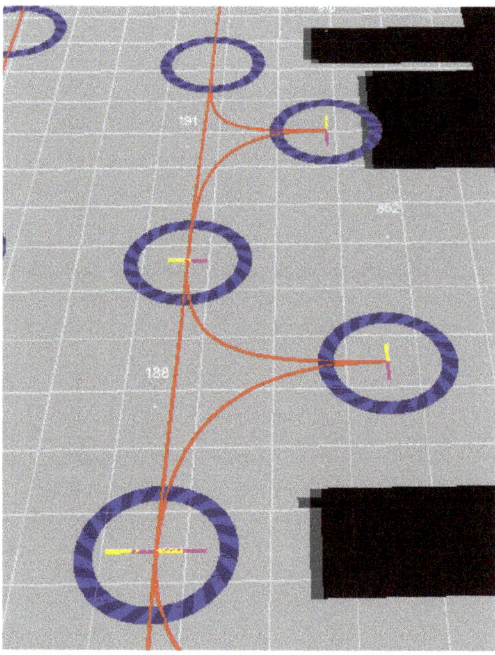

Figure 2. Cubic Bézier Curves Example.

4. Industrial Layout and Comparison Scenario Description

In this section, the TEA*, based on Robot Operating System (ROS), is compared with a state of the art alternative, namely the coordination algorithm presented in [46]. This algorithm relies on coordination diagrams for planning the coordinated motion of a fleet of AGVs. One of the contributions of [46] is the definition of a heuristic function that estimates the number of times a vehicle starts and stops during its path execution.

For the comparison of the two algorithms, a set of experiments were conducted using the same layout, the same number of vehicles and the same missions' list used by the authors of [11]. Each mission is defined by four tasks (Sn- Starting positions; Pn- Pick-up Positions; Dn- Drop-off Stations; Rn- Rest Positions). The layout dimensions are 80×110 m and 10 AGVs were used to generate the results of [11].

Figure 3 represents the input map of TEA* Algorithm. The graph was built using a graph editor, which was created with the Robot Operating System Visualization (RVIZ) platform and the Interactive Markers tool.

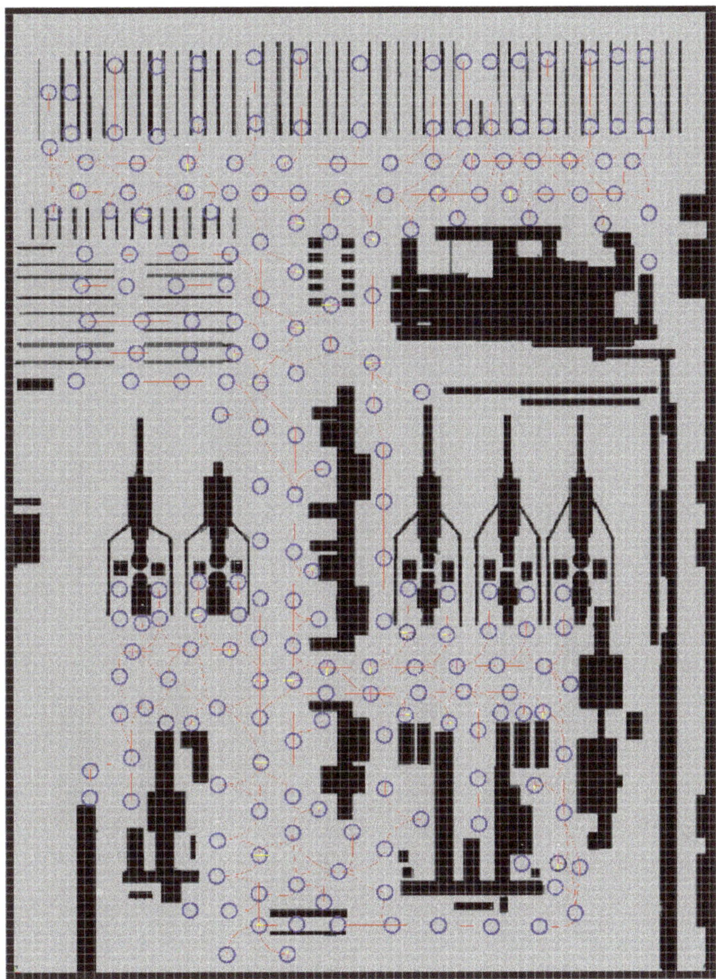

Figure 3. Layout Snapshot built in RVIZ.

5. Routing Algorithms Comparison

Considering the industrial scenario previously presented, Table 1 illustrates the mission execution time for each vehicle with the TEA* Algorithm, using 10 AGVs and 30 tasks. For each AGV, the time of advancement (T_{adv}) and the stopping time (T_{stop}) are reported. In Figure 4 is illustrated the final solution found for each vehicle. Two black circles represent possible collision points in different robot paths, but the third dimension of TEA* allows AGVs to share the same trajectory by passing at different times at the common points, i.e., avoiding possible collisions.

Comparing the results achieved with the TEA* (Table 1), with the results presented by the authors in [11] (Table 2), it is possible to conclude that TEA* is advantageous mainly considering the T_{adv} of the last vehicles (AGVs 6, 7, 8, 9, 10). Conversely, the T_{adv} of the AGVs 3, 4 and 5 surpasses the value of the same robots in Table 2. However, in these cases the $T_{average}$ and the T_{max} are lower in TEA*. Therefore, this algorithm presents itself as a better solution for multi-robot path planning.

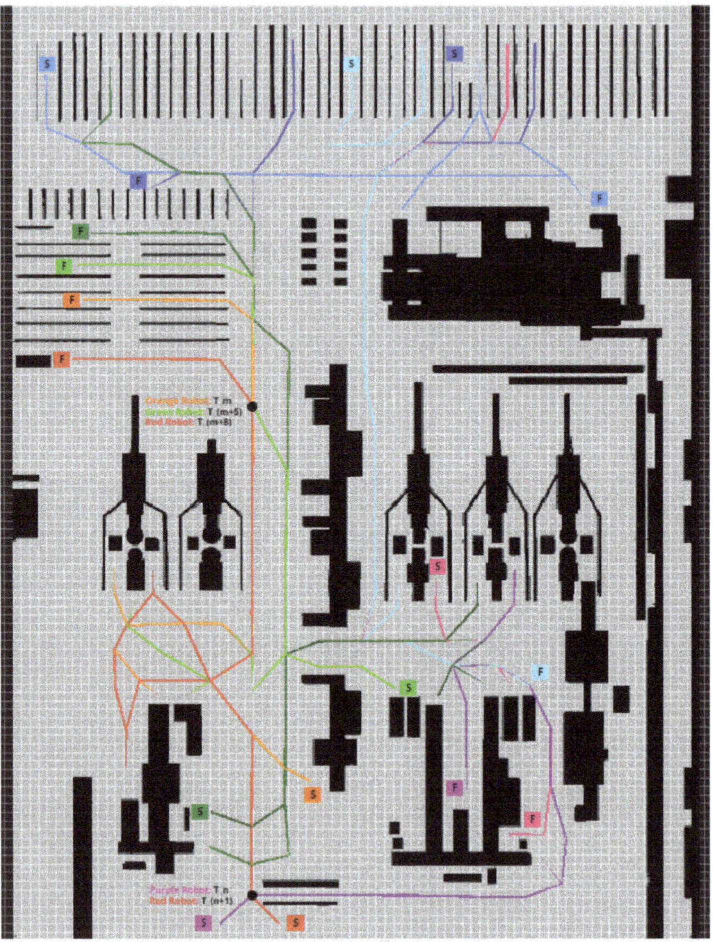

Figure 4. Routing Algorithm Results—10 AGVs.

The average time ($T_{average}$) to complete the missions list with TEA* is 116.7 s and in the case of [11] this time is approximately 121.78 s. The last vehicle in [11] takes 144.2 s to finish its tasks, while in TEA*, the last vehicle performs its tasks in 135.8 s.

To highlight the capability of TEA* Algorithm to optimize the routes even with considerable workload in the system, Table 3 presents the same industrial scenario but now with 20 AGVs and 60 tasks. Note that the difference in the average time between 10 and 20 AGVs are approximately 11 s. The fact that a vehicle considers the current and future positions of each vehicle as obstacles during the discrete-time, gives it the possibility to wait for some instants for the traveling of previous AGV, instead of calculating a longer deviation.

Table 1. TEA* Results—10 AGVs.

Vehicle	1	2	3	4	5
$T_{adv}(s)$	123.5	110.3	135.8	134.2	91.6
$T_{stop}(s)$	0	0	6	0	3
Vehicle	6	7	8	9	10
$T_{adv}(s)$	129.0	103.4	105.4	108.6	125.6
$T_{stop}(s)$	0	0	6	0	0
$T_{average}(s)$	116.7				
$T_{max}(s)$	135.8				
$T_{stop}(s)$	15				

Table 2. TRAFCON Results—10 AGVs, adapted from [11].

Vehicle	1	2	3	4	5
$T_{adv}(s)$	144.2	118.6	115.2	127	73.4
$T_{stop}(s)$	26.8	0	1.6	0	17
Vehicle	6	7	8	9	10
$T_{adv}(s)$	151	123	107	121.6	136.8
$T_{stop}(s)$	0	0	1.4	0	0
$T_{average}(s)$	121.8				
$T_{max}(s)$	151				
$T_{stop}(s)$	46.8				

For the sake of completeness, it is important to refer that the TEA* Algorithm has been designed to be integrated into dynamic industrial environments, thus allowing direct scaling concerning the number of robots to be used. The structure of the algorithm itself is already prepared for different exchanges of industrial scenarios.

Table 3. TEA* Results—20 AGVs.

Vehicle	1	2	3	4	5
$T_{adv}(s)$	123.5	110.3	135.8	134.2	91.6
$T_{stop}(s)$	0	0	6	0	3
Vehicle	**6**	**7**	**8**	**9**	**10**
$T_{adv}(s)$	129.0	103.4	105.4	108.6	125.6
$T_{stop}(s)$	0	0	6	0	0
Vehicle	**11**	**12**	**13**	**14**	**15**
$T_{adv}(s)$	144.3	166.6	107.7	129.6	121.5
$T_{stop}(s)$	9	3	0	12	0
Vehicle	**16**	**17**	**18**	**19**	**20**
$T_{adv}(s)$	161.4	139.5	179.3	81.5	154.7
$T_{stop}(s)$	3	0	3	3	3
$T_{average}(s)$	127.7				
$T_{max}(s)$	179.3				
$T_{stop}(s)$	51				

6. Task Scheduling Algorithm

As referred earlier, the problem of AGV coordination is not only closely related with the route planning of the AGVs, but also with the task scheduling. The performance of the routing algorithm can be improved. The AGV execution order affects the calculation of routes since the path positions over time for a given vehicle are obstacles for the following AGVs. If the execution order changes, the paths and respective task execution times for each vehicle are different.

6.1. Tabu Search Method

The Tabu Search Method is a 'meta-heuristic' adaptive method of local search in continuous exploration within a search space, moving from one solution to another, the Tabu Moves, diversifying the solutions found in this process of the search for an improved solution [47]. The best permissible movement is the one with the highest evaluation in the vicinity of the current solution regarding target function value and taboo restrictions. Thus, the 'meta-heuristic' Tabu Search is an iterative search algorithm characterized by dynamic memory and consisting of two parts: initialization and search.

Starting from an initial randomly generated solution or using a heuristic, the Tabu Search will evaluate a set of different mutations (neighborhood exploration) of the current solution in each iteration. The best mutation will be accepted, and the changes made saved in a Tabu List adopted to store the most used changes, which are classified as prohibited in later iterations. This strategy is necessary to avoid a return to solutions already checked previously.

Therefore, in this method, in each iteration, the evaluation function consists of validating a certain quantity of new solutions, where the best solution, based on the objective function, is accepted, even if its cost is higher than the cost of the current solution. Thus, the algorithm chooses the new solution that produces an improvement or the least deterioration in the cost function (an attempt to evade minimal locations). The Tabu Search algorithm runs until a stop criteria is reached.

Figure 5 presents the main blocks of the Tabu Search method.

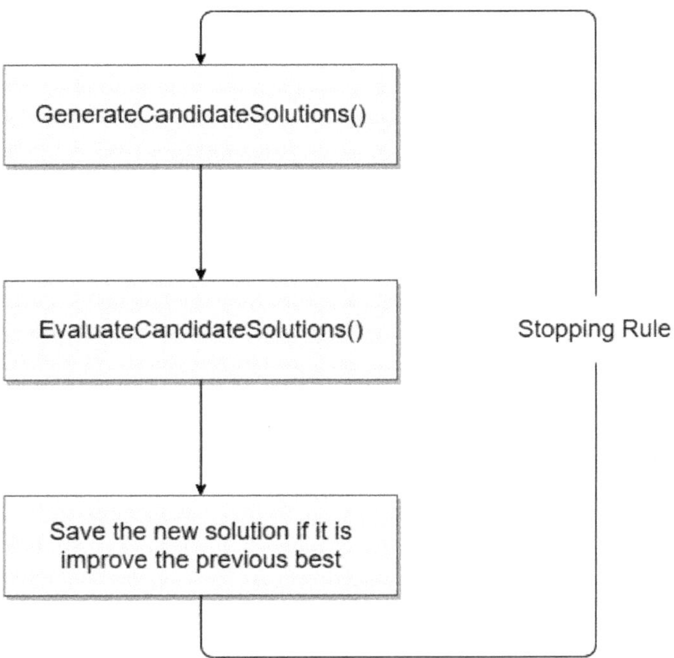

Figure 5. Tabu Search Diagram.

In the AGV scheduling problem presented, the final objective is the allocation of the n sub-tasks by the j available AGVs, in order to minimize the total time for the overall task completion. The main goal is to determine/distribute the best sequence of attendance of the sub-tasks by the AGVs to minimize the total time of execution.

For the problem presented, the initial solution (sequence of sub-tasks assigned to each AGV) is generated using the closest neighboring heuristic, as presented in Algorithm 2. In other words, each AGV (and taking into account the last sub-task performed, which influences its position on the map) is assigned to the next sub-task with lower cost (shorter travel time/distance).

This process is executed cyclically until all sub-tasks have been assigned to one, and only one, AGV.

Algorithm 2: CLOSEST NEIGHBOUR ALGORITHM—PSEUDO CODE

---

1 **while** $AllSubTasksUnallocated$ **do**
2   **for** $j = 1$ **to** $NumAGVs$ **do**
3     $NewAGV(j)SubTask \leftarrow NeighbourNext <$
    $SubTasksUnallocated, PreviousSubTaskAGV(j) >;$
4     $ListSubTasksAGV(j) \leftarrow< NewSubTaskAGV(j) >;$
5     $PreviousAGV(j)SubTask \leftarrow NewAGV(j)SubTask;$
6     $AllUnassignedSubTasks \leftarrow delete < NewSubTaskAGV(j) >;$

The Algorithm 3, describes the implementation of the Tabu Search used in the proposed approach.

6.2. TEA Algorithm with Tabu Search Method—Results*

The Tabu Search Method was implemented to find better vehicle configuration and to schedule the order in which vehicles execute their tasks. A configuration comprises the order in which vehicles should be processed by the TEA* Algorithm.

Algorithm 3: 'META-HEURISTIC' TABU SEARCH—PSEUDO CODE

```
1   s ← s0;
2   BestSolution ← s;
3   for k = 1 to TabuSearchMaxIteration do
4   |   CandidateList ← null;
5   |   for sCandidate in sNeighborhood do
6   |   |   if not containsTabuElements < Candidate, tabuList > then
7   |   |   |   candidateList ← candidateList + sCandidate;
8   |   sCandidate ← LocateBestCandidate < candidateList >;
9   |   tabuList ← addFeatureDifferences < sCandidate, sBest >;
10  |   s ← sCandidate;
11  |   if fitness < s > < fitness < sBest > then
12  |   |   sBest ← s;
13  |   UpdateTabuList < tabuList >;
14  return sBest
```

The AGV execution order affects the calculation of routes since the path positions over time for a given vehicle are obstacles for the following AGVs. If the execution order changes, each vehicle's paths and respective task execution times will be different.

The optimization goal is the minimization of the three following parameters:

1. Time of the last vehicle, denoted as T_{max} in seconds;
2. Average Time of the missions execution, denoted as $T_{average}$ in seconds;
3. Number of Stoppages, denoted as n_{stop};

That were aggregated in the following cost function (Equation (4)):

$$c = \gamma \times T_{last} + \psi \times T_{average} + \tau \times n_{stop} \tag{4}$$

Here, γ, ψ and τ are components that are weighting parameters. In the simulation experiments the following values were used, respectively 0.1, 0.7, 0.2. These were manually defined considering an iterative and experimental way, with the main goal of minimizing the cost function c.

The configuration that leads to the lowest cost function is chosen as the better solution. In our approach is not required to achieve the optimal solution, a near-optimal configuration that leads to an efficient TEA* execution is enough.

To obtain the initial configuration of the Tabu Search Method, a heuristic function was defined. It consists of the path computation for each vehicle without consider the other vehicles as obstacles (optimal solution) and ordering it by decreasing the order of execution task times. The objective is to process firstly the longer paths minimizing the number of stoppages. The candidate solutions are generated, changing two by two the vehicle execution order from the current solution.

Table 4 presents the results for the TEA* Algorithm using the AGV configuration solution found by the Tabu Search method. The average time for completing all tasks is lower, but the more significant improvement was the waiting time. In Table 1 using as initial configuration = $\{1, 2, 3, 4, 5, 6, 7, 8, 9, 10\}$ AGVs were stopped 15 s. Using the Tabu Search solution = $\{6, 2, 4, 3, 1, 10, 9, 7, 8, 5\}$, total waiting time was 6 s.

Considering 20 AGVs, besides the stoppage time to be higher, the task completion time is lower than TEA* results without scheduling module. In Table 3 the total time to complete all tasks was 179.3 s and in Table 5 this time was 159.1 s.

Table 4. TEA* Results with the Tabu Search Configuration—10 AGVs.

Vehicle	6	2	4	3	1
$T_{adv}(s)$	129.0	110.3	133.7	122.6	125.4
$T_{stop}(s)$	0	0	0	0	0
Vehicle	10	9	7	8	5
$T_{adv}(s)$	124.2	108.6	103.4	105.4	91.3
$T_{stop}(s)$	0	0	0	3	3
$T_{average}$	115.4				
$T_{max}(s)$	133.7				
$T_{stop}(s)$	6				

Table 5. TEA* Results with the Tabu Search Configuration—20 AGVs.

Vehicle	18	16	4	12	6
$T_{adv}(s)$	159.1	152.5	133.9	157.3	131.2
$T_{stop}(s)$	0	0	0	6	0
Vehicle	11	10	1	15	17
$T_{adv}(s)$	133.7	124.2	123.5	125.8	136.1
$T_{stop}(s)$	6	0	0	9	0
Vehicle	3	14	2	9	13
$T_{adv}(s)$	148.5	131.6	136.0	119.1	138.4
$T_{stop}(s)$	0	12	3	0	3
Vehicle	20	7	8	5	19
$T_{adv}(s)$	145.1	121.4	106.7	105.0	78.5
$T_{stop}(s)$	3	0	3	18	0
$T_{average}$	130.4				
$T_{max}(s)$	159.1				
$T_{stop}(s)$	63				

7. Conclusions

This article proposes a multi-AGV system that comprises a routing algorithm based on the search method A*. This algorithm is suitable for multi-robot applications, avoiding collisions and deadlocks and guaranteeing any industrial scenario required safety levels. To optimize the results achieved scheduling method (Tabu Search) minimizes a fitness function defined by several parameters calculated by TEA*. The two modules work cooperatively, sharing the TEA* information.

Our work's major contributions are: (i) Presentation of a promising approach for multi-AGV applications in warehouse environment, improving the flexibility and efficiency of the complete system; (ii) Validation of an on-line Multi-Robot Coordination Algorithm comparing it with a state-of-the-art alternative.

As future work, it will be interesting to validate the TEA* Algorithm in a real environment, with a real robotic system, by comparing it to the simulation results.

Author Contributions: The contributions of the authors of this work are pointed as follows: Conceptualization: J.S., L.F.R., P.C. and G.V.; Methodology: J.S., L.F.R., P.C. and G.V.; Software J.S., P.M.R. and P.C.; Validation: J.S., P.M.R. and P.C.; Writing—Review and Editing: J.S., P.M.R., L.F.R. and P.C.; Supervision: L.F.R., P.C. and G.V. All authors have read and agreed to the published version of the manuscript.

Funding: The research leading to these results has received funding from the European Union's Horizon 2020—The EU Framework Programme for Research and Innovation 2014–2020, under grant agreement No.101006798—Project `Mari4_YARD`.

Institutional Review Board Statement: Not applicable.

Informed Consent Statement: Not applicable.

Data Availability Statement: This study didn't report any new data.

Acknowledgments: The authors of this work would like to thank the members of INESC TEC for all the support rendered to this project.

Conflicts of Interest: The authors declare no conflict of interest.

Abbreviations

The following abbreviations are used in this manuscript:

AGV	Autonomous Guided Vehicle
AI	Artificial Intelligence
FORDPSO	Fractional Order Robotic Darwinian Particle Swarm Optimization
ILP	Integer Linear Programming
LTL	Linear Temporal Logic
MAPF	Multi-Agent Path-Finding
MDPI	Multidisciplinary Digital Publishing Institute
PSO	Particle Swarm Optimization
ROS	Robot Operating System
RVIZ	Robot Operating System Visualization
TEA*	Time Enhanced A*

References

1. Kalinovcic, L.; Petrovic, T.; Bogdan, S.; Bobanac, V. Modified Banker's algorithm for scheduling in multi-AGV systems. In Proceedings of the IEEE International Conference on Automation Science and Engineering, Hong Kong, China, 20–21 August 2011. [CrossRef]
2. Cirillo, M.; Pecora, F.; Andreasson, H.; Uras, T.; Koenig, S. Integrated motion planning and coordination for industrial vehicles. In Proceedings of the International Conference on Automated Planning and Scheduling (ICAPS), Portsmouth, NH, USA, 21–26 June 2014.
3. Erol, R.; Sahin, C.; Baykasoglu, A.; Kaplanoglu, V. A multi-agent based approach to dynamic scheduling of machines and automated guided vehicles in manufacturing systems. *Appl. Soft Comput. J.* **2012**, *12*, 1720–1732. [CrossRef]
4. Bräysy, O.; Gendreau, M. Vehicle routing problem with time windows, Part I: Route construction and local search algorithms. *Transp. Sci.* **2005**, *39*, 104–118. [CrossRef]
5. Möhring, R.H.; Köhler, E.; Gawrilow, E.; Stenzel, B. *Conflict-Free Real-Time AGV Routing*; Springer: Berlin/Heidelberg, Germany, 2005. [CrossRef]
6. Smolic-Rocak, N.; Bogdan, S.; Kovacic, Z.; Petrovic, T. Time windows based dynamic routing in multi-AGV systems. *IEEE Trans. Autom. Sci. Eng.* **2010**, *7*, 151–155. [CrossRef]
7. Ballal, P.; Giordano, V.; Lewis, F. Deadlock free dynamic resource assignment in multi-robot systems with multiple missions: A matrix-based approach. In Proceedings of the 14th Mediterranean Conference on Control and Automation (MED'06), Ancona, Italy, 28–30 June 2006. [CrossRef]
8. Guan, X.; Li, Y.; Xu, J.; Wang, C.; Wang, S. A literature review of deadlock prevention policy based on petri nets for automated manufacturing systems. *Int. J. Digit. Content Technol. Its Appl.* **2012**, *6*, 426–433. [CrossRef]
9. Santos, J.; Costa, P.; Rocha, L.F.; Moreira, A.P.; Veiga, G. Time enhanced A: Towards the development of a new approach for Multi-Robot Coordination. In Proceedings of the IEEE International Conference on Industrial Technology, Seville, Spain, 17–19 March 2015; pp. 3314–3319. [CrossRef]
10. Santos, J.; Costa, P.; Rocha, L.; Vivaldini, K.; Moreira, A.P.; Veiga, G. Validation of a time based routing algorithm using a realistic automatic warehouse scenario. In Proceedings of the Advances in Intelligent Systems and Computing, Lviv, Ukraine, 6–10 September 2016. [CrossRef]
11. Olmi, R.; Secchi, C.; Fantuzzi, C. Coordinating the motion of multiple AGVs in automatic warehouses. In Proceedings of the Workshop on Robotics and Intelligent Transportation Systems, Anchorage, Alaska, 8 May 2010.
12. Liu, C.; Kroll, A. A centralized multi-robot task allocation for industrial plant inspection by using A* and genetic algorithms. In *Lecture Notes in Computer Science (Including Subseries Lecture Notes in Artificial Intelligence and Lecture Notes in Bioinformatics)*; Springer: Berlin/Heidelberg, Germany, 2012. [CrossRef]

3. Tuncer, A.; Yildirim, M. Dynamic path planning of mobile robots with improved genetic algorithm. *Comput. Electr. Eng.* **2012**, *38*, 1564–1572. [CrossRef]

4. Kapanoglu, M.; Alikalfa, M.; Ozkan, M.; Yazıcı, A.; Parlaktuna, O. A pattern-based genetic algorithm for multi-robot coverage path planning minimizing completion time. *J. Intell. Manuf.* **2012**, *23*, 1035–1045. [CrossRef]

5. Zafar, M.N.; Mohanta, J.C. Methodology for Path Planning and Optimization of Mobile Robots: A Review. *Procedia Comput. Sci.* **2018**, *133*, 141–152. [CrossRef]

6. Kala, R. Multi-robot path planning using co-evolutionary genetic programming. *Expert Syst. Appl.* **2012**, *39*, 3817–3831. [CrossRef]

7. Jones, E.G.; Dias, M.B.; Stentz, A. Time-extended multi-robot coordination for domains with intra-path constraints. *Auton. Robot.* **2011**, *30*, 41–56. [CrossRef]

8. Fauadi, M.H.F.B.M.; Yahaya, S.H.; Murata, T. Intelligent combinatorial auctions of decentralized task assignment for AGV with multiple loading capacity. *IEEJ Trans. Electr. Electron. Eng.* **2013**, *8*, 371–379. [CrossRef]

9. Sun, D.; Kleiner, A.; Nebel, B. Behavior-based multi-robot collision avoidance. In Proceedings of the IEEE International Conference on Robotics and Automation, Hong Kong, China, 31 May–7 June 2014. [CrossRef]

10. Zhang, Y.; Gong, D.W.; Zhang, J.H. Robot path planning in uncertain environment using multi-objective particle swarm optimization. *Neurocomputing* **2013**, *103*, 172–185. [CrossRef]

11. Wang, D.; Wang, H.; Liu, L. Unknown environment exploration of multi-robot system with the FORDPSO. *Swarm Evol. Comput.* **2016**, *26*, 157–174. [CrossRef]

12. Simmons, R.; Apfelbaum, D.; Burgard, W.; Fox, D. Coordination for multi-robot exploration and mapping. In Proceedings of the Aaai/Iaai, Austin, TX, USA, 1–3 August 2000.

13. Berhault, M.; Huang, H.; Keskinocak, P.; Koenig, S.; Elmaghraby, W.; Griffin, P.; Kleywegt, A. Robot Exploration with Combinatorial Auctions. In Proceedings of the IEEE International Conference on Intelligent Robots and Systems, Las Vegas, NV, USA, 27 October–1 November 2003. [CrossRef]

14. Ulusoy, A.; Smith, S.L.; Ding, X.C.; Belta, C.; Rus, D. Optimality and robustness in multi-robot path planning with temporal logic constraints. *Int. J. Robot. Res.* **2013**, *32*, 889–911. [CrossRef]

15. Yu, J.; Lavalle, S.M. Planning optimal paths for multiple robots on graphs. In Proceedings of the IEEE International Conference on Robotics and Automation, Karlsruhe, Germany, 6–10 May 2013. [CrossRef]

16. Chaimowicz, L.; Sugar, T.; Kumar, V.; Campos, M.F. An architecture for tightly coupled multi-robot cooperation. In Proceedings of the IEEE International Conference on Robotics and Automation, Seoul, Korea, 21–26 May 2001. [CrossRef]

17. Guo, Y.; Parker, L.E. A distributed and optimal motion planning approach for multiple mobile robots. In Proceedings of the IEEE International Conference on Robotics and Automation, Washington, DC, USA, 11–15 May 2002. [CrossRef]

18. Jose, K.; Pratihar, D.K. Task allocation and collision-free path planning of centralized multi-robots system for industrial plant inspection using heuristic methods. *Robot. Auton. Syst.* **2016**, *80*, 34–42. [CrossRef]

19. Zhang, X.; Yan, M.; Ju, Y. A tabu search based flocking algorithm of motion control for multiple mobile robots. In Proceedings of the 2012 5th International Conference on Intelligent Computation Technology and Automation (ICICTA 2012), Zhangjiajie, China, 12–14 January 2012. [CrossRef]

20. Fierro, R.; Das, A.K.; Kumar, V.; Ostrowski, J.P. Hybrid control of formations of robots. In Proceedings of the IEEE International Conference on Robotics and Automation, Seoul, Korea, 21–26 May 2001. [CrossRef]

21. Stentz, A. Optimal and efficient path planning for partially-known environments. In Proceedings of the IEEE International Conference on Robotics and Automation, San Diego, CA, USA, 8–13 May 1994. [CrossRef]

22. Vivaldini, K.; Rocha, L.F.; Martarelli, N.J.; Becker, M.; Moreira, A.P. Integrated tasks assignment and routing for the estimation of the optimal number of AGVS. *Int. J. Adv. Manuf. Technol.* **2016**, *82*, 719–736. [CrossRef]

23. Nilsson, N.J. *Principles of Artificial Intelligence*; Morgan Kaufmann: Burlington, MA, USA, 2014.

24. Mitchell, R.; Michalski, J.; Carbonell, T. *An Artificial Intelligence Approach*; Springer: Berlin/Heidelberg, Germany, 2013.

25. Atzmon, D.; Stern, R.; Felner, A.; Wagner, G.; Barták, R.; Zhou, N.F. Robust multi-agent path finding and executing. *J. Artif. Intell. Res.* **2020**. [CrossRef]

26. Li, J.; Sun, K.; Ma, H.; Felner, A.; Satish Kumar, T.K.; Koenig, S. Moving agents in formation in congested environments. In Proceedings of the International Joint Conference on Autonomous Agents and Multiagent Systems (AAMAS), Auckland, New Zealand, 9–13 May 2020.

27. Atzmon, D.; Stern, R.; Felner, A.; Sturtevant, N.R.; Koenig, S. Probabilistic robust multi-agent path finding. In Proceedings of the International Conference on Automated Planning and Scheduling (ICAPS), Nancy, France, 14–19 June 2020.

28. Liu, M.; Ma, H.; Li, J.; Koenig, S. Task and path planning for multi-agent pickup and delivery. In Proceedings of the International Joint Conference on Autonomous Agents and Multiagent Systems (AAMAS), Montreal, QC, Canada, 13–17 May 2019.

29. Surynek, P. A novel approach to path planning for multiple robots in bi-connected graphs. In Proceedings of the IEEE International Conference on Robotics and Automation, Kobe, Japan, 12–17 May 2009. [CrossRef]

30. Wagner, G.; Kang, M.; Choset, H. Probabilistic path planning for multiple robots with subdimensional expansion. In Proceedings of the IEEE International Conference on Robotics and Automation, Saint Paul, MN, USA, 14–18 May 2012. [CrossRef]

31. Barták, R.; Švancara, J.; Škopková, V.; Nohejl, D. Multi-agent path finding on real robots: first experience with ozobots. In *Lecture Notes in Computer Science (Including Subseries Lecture Notes in Artificial Intelligence and Lecture Notes in Bioinformatics)*; Springer: Cham, Switzerland, 2018. [CrossRef]

42. Li, D.; Ouyang, B.; Wu, D.; Wang, Y. Artificial intelligence empowered multi-AGVs in manufacturing systems. *arXiv* **2019**, arXiv:1909.03373.
43. Sasi Kumar, G.; Shravan, B.; Gole, H.; Barve, P.; Ravikumar, L. *Path Planning Algorithms: A Comparative Study*; Space Transportation Systems Division: Houston, TX, USA, 2011.
44. Moreira, A.P.; Costa, P.J.; Costa, P. Real-time path planning using a modified A* algorithm. In Proceedings of the 9th Conference on Mobile Robots and Competitions, Castelo Branco, Portugal, 7 May 2009; pp. 141–146.
45. Petrinec, K.; Kovačić, Z. The application of spline functions and bézier curves to AGV path planning. In Proceedings of the IEEE International Symposium on Industrial Electronics, Dubrovnik, Croatia, 20–23 June 2005. [CrossRef]
46. Secchi, C.; Olmi, R.; Fantuzzi, C.; Casarini, M. TRAFCON—Traffic control of AGVs in automatic warehouses. In *Springer Tracts in Advanced Robotics*; Springer: Cham, Switzerland, 2014. [CrossRef]
47. Glover, F. Future paths for integer programming and links to artificial intelligence. *Comput. Oper. Res.* **1986**, *13*, 533–549. [CrossRef]

 **robotics**

Article

Human–Robot Interaction in Industrial Settings: Perception of Multiple Participants at a Crossroad Intersection Scenario with Different Courtesy Cues

Carla Alves [1], André Cardoso [1], Ana Colim [1,2,*], Estela Bicho [1], Ana Cristina Braga [1], João Cunha [2], Carlos Faria [2] and Luís A. Rocha [2]

1   Algoritmi Centre, School of Engineering, University of Minho, 4800-058 Guimaraes, Portugal; id9067@alunos.uminho.pt (C.A.); andre.cardoso@dps.uminho.pt (A.C.); estela.bicho@dei.uminho.pt (E.B.); acb@dps.uminho.pt (A.C.B.)

2   DTx—Digital Transformation CoLab, 4800-058 Guimaraes, Portugal; joao.cunha@dtx-colab.pt (J.C.); carlos.faria@dtx-colab.pt (C.F.); luis.rocha@dtx-colab.pt (L.A.R.)

*   Correspondence: ana.colim@dtx-colab.pt

Abstract: In environments shared with humans, Autonomous Mobile Robots (AMRs) should be designed with human-aware motion-planning skills. Even when AMRs can effectively avoid humans, only a handful of studies have evaluated the human perception of mobile robots. To establish appropriate non-verbal communication, robot movement should be legible and should consider the human element. In this paper, a study that evaluates humans' perceptions of different AMR courtesy behaviors at industrial facilities, particularly at crossing areas, is presented. To evaluate the proposed kinesic courtesy cues, we proposed five tests (four proposed cues—stop, deceleration, retreating, and retreating and moving aside—and one control test) with a set of participants taken two by two. We assessed three different metrics, namely, the participants' self-reported trust in AMR behavior, the legibility of the courtesy cues in the participants' opinions, and the behavioral analysis of the participants related to each courtesy cue tested. The retreating courtesy cue, regarding the legibility of the AMR behavior, and the decelerate courtesy cue, regarding the behavioral analysis of the participants' signs of hesitation, are better perceived from the forward view. The results obtained regarding the participants' self-reported trust showed no significant differences in the two participant perspectives.

Keywords: human–robot interaction; AMR navigation; AMR safety; human perception; courtesy cues; forward and backward scenarios; HTA questionnaire; shop floor configuration

Citation: Alves, C.; Cardoso, A.; Colim, A.; Bicho, E.; Braga, A.C.; Cunha, J.; Faria, C.; Rocha, L.A. Human–Robot Interaction in Industrial Settings: Perception of Multiple Participants at a Crossroad Intersection Scenario with Different Courtesy Cues. *Robotics* **2022**, *11*, 59. https://doi.org/10.3390/robotics11030059

Academic Editor: Carlo Alberto Avizzano

Received: 22 February 2022
Accepted: 9 May 2022
Published: 13 May 2022

Publisher's Note: MDPI stays neutral with regard to jurisdictional claims in published maps and institutional affiliations.

1. Introduction

The implementation of collaborative robots is seen as one of the technologies enabling Industry 5.0. This new industrial paradigm prioritizes essential needs and interests by placing humans at the core of the industrial production processes. It recognizes the power of the role of industry in achieving social and environmental objectives without setting aside the role of human workers in this process [1]. In fact, along with this new industrial paradigm, robots are no longer only programmable machines but are expected to be recognized as co-workers, side by side with human workers [2]. This relationship should increase production flexibility and efficiency while supporting the human workers in their tasks [3]. One of these technologies already being introduced on the shop floor is Industrial AMRs [4].

AMRs have evolved from Automated Guided Vehicles (AGVs), which are restricted to predefined paths using magnetic/electrical wires, among other sensors [4,5]. Compared to AGVs, AMRs are more flexible, collaborative, and cost-efficient [3]. This type of robot can move autonomously without the help of external workers [6] and they can detect obstacles and recalculate a new route around them [7]. The autonomy of AMRs implies continuous decision-making on how to behave according to their environment, with predefined rules and constraints [4].

Cooperation between humans and robots sharing a workspace is becoming increasingly common [8]. Human–Robot Collaboration (HRC) is the process wherein human and robot agents work together to achieve shared goals. For any level of collaboration, human safety has been a primary concern ever since robots were first created [8]. Beyond physical safety, other aspects also need to be considered when humans and robots interact, such as psychological aspects and mental stress [3]. In fact, when a robot comes close to a human, the robot may generate negative feelings, such as stress, mistrust, and anxiety [9]. This may be linked with human nature. For instance, instinct may lead humans to take evasive action when they perceive a threat due to approaching an unfamiliar object [10]. Therefore, foreseeing the humans' acceptance, trust, and comfort requires us to take into account the robots' appearance, movement, and behavior [11,12]. Because most industrial AMRs have non-humanoid features, the way to promote appreciation lies in non-verbal communication linked to their movement and behavior [13]. Motion is a way of communication and not only an instrument to reach a goal position, and predictability is of much importance for efficient human–robot interaction and collaboration [14]. Communication is a key factor in HRC activities to achieve a common goal [15]. Friendly and comprehensible AMR movements and behaviors are key factors for proper communication with the human worker [16].

AMRs can emulate the social behavior of humans through kinesic courtesy cues. In human–human interactions, kinesic courtesy cues promote social affiliation (e.g., physical distance from others, postural orientation, smooth social encounters, and acceptance of others) [13,17]. In the specific case of AMRs, legible kinesic cues can give, to humans, information about granting them the privilege of first passage at a crossroads. To be legible, AMRs' kinesic courtesy cues need to be predictable and resemble human behavior [13]. Lichtenthäler et al. [14] showed that a good strategy for an AMR is moving as far as is possible straight towards its goal and reacting as smoothly as possible to a human. Kaiser et al. [13] showed that when robots present legible behaviors, they are better appreciated by humans.

The assessment of humans' perception of robot behavior is a non-trivial problem. In industrial HRC scenarios, cognitive ergonomics deals with this issue. It is concerned with principles of interaction acceptability by minimizing mental stress and psychological discomfort, which could be felt by workers sharing a workspace with robots [18]. There is no single best way to assess these psychological parameters because they depend on the purpose of the assessment. According to Gualtieri et al. [19], there is a set of cognitive variables related to HRC, namely, trust, usability, frustration, perceived enjoyment, satisfaction, and acceptance. There are three categories of measurements available to measure these types of variables: (i) performance measures, (ii) subjective measures, and (iii) psychophysiological measures [20]. Performance measures are conducted based on reaction time and mistakes. Subjective measures assess the workers' opinions, providing information on how they assess aspects of their interaction within the workspace. Psychophysiological measures include direct measurement of cognitive variables, namely, heart rate variability (HRV), galvanic skin response (GSR), and eye blink rate [20]. These technologies are more feasible and capable of providing human cognitive status assessment and interpretation [8]. However, subjective measures are more often used because they show practical advantages, e.g., ease of implementation and non-intrusiveness. Additionally, previous research supports their capacity to provide sensitive measures of the cognitive status of workers [21].

Hetherington et al. [22] highlighted the need to carry out in-person experiments with mobile robots, in open spaces, and to apply courtesy cues in scenarios that include several participants at the same time. These authors even wondered what the impact would be of courtesy cues from the perspective of participants who have a view of the robot from behind. Kaiser et al. [13] tested the legibility of two kinesic courtesy cues common in human interaction with an autonomous mobile robot in two different situations, but not simultaneously: one participant and robot share the same trajectory next to each other or moving from opposite ends. The authors also pointed out the need to explore other courtesy cues.

Objectives

The current exploratory study intends to go a step further, recognizing the im-portance of legible movements in the communication processes of AMRs. This study leads us to create an experimental protocol on a real-industry in-person scenario and contribute with some specific conditions not previously considered. Specifically, the aim is to investigate how different kinesic courtesy cues (stop, decelerate, retreat, and retreat and move to the left) would be understood in the view of two participants with different perspectives of the robot (one with a frontward view and the other with a backward view) at an industrial crossroad under the same test conditions, i.e., within one simultaneous scenario.

2. Materials and Methods

The following subsections provide the sample characterization, a description of the AMR used and its operating conditions, the hypotheses and measures that we intended to analyze, and an explanation of the experimental procedure and apparatus.

2.1. Participants

The participants collaborated voluntarily and signed an informed consent form in agreement with the Committee of Ethics for Research in Social and Humans Sciences of the University of Minho (approval number CEICSH 095/2019), in agreement with the Declaration of Helsinki.

A total number of 34 participants were recruited, with 13 females (38.2%) and 21 males (61.8%). To conduct each trial, two participants were required simultaneously. Participant A has a backward perspective of the robot, where the robot is moving away from the participant, and Participant B has a forward perspective of the robot, where the robot is moving toward the participant. Regarding the characterization of the participants' ages, the average age was 29.8, with a standard deviation of 7.5, in the range of 21–46 years old.

2.2. Material and Experimental Setup
2.2.1. MiR 200 Specifications, Navigation, Control, and Safety

In this user study, an AMR, the MiR 200, and its automatic battery charging station, the MiR Charge 24 V, were used (Figure 1). The main specifications of the MiR 200 used for conducting the experiments are: weight (without load) of 65 kg; maximum speed forwards of 1.1 m/s; maximum speed backwards of 0.3 m/s; battery running time 10 h (or 15 km continuous driving); charging time with charge station up to 3.0 h ((0–80%): 2.0 h); charging time with cable up to 4.5 h ((0–80%): 3.0 h) [23].

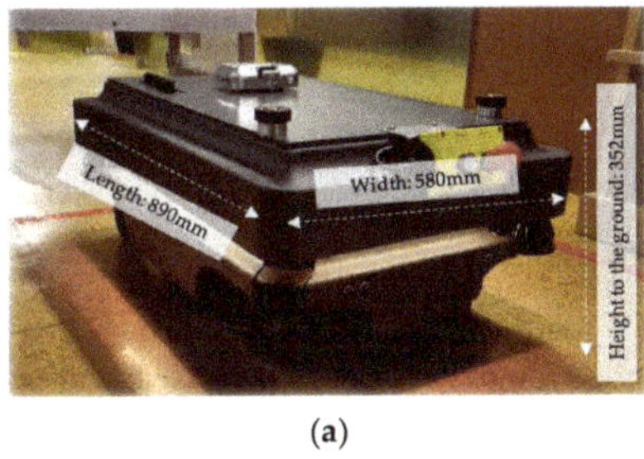

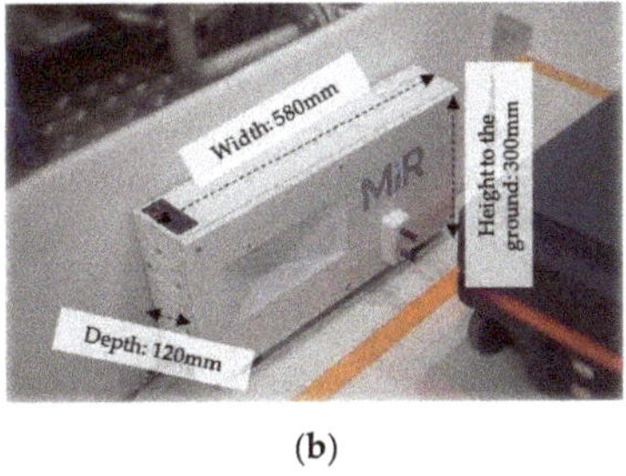

(a)   (b)

Figure 1. Principal dimensions of: (**a**) the MiR 200 robot; (**b**) the MiR Charge 24 V.

The MiR 200 is a nonholonomic wheeled mobile robot (WMR) with a rectangular configuration, controlling its movement speed based on wheel odometry, with six wheels in total: one omnidirectional swivel wheel in each corner; and two driving wheels (differential control) in the center of the platform to ensure the stability of the mobile robot when it rotates [24–28]. The robot adjusts how much power is sent to each motor based on sensory input. The robot is equipped with two ISO 13849-certified SICK S300 safety laser scanners, one in the front left corner and another one in the rear right corner, offering 360° visual protection around the robot (Figure 2a); two Intel RealSense™ D435 3D cameras on the front of the robot for the detection of objects vertically up to 1800 mm at a distance of 1950 mm in front of the robot (Figure 2b), and with an angle of 118° in the horizontal field of view (FoV) at 180 mm height from the floor (Figure 2c); and four ultrasound sensors, two placed at the front of the robot and two placed at the rear of the robot [23,27,29].

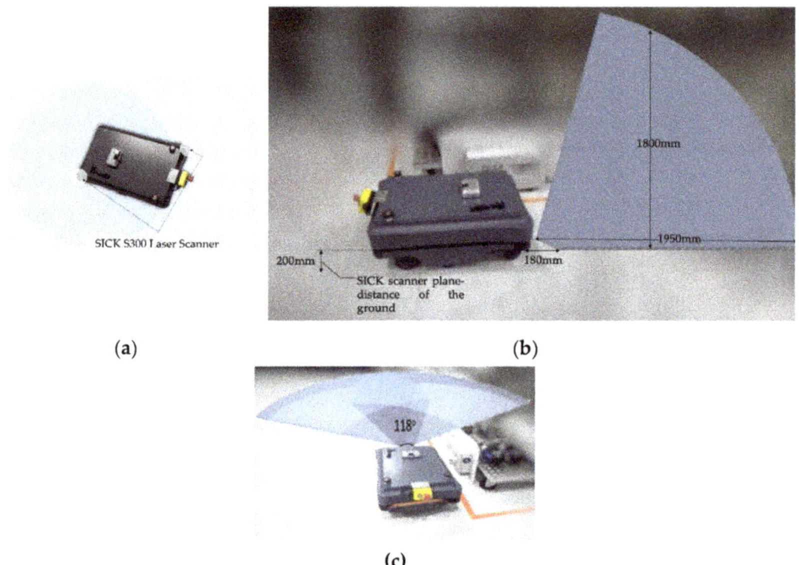

Figure 2. (**a**) Top view of the SICK laser scanners; (**b**) Configuration of the 3D cameras and SICK laser scanners, side view; (**c**) FoV of 118°.

The SICK safety laser scanners provide the sensorial information for the collision avoidance function. This function prevents the robot from colliding with a person or an object by stopping it before a collision happens. To that end, the safety laser scanners are programmed with two sets of protective fields, each one individually configured to contour around the robot. One set is used when the robot is driving forward, and the other set is used when the robot is driving backward. Based on the speed, the robot activates the corresponding protective field. If an obstacle is detected, whether person or object, within the active protective field, the robot enters a protective stop automatically (Figure 3) until the protective field is cleared of obstacles for at least two seconds. The protective stop is a state of the robot where a robot status light turns red, and it is not possible to move the robot or send it on missions until it is brought out of the protective stop [27].

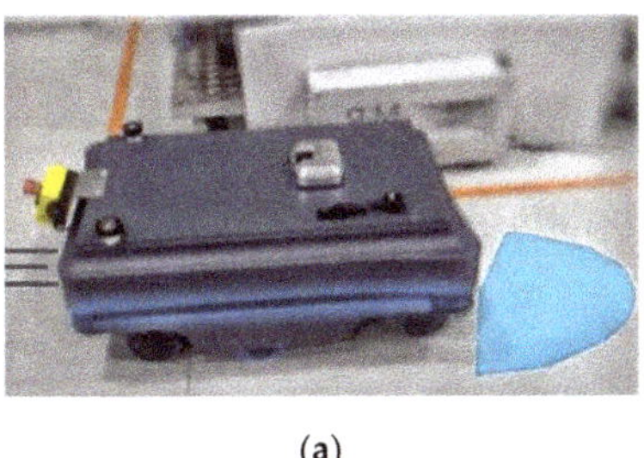
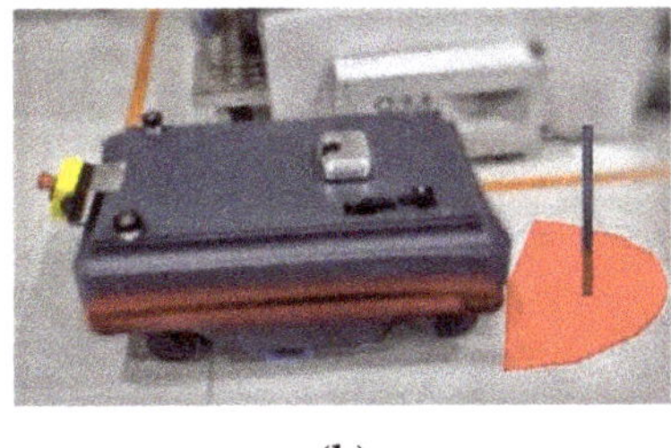

(a)   (b)

Figure 3. MiR 200 protective field mechanism: (**a**) The robot drives when its path is clear; (**b**) the robot activates the proactive stop when an obstacle is detected within its protective field.

According to [27], the velocity and the protective field ranges when the robot is driving forward, i.e., in front of the robot, and backward, are different for specific cases. Each case describes a velocity interval in which the robot may operate. Figure 4a shows a representation of the protective field ranges according to five different forward robot velocities. For example, when the robot is moving at a velocity between −0.14 to 0.20 m/s (1) the protective field range is from 0 to 20 mm. The other velocity scenarios (2 to 5) are 0.21 to 0.40 m/s; 0.41 to 0.80 m/s; 0.81 to 1.10 m/s; and 1.11 to 2.00 m/s. The protective field range are respectively 0–120 mm; 0–290 mm; 0–430 mm; and 0–720 mm. Figure 4b shows the similar representation for when the robot is driving backward. In this case, the four velocity scenarios are −0.14 to 1.80 m/s; −0.20 to −0.15 m/s; −0.40 to −0.21 m/s; and −1.50 to −0.41 m/s. The protective field ranges are (1 to 4), respectively, 0–30 mm; 0–120 mm; 0–290 mm; and 0–430 mm.

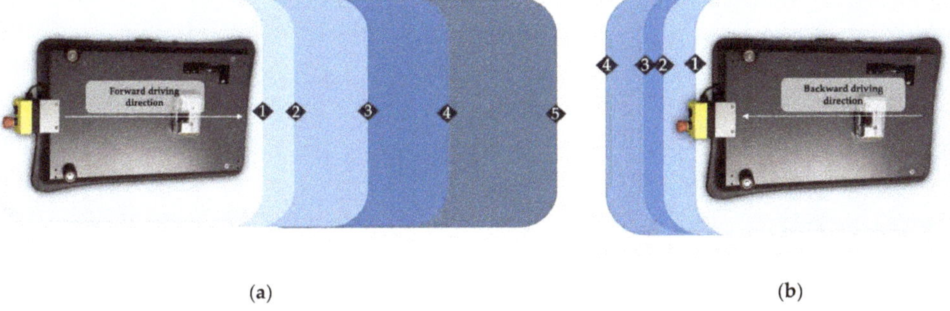

(a)   (b)

Figure 4. Range of the robot's active protective field that changes with the robot's speed, represented in millimeters: (**a**) Forward driving direction; (**b**) Backward driving direction.

To execute the experiments, a 2D map of the test area was created through the cartographer algorithm available on the robot platform (Figure 5). The robot localization within this map is determined by an adaptive Monte Carlo localization (AMCL) navigational system combining wheel odometry, information from the inertial measurement unit (IMU) encoders, and laser scanner data [27–31].

The kinesic courtesy cues were implemented on the robot through commands programmed in the software interface supplied by the respective vendor. During the tests, the desired value for the robot's linear speed was set to 0.6 m/s. According to Lauckner et al. [31], speeds slower than 0.6 m/s are perceived as too slow. In terms of obstacle detection, the higher the speed (forward and backward speed), the larger the protective safety range [27].

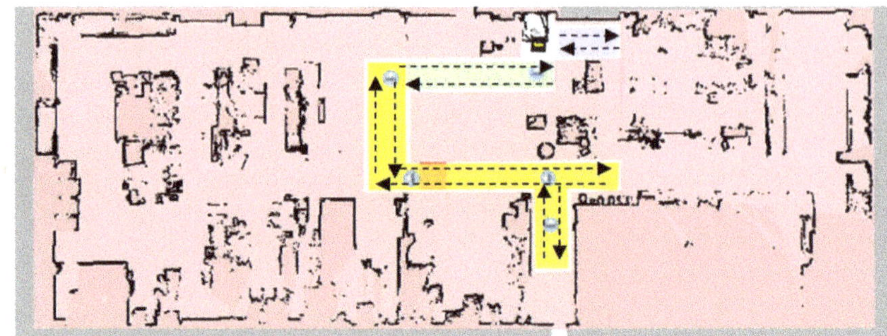

Figure 5. Clean and edited map corresponding to the total industrial area used for the tests: the arrows indicate the area in which the robot was allowed to circulate.

2.2.2. Experimental Setup

The experiments were conducted in an open space within an industrial environment in a crossroad-like configuration with simultaneous forward and backward scenarios (Figure 6).

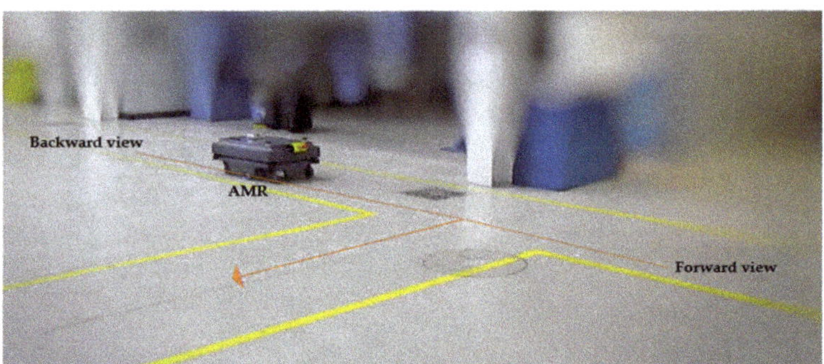

Figure 6. Industrial environment: A crossroad-like configuration with simultaneous forward and backward scenarios.

A control courtesy cue, i.e., a control condition where the robot does not stop, and another four courtesy cues—stop, decelerate, retreat, and retreat and move to the left—were programmed into the MiR 200. For each courtesy cue, the MiR 200 moved towards the crossing area with linear speed ($v = 0.6$ m/s) and then executed specific movements to communicate to the participants that it was yielding the right of way at the intersection. The crossing area is an interaction and decision area where Participants A and B outpace the MiR 200, localized in the experimental apparatus between the end of the horizontal paths for Participant A, the MiR 200, and Participant B and the beginning of the vertical path. The four courtesy cues (Figure 7) tested were the following:

(i)   "**stop**": The AMR stopped suddenly before the crossing area, made a two-second stop, and returned to its trajectory to the final position.

(ii)   "**decelerate**": The AMR started to slow down its linear speed ($v = 0.6$ m/s) to $v = 0.2$ m/s at a distance of 1.0 m (represented by X_d in Figure 7) before stopping before the crossing area. Then, it stopped for two seconds and returned to its trajectory to the final position.

(iii) "**retreat**": The robot stopped suddenly before the crossing area, then retreated 1.0 m (X_r in Figure 7), stopped for two seconds, and returned to its trajectory to the final position.

(iv) "**retreat and move to the left**": The robot stopped suddenly before the crossing area, then retreated 1.0 m, and moved to the left by 0.2 m (X_{left} in Figure 7) relative to the central point of the crossing area. Then, it stopped for two seconds and returned to its trajectory to the final position.

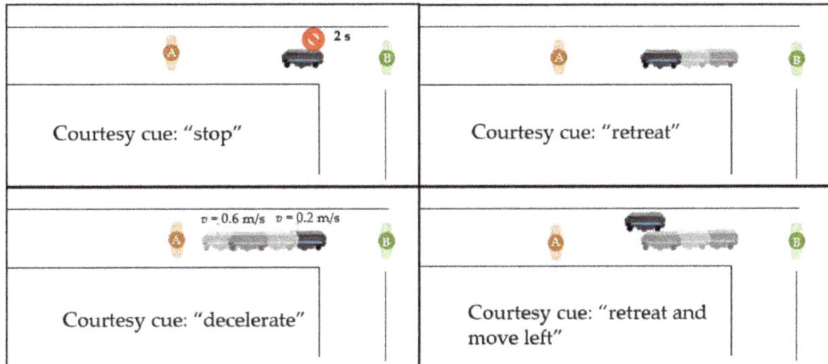

Figure 7. Scheme of the four courtesy cues tested.

Participant A and the MiR 200 shared the trajectory from the position start, and Participant B shared the trajectory with the MiR 200 from the crossing area, both passing through the intersection. When the participants reached the crossing area, they were asked to decide whether (or not) to overtake the robot, according to the kinesic courtesy cue of the test in progress (Figure 8).

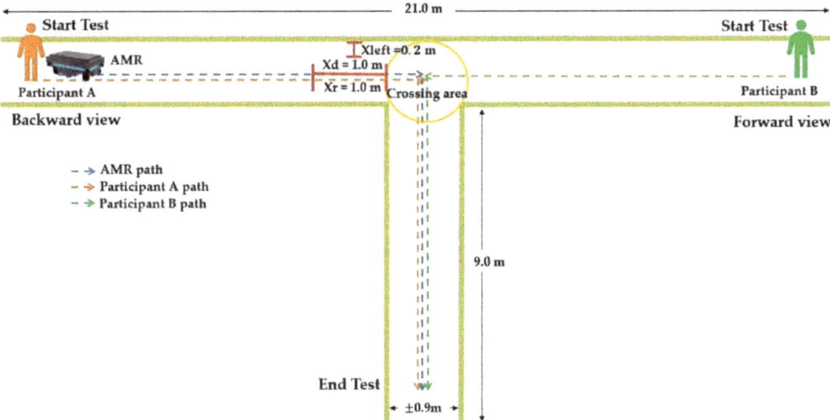

Figure 8. Experimental apparatus of the interaction's HRC kinesic cue study conditions: X_d decelerate distance, X_r retreat distance, X_{left} move left distance; distances are represented in meters.

2.3. Procedure

The participants were informed about the scope of the study and the general instructions of the experiment; precisely, that the MiR 200 works autonomously, i.e., the robot moves on a preprogrammed map and trajectory (Figure 5)—the robot does not collide with any participant or any object due to its sensors that monitor the environment around the robot [27]. Each participant completed five tests: one for the control courtesy cue condition and the rest related to the other four courtesy conditions (see Section 2.2). The total time for performing the five test conditions was 20 min. Participants were instructed to start the test after hearing a "Beep" and seeing a green light. To ensure that they would encounter the robot at a specific point on the navigation map (where the robot would show the courtesy cue behavior), participants were also instructed to walk at the pace imposed by the evaluators (about 1 m/s). They could also abandon the test if they felt uncomfortable at any moment. After each courtesy cue condition, an appreciation of the subjects' perception, through an adapted version of the Human Trust in Automation questionnaire (HTA), was obtained. To reduce hysteresis phenomena, i.e., different responses to identical inputs [32,33], the different kinesic courtesy cues were randomly assigned among the participants.

2.4. Measures

2.4.1. Perceived Trust and Mistrust Assessment

To measure the participants' perceived trust in the AMR behavior, we applied an adapted version of the HTA questionnaire [34]. Comparatively to the original questionnaire, we replaced the word "system" by "robot". Trust is an important factor in HRC. It determines humans' use of autonomy, and improper trust can lead to either over-reliance or under-reliance on the robot [35]. The HTA is a validated questionnaire composed of 12 statements, assessed by a 7-point Likert scale (between 1 = "Totally disagree" and 7 = "Totally agree"). The subjects answered at the end of each interaction with the AMR presenting the different courtesy cues. The first five statements have a negative connotation while the last seven statements have a positive connotation. We assessed the responses to the negative connotation statements as mistrust and the others as trust in the robot. To assess the condition of normality of the results, the Kolmogorov–Smirnov test was applied. Additionally, a one-way ANOVA was applied to compare the results of trust and mistrust with the conditions of different courtesy cues.

2.4.2. Legibility Assessment

To measure the legibility of the implemented kinesic courtesy cues tested, we asked the participants to select one of eight options after the end of each trial. The options available were: (i) the robot stopped; (ii) the robot decelerated and stopped; (iii) the robot stopped and retreated; (iv) the robot stopped, then retreated, and then moved forward and moved to the left/right; (v) the robot followed its path without stopping; (vi) the robot stopped and nudged; (vii) the robot stopped and tilted to one side; (viii) none of the above options. When the participant answered correctly, we considered the answer "true". When the participant answered wrongly but pointed out one of the courtesies used in the test (i, ii, iii, iv, or v), we considered it "false". If the participant answered wrongly and pointed out one courtesy that was not in the scope of the experiment (vi, vii, or viii), we considered it "false out of the test". To assess the association between the levels of courtesy kinesic cues and the answer to control questions, a Pearson's chi-square test was applied.

2.4.3. Behavioral Analysis

Based on video recordings of each experimental trial, we assessed the lack of hesitation with which the participants moved through the crossroad. For this purpose, two raters assessed each video individually and inferred the participants' lack of hesitation. To assess this condition, we assumed that a person's movement showed signs of hesitation if one of the following situations was observed: slowed down, stopped, moved to the side, retreated, granted the robot the right to pass, visually checked the robot, moved first but tentatively, seemed somewhat forced by the robot to pass first, passed the bottleneck jointly with the robot, or both got stuck in the crossroad [13]. To assess the association between the robot's kinesic courtesy cues and the observable participant's signs of hesitation, Pearson's chi-square test was applied.

3. Results

We report our findings in three subsections. First, we tested the effect of the four kinesic courtesy cues on subjects' trust and mistrust. Second, we assessed the legibility of the kinesic courtesy cues via participants' self-reporting. Third, we compared the hesitation behavior of the participants when subjected to each kinesic courtesy cue from the point of view of the participants concerning their relative position (behind/in front) to the robot. We started from the assumption that an AMR without skills in non-verbal communication will be hard to read by its human counterparts.

3.1. Perceived Trust and Mistrust

1. To assess perceived trust, the HTA was applied. A one-way ANOVA was applied, and the application condition of homogeneity of variance was verified ($p > 0.05$). Additionally, the Kolmogorov–Smirnov test was applied to assess the condition of normality. The normality of the residuals was confirmed ($p > 0.05$). The results of trust and distrust in the human counterpart are shown in Table 1.

Table 1. Results of one-way ANOVA related to the subjects' perceived trust and distrust of the AMR from both points of view (forward and backward).

	Forward	Backward
Trust	$F(4,80) = 1.082, p = 0.371$	$F(4,80) = 0.486, p = 0.746$
Mistrust	$F(4,80) = 0.564, p = 0.689$	$F(4,80) = 0.966, p = 0.431$

No statistically significant difference was found in the mean values of trust and distrust from both points of view regarding the kinesic courtesy cues. The graphs in Figure 9 illustrate these results.

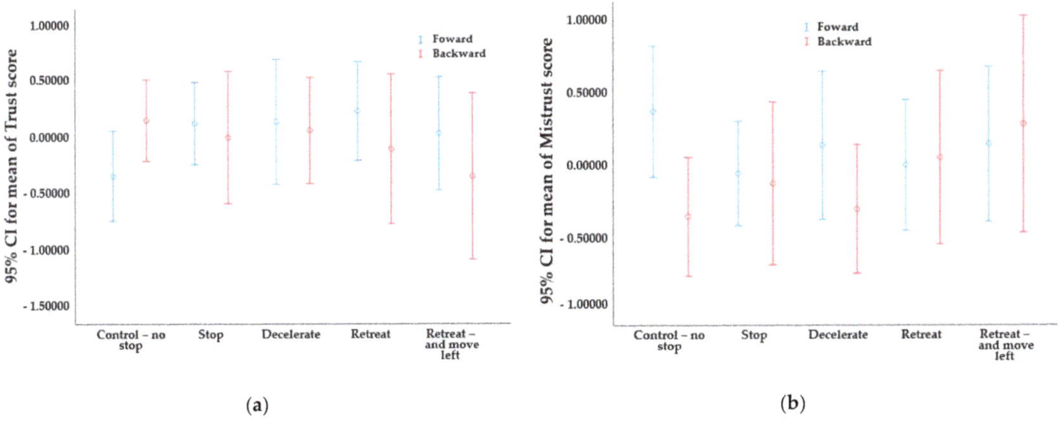

Figure 9. Profile plots for trust (**a**) and mistrust (**b**) scores by kinesic courtesy cue and by point of view.

3.2. Legibility Assessment

Pearson's chi-square test of independence revealed a statistically significant association between the levels of courtesy kinesic cues and the answer to the control question for the forward point of view ($X^2(8) = 16.316, p = 0.038$). The graph in Figure 10a shows that the courtesy cue that presented better legibility for the participants was retreat (with the same percentage as the control situation) (64.7%). From this perspective, the courtesy cue less understood by the participants was "retreat and move left" (11.8%).

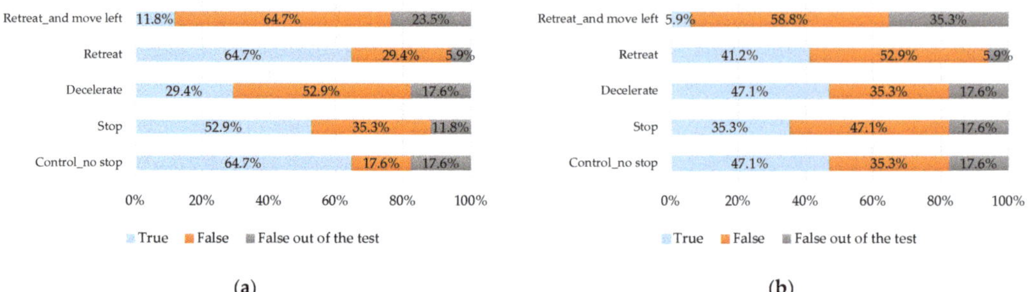

(a) **(b)**

Figure 10. Legibility of the robot courtesy kinesic cues: (**a**) Forward view; (**b**) Backward view.

For the backward view, Pearson's chi-square test did not reveal a statistically significant association between the levels of courtesy kinesic cues and the answer to the control question for the forward point of view ($X^2(8) = 11.308$, $p = 0.186$). Figure 10b illustrates this result, and it can be seen that the courtesy cue better understood by the participants was the decelerate (also with the same percentage as the control situation) (47.1%). Regarding the courtesy with less legibility for the participants, retreat and move left was the courtesy cue with the lowest percentage of right answers (5.9%).

3.3. Behavioral Analysis

Pearson's chi-square test revealed a statistically significant association ($X^2(4) = 12.143$, $p = 0.016$) between the robot courtesy cue and observable signs of hesitation in the participant from the forward view. The graph in Figure 11a shows that the courtesy cue for which the participants presented lesser signs of hesitation was decelerate (82.4%). On the contrary, the control condition was the one that presented a greater percentage of participant hesitation (70.6%).

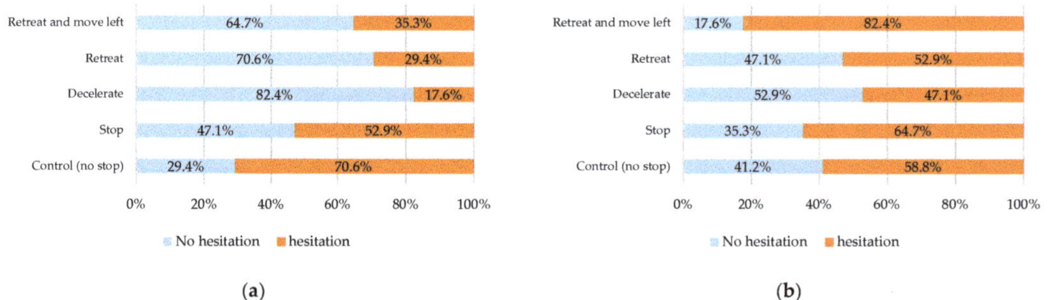

(a) **(b)**

Figure 11. Observable signs of hesitation in the participants' behavior while encountering the AMR related to each kinesic courtesy cue condition: (**a**) Forward view; (**b**) Backward view.

Related to the backward view, Pearson's chi-square test did not detect a significant association ($X^2(4) = 5.251$, $p = 0.263$). Figure 11b illustrates that decelerate was the courtesy cue with a lesser percentage of participant hesitation (52.9%). On the contrary, retreat and move left was the courtesy with a greater percentage of participant hesitation behavior when encountering the AMR.

Additionally, an exact Fisher test was conducted to evaluate the association between the type of courtesy cue and the behavior of participants. The result ($p = 0.014$) revealed a statistically significant association in the sense that a higher percentage of participants showed hesitant behavior when they saw the robot from behind (61.2%) than when they approached from the front (41.2%).

4. Discussion

In this exploratory in-person experiment, we implemented four courtesy cues (stop, decelerate, retreat, and retreat and move to the left) and one control courtesy cue (no stop) on an AMR to investigate how these different kinesic courtesy cues would be understood in the view of two participants with different perspectives of the robot. That is, we intended to understand how participants' behaviors were influenced by the courtesy cues of the robot, in a simultaneous scenario, with a set of participants taken two by two. One participant had a front-facing view of the robot, while the other had a back-facing view. The study being simultaneous becomes important when the behavior of one participant affects the behavior of the other. In environments where several participants simultaneously share the space with each other and with the robot, it is very likely that there will be different views. Different views imply different perceptions of the robot's behavior and, therefore, different behaviors on the part of the participants towards the robot. In turn, this also has implications for the perception and behavior of other participants.

We tested three different metrics in our research, namely, the participants' self-reported trust in AMR behavior, the legibility of the courtesy cues in the participants' opinions, and behavioral analysis of the participants related to each courtesy cue tested. These metrics were assessed via an experimental protocol that consisted of two participants interacting with an AMR at an industrial crossroad.

The results of the participants' self-reported trust showed no significant differences from the two participants' perspectives between the control situation and the four kinesic courtesy cues implemented on the robot. This may be related to the fact that we measured trust right after the interaction with the robot. As Hancock et al. [36] pointed out, this is an issue that needs to be addressed in HRC because the process of trust development is not clear and needs to be further studied. Kaiser et al. [13], in a study where two kinesic courtesy cues were investigated (robot stop and robot stop and move to the side), found that an AMR that presented polite behaviors was better accepted by its human counterpart in an interaction, regardless of the specific courtesy cue. Here, we measured trust because this parameter directly affects people's acceptance of the robot [37], but we did not find the same results. The fact of having two participants interacting simultaneously with the robot may have influenced our results, leading to different conclusions. Additionally, trust can be dynamically influenced by different factors, namely, the robotic system itself, the surrounding operational environment, and the respective natures and characteristics of the human team members [38].

Regarding the legibility of the robot behavior, we only found a statistically significant association in the participants with the forward view. From this point of view, the results point out that the users better perceived the robot behavior when it presented a retreating courtesy cue, granting the human the right to pass first at the crossroad. Hetherington et al. [22] presented results that are in agreement with our results. They explored the common human–robot interaction at a doorway or bottleneck in a structured environment and found that a robot's retreating cue was the most socially acceptable cue and, therefore, the most legible.

Related to the behavioral analysis of the participants' signs of hesitation, we also only found a statistically significant association in the forward view. When a human interacts with a robot and presents lesser patterns of hesitation, these interactions lead to less cognitive effort to decide how to interact [13]. The results show that the decelerate courtesy cue was the one with which the participants presented a lesser percentage of hesitation. These results are in accordance with Dondrup et al. [39]. These authors showed that the robot's deceleration within the pedestrians' personal space resulted in less disruption to their movement.

It was expected that the lower the understanding of the robot behavior, the higher the participants' hesitancy [40]. Our results from the front view related to the retreat courtesy cue show that more than half of the participants understood the robot behavior (64.7%), and more than half of the participants showed no signs of hesitation (70.6%). However, the other courtesy cues are not in accordance with this. For instance, the decelerate and the retreat and move left courtesy cues presented a lower percentage of legibility and, also, a lower percentage of hesitance in the participants. This may indicate that a robot behavior understood by humans may be not enough to present good communication of the robot's intentions. For that reason, further research should be carried out in this direction. Unlike what was expected in the four courtesy cues tested, the control situation was designed for the robot to not be polite and to thereby increase the participant's hesitancy as the robot does not stop at the crossroad. Our results show that the control situation was understood by the majority of the participants (64.7%) and resulted in higher hesitance (70.6%), as expected. The results related to the forward view of the robot show that less than half of the participants did not understand the robot behavior, and this led to greater signs of hesitation in the participants in all courtesy cues tested. This shows that further research involving different perspectives of the robot needs to be carried out, to understand how AMRs should behave in order to be accepted and understood by all humans that interact with them.

Limitations and Future Work

This article addresses human perceptions of robot actions in a shared environment at an industrial crossing area. While this is a relevant topic that provides insight into human–robot interaction and human–robot collaboration research, some limitations that affected our research work should be noted. We can point to restraints regarding the low number of participants and the low variability in terms of the representativeness of the participants (the participants were all recruited from the academic community). Another limitation is related to the fact that there was no full human–robot interaction in play: the robot behavior was indifferent to human presence, because the behaviors' activation was hardcoded. However, if proper human recognition were to be implemented, the outcome should be similar if full human–robot interaction was implemented. Finally, it should be noted that to carry out this exploratory study in an industrial environment and via in-person experiments, the parameters (such as linear speed and test conditions for courtesy cues) were adapted from other related works [22,31] using types of mobile robots other than the robot model under study (MiR 200), which may have conditioned the results obtained.

These limitations pave the way to further research. That is, they illustrate the need to develop an algorithm for the robot's movement to be completely autonomous and for it to show courtesy cues when it finds a person at any time during the execution of its tasks, as suggested by Kaiser et al. [13]. This exploratory study already includes a scenario for analysis of the forward and backward view configurations of a group of two simultaneous participants with the robot. For that reason, it might be important to understand which parameters intrinsic to the robot and to courtesy cues must be applied in a scenario with more than two participants. We intend for future research to be applied on a shop floor where the AMR has to share the same trajectory with more than two workers, taking into account that the type of robot under study is intended for application in dynamic environments (e.g., industrial sector, logistics, hospitals) with a significant number of people in circulation [7,23]. On the other hand, by increasing the number of participants, the results, especially concerning trust and distrust analysis in HRC, become more difficult to analyze using only standard interviews and questionnaires [41]. Therefore, the application of psychophysiological measures (e.g., eye-tracking systems) should be a relevant measurement approach for this type of variable [42]. Another question that arose during the experiments and that should be addressed in further research is how the presence of a second co-actor (Participant A) affects the legibility of movements and courtesy cues perceived by the first co-actor (Participant B) and vice versa.

Author Contributions: C.A., conceptualization, methodology, investigation, writing, review, and validation; A.C. (André Cardoso), conceptualization, methodology, investigation, writing, review, and validation; A.C. (Ana Colim), investigation support, review, validation, and supervision; E.B., investigation support, validation, supervision, and review; A.C.B., statistical analysis, validation, supervision, and review; J.C., writing review; C.F., writing review; L.A.R., investigation support. All authors have read and agreed to the published version of the manuscript.

Funding: This work was supported by NORTE-06-3559-FSE-000018, integrated in the invitation NORTE-59-2018-41, aimed at the Hiring of Highly Qualified Human Resources, co-financed by the Regional Operational Programme of the North 2020, thematic area of Competitiveness and Employment, through the European Social Fund (ESF). This work was also supported by FCT–Fundação para a Ciência e Tecnologia within the R&D Units Project Scope: UIDB/00319/2020.

Institutional Review Board Statement: The study was conducted according to the guidelines of the Declaration of Helsinki and approved by the Committee of Ethics for Research in Social and Humans Sciences of the University of Minho (approval number CEICSH 095/2019).

Informed Consent Statement: Informed consent was obtained from all subjects involved in the study.

Data Availability Statement: Not applicable.

Acknowledgments: The authors would like to acknowledge PIEP (Innovation in Polymer Engineering) for the use of their facilities.

Conflicts of Interest: The authors declare no conflict of interest.

References

1. European Commission. *Industry 5.0—Towards a Sustainable, Human-Centric and Resilient European Industry*; Publications Office of the European Union: Luxembourg, 2021. [CrossRef]
2. Nahavandi, S. Industry 5.0—A Human-Centric Solution. *Sustainability* **2019**, *11*, 4371. [CrossRef]
3. Berx, N.; Pintelon, L.; Decré, W. Psychosocial Impact of Collaborating with an Autonomous Mobile Robot: Results of an Exploratory Case Study. In Proceedings of the HRI'21: ACM/IEEE International Conference on Human-Robot Interaction, Boulder, CO, USA, 8–11 March 2021. [CrossRef]
4. Fragapane, G.; de Koster, R.; Sgarbossa, F.; Strandhagen, J.O. Planning and control of autonomous mobile robots for intralogistics: Literature review and research agenda. *Eur. J. Oper. Res.* **2021**, *294*, 405–426. [CrossRef]
5. Saeidi, H.; Wang, Y. Incorporating Trust and Self-Confidence Analysis in the Guidance and Control of (Semi)Autonomous Mobile Robotic Systems. *IEEE Robot. Autom. Lett.* **2019**, *4*, 239–246. [CrossRef]
6. Rubio, F.; Valero, F.; Llopis-Albert, C. A review of mobile robots: Concepts, methods, theoretical framework, and applications. *Int. J. Adv. Robot. Syst.* **2019**, *16*, 1–22. [CrossRef]
7. Liaqat, A.; Hutabarat, W.; Tiwari, D.; Tinkler, L.; Harra, D.; Morgan, B.; Taylor, A.; Lu, T.; Tiwari, A. Autonomous mobile robots in manufacturing: Highway Code development, simulation, and testing. *Int. J. Adv. Manuf. Technol.* **2019**, *104*, 4617–4628. [CrossRef]
8. Chen, Y.; Yang, C.; Song, B.; Gonzalez, N.; Gu, Y.; Hu, B. Effects of Autonomous Mobile Robots on Human Mental Workload and System Productivity in Smart Warehouses: A Preliminary Study. *Proc. Hum. Factors Ergon. Soc. Annu. Meet.* **2020**, *64*, 1691–1695. [CrossRef]
9. Toyoshima, A.; Nishino, N.; Chugo, D.; Muramatsu, S.; Yokota, S.; Hashimoto, H. Autonomous Mobile Robot Navigation: Consideration of the Pedestrian's Dynamic Personal Space. In Proceedings of the 2018 IEEE 27th International Symposium on Industrial Electronics (ISIE), Cairns, QLD, Australia, 13–15 June 2018; pp. 1094–1099. [CrossRef]
10. Sunada, K.; Yamada, Y.; Hattori, T.; Okamoto, S.; Hara, S. Extrapolation simulation for estimating human avoidability in human-robot coexistence systems. In Proceedings of the 2012 IEEE RO-MAN: The 21st IEEE International Symposium on Robot and Human Interactive Communication, Paris, France, 9–13 September 2012; pp. 785–790. [CrossRef]
11. Shen, Z.; Elibol, A.; Chong, N.Y. Understanding nonverbal communication cues of human personality traits in human-robot interaction. *IEEE/CAA J. Autom. Sin.* **2020**, *7*, 1465–1477. [CrossRef]
12. Stroessner, S.J.; Benitez, J. The Social Perception of Humanoid and Non-Humanoid Robots: Effects of Gendered and Machinelike Features. *Int. J. Soc. Robot.* **2019**, *11*, 305–315. [CrossRef]
13. Kaiser, F.G.; Glatte, K.; Lauckner, M. How to make nonhumanoid mobile robots more likable: Employing kinesic courtesy cues to promote appreciation. *Appl. Ergon.* **2019**, *78*, 70–75. [CrossRef]
14. Lichtenthäler, C.; Kirsch, A. Towards Legible Robot Navigation—How to Increase the Intend Expressiveness of Robot Navigation Behavior. In Proceedings of the International Conference on Social Robotics—Workshop Embodied Communication of Goals and Intentions, Bristol, UK, 27–29 October 2013.
15. Gildert, N.; Millard, A.; Pomfret, A.; Timmis, J. The Need for Combining Implicit and Explicit Communication in Cooperative Robotic Systems. *Front. Robot. AI* **2018**, *5*, 65. [CrossRef]
16. Sisbot, E.A.; Marin-Urias, L.F.; Alami, R.; Simeon, T. A Human Aware Mobile Robot Motion Planner. *IEEE Trans. Robot.* **2007**, *23*, 874–883. [CrossRef]
17. Leroy, T.; Christophe, V.; Delelis, G.; Corbeil, M.; Nandrino, J.L. Social Affiliation as a Way to Socially Regulate Emotions: Effects of Others' Situational and Emotional Similarities. *Curr. Res. Soc. Psychol. Univ. Iowa* **2010**, *16*.

18. Gualtieri, L.; Rauch, E.; Vidoni, R. Emerging research fields in safety and ergonomics in industrial collaborative robotics: A systematic literature review. *Robot. Comput. Integr. Manuf.* **2021**, *67*, 101998. [CrossRef]

19. Gualtieri, L.; Fraboni, F.; De Marchi, M.; Rauch, E. Evaluation of Variables of Cognitive Ergonomics in Industrial Human-Robot Collaborative Assembly Systems. In *Proceedings of the 21st Congress of the International Ergonomics Association (IEA 2021)*; Springer: Cham, Switzerland, 2021; pp. 266–273. [CrossRef]

20. Fista, B.; Azis, H.A.; Aprilya, T.; Saidatul, S.; Sinaga, M.K.; Pratama, J.; Syalfinaf, F.A.; Steven; Amalia, S. Review of Cognitive Ergonomic Measurement Tools. *IOP Conf. Ser. Mater. Sci. Eng.* **2019**, *598*, 012131. [CrossRef]

21. Rubio, S.; Diaz, E.; Martin, J.; Puente, J.M. Evaluation of Subjective Mental Workload: A Comparison of SWAT, NASA-TLX, and Workload Profile Methods. *Appl. Psychol.* **2004**, *53*, 61–86. [CrossRef]

22. Hetherington, N.J.; Lee, R.; Haase, M.; Croft, E.A.; Van der Loos, H.F.M. Mobile Robot Yielding Cues for Human-Robot Spatial Interaction. In Proceedings of the 2021 IEEE/RSJ International Conference on Intelligent Robots and Systems (IROS), Prague, Czech Republic, 27 September–1 October 2021; pp. 3028–3033. [CrossRef]

23. Mobile Industrial Robots A/S (MiR). MiR a Better Way. 2019. Available online: https://epl-si.com/produto/mir200/ (accessed on 24 January 2022).

24. Tzafestas, S.G. Mobile Robot Control and Navigation: A Global Overview. *J. Intell. Robot. Syst.* **2018**, *91*, 35–58. [CrossRef]

25. Recker, T.; Heilemann, F.; Raatz, A. Handling of large and heavy objects using a single mobile manipulator in combination with a roller board. *Procedia CIRP* **2020**, *97*, 21–26. [CrossRef]

26. Recker, T.; Zhou, B.; Stüde, M.; Wielitzka, M.; Ortmaier, T.; Raatz, A. LiDAR-Based Localization for Formation Control of Multi-Robot Systems. In *Annals of Scientific Society for Assembly, Handling and Industrial Robotics 2021*; Schüppstuhl, T., Tracht, K., Raatz, A., Eds.; Springer International Publishing: Cham, Switzerland, 2022; pp. 363–373.

27. Mobile Industrial Robots A/S (MiR). *User Guide (En) MiR 200*, 3rd ed.; Mobile Industrial Robots A/S: Odense, Denmark, 2020; pp. 1–173

28. Siegwart, R.; Nourbakhsh, L.R. *Introduction to Autonomous Mobile Robots*; MIT Press: Cambridge, MA, USA, 2004. [CrossRef]

29. Mobile Industrial Robots A/S. MiR Robot Safety. Available online: https://www.mobile-industrial-robots.com/insights/amr-safety/mir-robot-safety/ (accessed on 26 January 2022).

30. Wadsten, J.; Klemets, R.E. Automated Deliverance of Goods by an Automated Guided Vehicle—Case study of the testing and implementation of an AGV within the production at Volvo Group AB, Tuve Gothenburg. Bachelor Thesis, Mechanical Engineering, Chalmers University of Technology, Gothenburg, Sweden, 14 June 2019.

31. Lauckner, M.; Kobiela, F.; Manzey, D. 'Hey Robot, Please Step Back!'—Exploration of a Spatial Threshold of Comfort for Human-Mechanoid Spatial Interaction in a Hallway Scenario. In Proceedings of the 23rd IEEE International Symposium on Robot and Human Interactive Communication, Edinburgh, UK, 25–29 August 2014; pp. 780–787. [CrossRef]

32. Farrell, P.S.E. The Hysteresis Effect. *Hum. Factors J. Hum. Factors Ergon. Soc.* **1999**, *41*, 226–240. [CrossRef]

33. Bicho, E.; Schoner, G.; Vaz, F. Modelo Dinâmico Neuronal Para a Percepção Categorial Da Fala. *Electrónica Telecomunicações* **1999**, *2*, 617.

34. Jian, J.-Y.; Bisantz, A.; Drury, C.G. Foundations for an Empirically Determined Scale of Trust in Automated Systems. *Int. J. Cogn. Ergon.* **2000**, *4*, 53–71. [CrossRef]

35. Sadrfaridpour, B.; Saeidi, H.; Wang, Y. An integrated framework for human-robot collaborative assembly in hybrid manufacturing cells. In Proceedings of the 2016 IEEE International Conference on Automation Science and Engineering (CASE), Fort Worth, TX, USA, 21–25 August 2016; pp. 462–467. [CrossRef]

36. Hancock, P.A.; Billings, D.R.; Schaefer, K.E.; Chen, J.Y.C.; De Visser, E.J.; Parasuraman, R. A Meta-Analysis of Factors Affecting Trust in Human-Robot Interaction. *Hum. Factors J. Hum. Factors Ergon. Soc.* **2011**, *53*, 517–527. [CrossRef] [PubMed]

37. Freedy, A.; DeVisser, E.; Weltman, G.; Coeyman, N. Measurement of trust in human-robot collaboration. In Proceedings of the 2007 International Symposium on Collaborative Technologies and Systems, Orlando, FL, USA, 25 May 2007; pp. 106–114. [CrossRef]

38. Park, E.; Jenkins, Q.; Jiang, X. Measuring trust of human operators in new generation rescue robots. *Proc. JFPS Int. Symp. Fluid Power* **2008**, *2008*, 489–492. [CrossRef]

39. Dondrup, C.; Lichtenthäler, C.; Hanheide, M. Hesitation signals in human-robot head-on encounters: A Pilot Study. In Proceedings of the HRI'14: ACM/IEEE International Conference on Human-Robot Interaction, Bielefeld, Germany, 3–6 March 2014; pp. 154–155. [CrossRef]

40. Brooks, C.; Szafir, D. Visualization of Intended Assistance for Acceptance of Shared Control. In Proceedings of the 2020 IEEE/RSJ International Conference on Intelligent Robots and Systems (IROS), Las Vegas, NV, USA, 25–29 October 2021; pp. 11425–11430. [CrossRef]

41. Evans, D.C.; Fendley, M. A multi-measure approach for connecting cognitive workload and automation. *Int. J. Hum.-Comput. Stud.* **2017**, *97*, 182–189. [CrossRef]

42. Jenkins, Q.; Jiang, X. Measuring Trust and Application of Eye Tracking in Human Robotic Interaction. In Proceedings of the IIE Annual Conference and Expo 2010, Cancun, Mexico, 5–9 June 2010.

Article

Infrastructure-Aided Localization and State Estimation for Autonomous Mobile Robots

Daniel Flögel [1], Neel Pratik Bhatt [2] and Ehsan Hashemi [3,*]

[1] Institute for Regulation and Control Systems, Karlsruhe Institute of Technology (KIT), 76131 Karlsruhe, Germany
[2] Mechanical and Mechatronics Engineering Department, University of Waterloo, 200 University Ave W, Waterloo, ON N2L 3G1, Canada
[3] Mechanical Engineering Department, University of Alberta, 9211-116 Street NW, Edmonton, AB T6G 1H9, Canada
* Correspondence: ehashemi@ualberta.ca

Abstract: A slip-aware localization framework is proposed for mobile robots experiencing wheel slip in dynamic environments. The framework fuses infrastructure-aided visual tracking data (via fisheye lenses) and proprioceptive sensory data from a skid-steer mobile robot to enhance accuracy and reduce variance of the estimated states. The slip-aware localization framework includes: the visual thread to detect and track the robot in the stereo image through computationally efficient 3D point cloud generation using a region of interest; and the ego motion thread which uses a slip-aware odometry mechanism to estimate the robot pose utilizing a motion model considering wheel slip. Covariance intersection is used to fuse the pose prediction (using proprioceptive data) and the visual thread, such that the updated estimate remains consistent. As confirmed by experiments on a skid-steer mobile robot, the designed localization framework addresses state estimation challenges for indoor/outdoor autonomous mobile robots which experience high-slip, uneven torque distribution at each wheel (by the motion planner), or occlusion when observed by an infrastructure-mounted camera. The proposed system is real-time capable and scalable to multiple robots and multiple environmental cameras.

Keywords: indoor localization; state estimation; covariance intersection; uncertainty-aware state observer

Citation: Flögel, D.; Bhatt, N.P.; Hashemi, E. Infrastructure-Aided Localization and State Estimation for Autonomous Mobile Robots. *Robotics* **2022**, *11*, 82. https://doi.org/10.3390/robotics11040082

Academic Editors: António Paulo Moreira, Félix Vilariño and Pedro Neto

Received: 11 July 2022
Accepted: 16 August 2022
Published: 18 August 2022

Publisher's Note: MDPI stays neutral with regard to jurisdictional claims in published maps and institutional affiliations.

1. Introduction

Navigating mobile robots in dynamic environments with human presence makes visual odometry challenging due to occlusion and dynamic features. This necessitates multi-modal (e.g., camera, LiDAR, inertial) data fusion to identify and remove the dynamic features for feature-based localization [1,2], address disturbance and model mismatch challenges for LiDAR based localization [3,4], or tackle perceptually degraded conditions through distributed estimation [5,6]. In this regard, multi-modal state estimation approaches for mobile robots [7,8] are revolutionizing accurate navigation for indoor applications (e.g., warehouse robotics or service robots using on-board sensors) where the loss of reception and low bandwidth of commercial Global Navigation Satellite Systems (GNSS), inhibit reliable robot state measurements.

One of main challenges for the the existing multi-modal state estimators that utilize on-board inertial measurement unit (IMU) data and visual odometry through monocular/stereo cameras is the wheel slip in the longitudinal and lateral directions. This is due to: (i) Model uncertainties caused by the wheel force saturation in the robot dynamical model (by various robot payloads, changing surface conditions, or harsh cornering scenarios) impacting estimation error and update frequency in real-time [9–11]; and (ii) The real-time performance of state estimators for safe motion planning and controls in a scene with dynamic features [12,13]. Infrastructure-aided state estimation approaches which leverage visual/radar data measured by fixed sensors and communication with the robot are proposed in the literature to deal with perceptually degraded conditions and dynamic features for navigation of mobile autonomous systems [14–16]. This is cost effective as it reduces the number of on-board sensors, specially for large-scale networked robotic systems. In [17], cameras installed on the ceiling detect multiple robots with unique markers and determine their position and heading states based on the distance to fixed markers on the ground and known marker sizes. A stationary fisheye camera installed on the ceiling is used in [14] for indoor robot localization, in which the pose is determined based on the azimuth and elevation of the line of view (to the center of the segmentation). Multiple fixed surveillance cameras are used in [18] to detect the robot and static objects to construct a 2D map. Pose data from low-cost cameras mounted on ceiling is fused with on-board LiDAR odometry data for robot state estimation in [19] where the fusion of camera and odometry is performed in a map with an adaptive Monte Carlo approach. The existing infrastructure-aided localization approaches require visual markers or free line of sight to the robot [17,19], heavily rely on robot model, and are challenged by occluded scenes and model uncertainties due to the wheel slip.

In order to compensate for the wheel longitudinal/lateral slip in robots with nonholonomic constraints, kinematic- or dynamic-based slip estimation/compensation methods have been adopted in the literature [20,21] using on-board sensory data. The dynamic-based approaches require wheel stiffness properties and vertical forces that may change due to various payloads and road surface conditions [22]. Kinematic-based methods, on the other hand, use wheel odometry and inertial data to estimate the slip with upper bounded mean square estimation error (MSE) through nonlinear or stochastic observers [12,23,24]. A high gain observer is designed in [25] to deal with unknown model parameters. To avoid model complexities due to tire force nonlinearities (and the combined-slip effect), an empirical parameterized kinematic model is proposed in [26] for robot state estimation. An event-based Kalman observer is designed in [27] to fuse IMU data and wheel odometry for heading and speed estimation. However, the information from on-board state observers has not been used for fusion with infrastructure sensing units to enhance reliability of the pose estimation. In addition, the computational efficiency and accuracy are main challenges for the existing infrastructure-mounted visual tracking and localization methods that use low-cost wide-angle lenses.

To address computational time and accuracy challenges of the existing visual and kinematic/dynamic model based localization methods (to be executed on embedded systems and robot's on-board processing units), this paper develops and experimentally verifies a cooperative state estimator using: (i) Proprioceptive data from low-cost odometry sensors of a skid-steer mobile robot; and (ii) Region of Interest (ROI)-based processing and visual tracking on the 3D point clouds obtained from fixed sensing units. The main contributions of the paper are summarized as:

- Design of a computationally efficient ROI-based pose estimator using 3D point clouds from a stationary stereo camera with a wide-angle (fisheye) lens.
- Developing an infrastructure-aided localization framework which is scalable for large systems with multiple robots using communication between a slip-aware onboard observer and the stationary sensing unit.

2. Background and System Overview

The localization framework includes visual tracking through forming an ROI for computationally enhanced processing at the edge (e.g., embedded Jetson Xavier) and a slip-aware state observer at the robot using proprioceptive data. The visual tracking is through a fixed low-cost stereo camera, Intel Realsense T265. As illustrated in Figure 1 the system has independent visual tracking thread and ego motion thread.

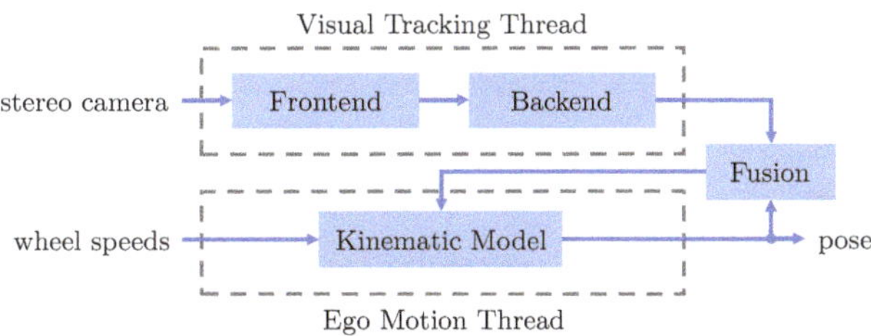

Figure 1. The slip-aware localization framework overview.

The state vector is defined by $\xi(t) = [x(t),\, y(t),\, \theta(t)]^\top$ for the proposed framework, where the longitudinal position, lateral position, and heading of the robot in the reference fixed frame $\{W\}$ is denoted by x, y, and θ, respectively. The local robot body frame is denoted by $\{b\}$, which is at the geometrical center of the robot and is depicted in Figure 2. The reference coordinate system $\{W\}$ is derived from $\{b\}$ at time zero t_0.

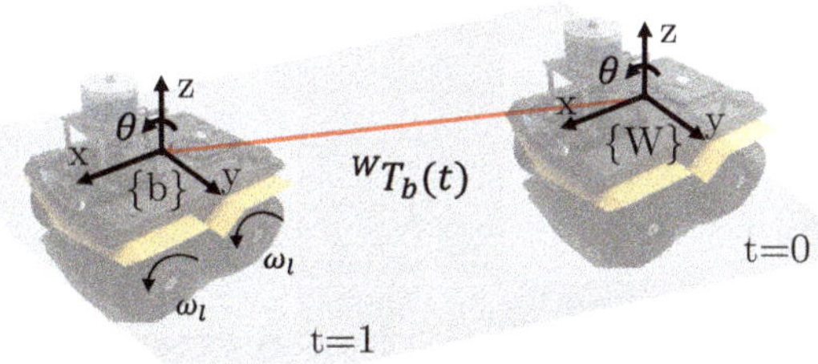

Figure 2. The mobile robot platform and the coordinates.

The visual tracking thread estimates the robot pose $\hat{\xi}^v$ based on the captured images of the stationary stereo camera in the environment. The occlusion cases, in which visual-based pose estimates are intermittent (or not available), will be addressed by the Covariance Intersection (CI) fusion with the estimated states $\hat{\xi}^p$ from the slip-aware motion model. The updated pose by CI is then used as a corrected pose for the relative motion prediction in the next sample time. The robot pose is a time-varying transformation $^W T_b(t) = \begin{bmatrix} ^W R_b & ^W p_b \\ 0 & 1 \end{bmatrix}$ where the rotation matrix $^W R_b$ with $\theta(t)$ is about the Z-axis of the $\{W\}$, and the position vector $^W p_b = [x,\, y,\, 0]^\top$ with x, y is the longitudinal and lateral robot position in the reference frame $\{W\}$.

2.1. Visual Tracking Thread

The visual tracking thread includes frontend and backend modules as illustrated in Figure 3. The frontend performs image processing and object detection. In the image processing step, the stereo image pair is undistorted and rectified. The object detection generates a boundingbox for the robot within the rectified stereo images. The area in the images enclosed with the boundingbox is termed as region of interest. The undistorted and rectified images, and image coordinates of the corresponding bounding box are used in the backend to localize the robot using the 3D position of points on the robot.

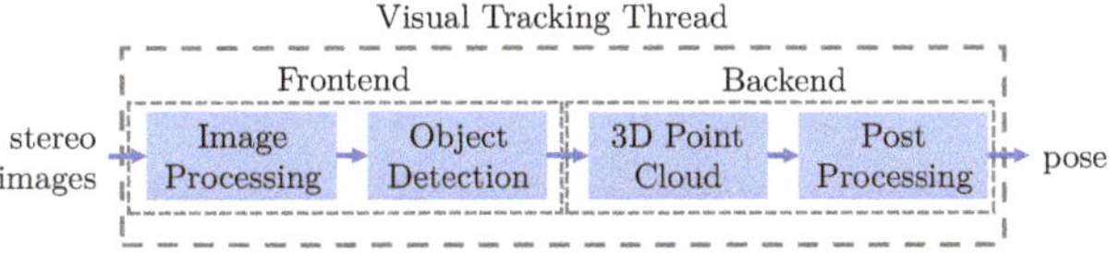

Figure 3. The visual tracking thread with ROI.

With the assumption of the pinhole model and known extrinsic parameters, the constraints for the projection of point clouds in $\{W\}$ onto the two image planes are derived. These constraints are described with epipolar geometry, and determine the area in the image planes where the same point in $\{W\}$ is mapped on. Figure 4 illustrates the epipolar geometry for two non rectified images. The projection of the point m into the camera centers C_1 and C_2 defines the epipolar plane which intersects the image plane P_1 and P_2 forming epipoles e_1 and e_2 for the left/right images. The homogeneous transformation $T = [R, p] \in SO(3)$ with the rotation matrix R and translation vector p between the camera centers describes the extrinsic parameters [28].

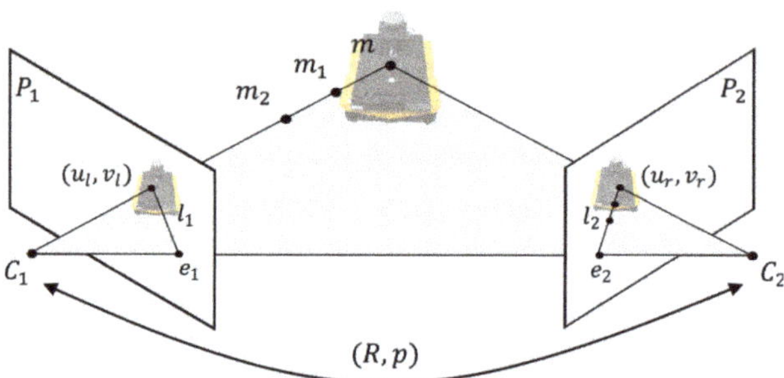

Figure 4. Epipolar geometry for non-rectified stereo images.

The position of a point in $\{W\}$ is determined with the intersection of the projection ray in 3D from the left and right image plane for the same mapped world point. The mapping of the x,y and z coordinate of a point from $\{W\}$ onto the left and right rectified image (Figures 5 and 6) plane as $\bar{u} = [u, v, 1]^\top$ is described as $z\bar{u} = K_j\bar{x}, j \in \{l, r\}$ (l, r denotes the left and right sides, respectively) where $\bar{x} = [x, y, z, 1]^\top$ and

$$K_l = \begin{bmatrix} f & 0 & c_{x1} & 0 \\ 0 & f & c_y & 0 \\ 0 & 0 & 1 & 0 \end{bmatrix}, \ K_r = \begin{bmatrix} f & 0 & c_{x2} & b \cdot f \\ 0 & f & c_y & 0 \\ 0 & 0 & 1 & 0 \end{bmatrix} \tag{1}$$

are the extended camera matrix for the left and right image planes. The images have the same focal length in X and Y direction as well as the same principal point in Y direction, they are geometrically shifted with the baseline b in X direction. The radial distance r for perspective pinhole projection between the principal point and image coordinates of incoming ray of the point m is $r = \sqrt{u^2 + v^2}$ and the angle Ψ between the principal axis and the ray is $\Psi = \tan^{-1}(r)$. The radial fisheye distortion factor Ψ_d is modeled [29] as $\Psi_d = \Psi(1 + k_1\Psi^2 + k_2\Psi^4 + k_3\Psi^6 + k_4\Psi^8)$ with the individual lens distortion parameter $k_i, i \in \{1, \dots, 4\}$. The distorted image coordinates u' and v' are

$$u' = \frac{\Psi_d}{r}u \quad v' = \frac{\Psi_d}{r}v, \tag{2}$$

which are then converted into undistorted image coordinates

$$u = f_x(u' + \alpha v') + c_x, \ v = f_y v' + c_y, \tag{3}$$

Subsequent to this, a Yolov4 object detector [30] is used for 2D detection of the robot in the undistorted left image. The Yolov4 model is trained on a custom collected dataset of the robot for identification of the robot as a class label since the state-of-the-art COCO class labels have no training data corresponding to the robot.

Remark 1. *The output of the Yolov4 custom training detector at $k-$th step is a bounding box $\mathcal{B}_d(k)$ around the robot in the image yielding the extents of the box in the horizontal and vertical directions* *of the image. This enables an ROI which will be used to extract a frustum of the point cloud representing the robot. Point cloud processing will then be applied exclusively to the ROI-based frustum, i.e., interior $\text{Int}(\mathcal{B}_d(k))$. This bounding box-informed frustum significantly reduces the computational cost compared to processing the point cloud as a whole.*

2.2. Point Cloud Computation and Post Processing

The feature extraction is restricted to the ROI, $\text{Int}(\mathcal{B}_d(k))$, and is scalable for visual tracking in multi-robot settings. The robot is depicted inside the ROI in the left and right image plane of the undistorted and rectified images. The aim is to find the image coordinates u_l and v (in the left image) and u_r (in the right image plane) of the world point m, as see in Figures 5 and 6.

Figure 5. Unrectified stereo images with fisheye distortion.

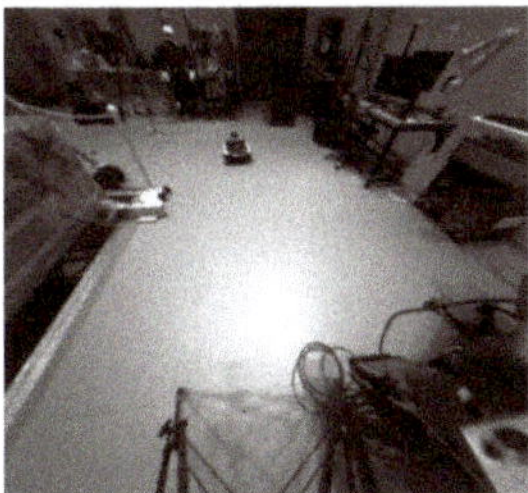

Figure 6. Rectified and undistorted stereo images.

For feature extraction, ORB features [31,32] were used, where the extracted features are matched within the stereo image pair and between subsequent captured image pairs. It is assumed that the remaining image coordinates represent the same point on the robot platform, then, these points' 3D coordinates are reconstructed. Based on the epipolar geometry, the depth $z = \frac{bf}{u_l - u_r}$ is computed for each match with the horizontal image coordinates u_l and u_r of the left and right stereo image and the baseline b, as well as the focal length f of the camera, then the depth is used for $x = \frac{u_l}{f}z$ and $y = \frac{v_l}{f}z$ with the vertical image coordinate v_l of the left stereo image plane as illustrated in Figure 7. The coordinates are computed for every match and transformed into $\{W\}$. All points lead to a point cloud assumed to be derived from the surface of the robot. The point cloud is processed with the PCL library [33,34] and a statistical outlier filter. The filter rejects points that are further away from their neighbors compared to the average of the point cloud. The input parameters are the number of neighbors to calculate the average distance for a given point and a ratio to set the threshold based on the standard deviation across the point cloud.

The 2D projection of the point cloud is used to enhance the reliability of the 3D point clouds for navigating the robot far from the stationary sensing unit (i.e., the stereo visual node). The Euclidean center of the 2D points (which is less sensitive to outliers) is considered as an estimate of the position, i.e., $\hat{\xi}^p(k)$ at time step t_k and will be corrected using the slip-aware motion model, which is described in the next section.

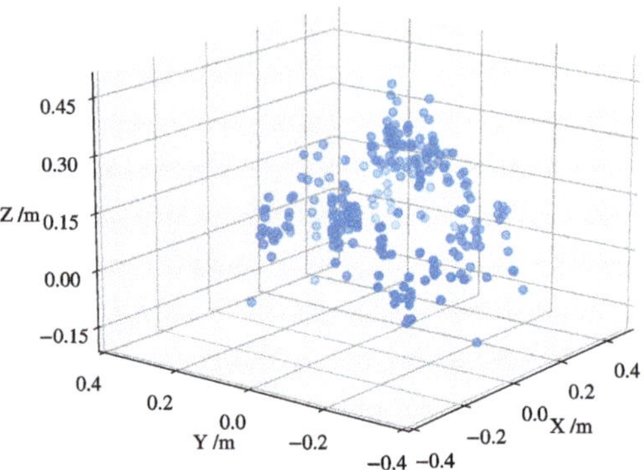

Figure 7. Robot point cloud post processed by the statistical outlier filter. The robot is in the closest position to the stereo camera. The outer dimension of the points is used for a Euclidean distance based sparsity as a measure to be close to the actual geometry of the *Jackal* robot.

For orientation estimation, a linear regressor is used for a moving horizon N_h of the estimated states. The angle between the estimated linear function and the world frame's longitudinal axis is then considered as the orientation of the robot. To cope with situations when the robot is not driving or turning with zero radius, a plausibility check is applied. The plausibility check rejects estimates if the linear regression is too short or the distance between the position estimates and the line is greater than a threshold.

3. Infrastructure-Aided State Estimation

A kinematic model is introduced and parametrized to predict the motion in presence of wheel skidding and slipping. A covariance intersection (CI) method is then used to update the prediction.

3.1. Slip-Aware Motion Model

The autonomous mobile robot used to evaluate the localization approach is the skid-steer *Clearpath's Jackal* robot, which is subject to the large wheel longitudinal slip in various cornering scenarios. The kinematic motion model in the following predicts the robot states using the heading and wheel rotational speed in the robot body frame $\{b\}$. The robot's motion is defined based on the instantaneous center of rotation (IC) as shown in Figure 8, assuming that the robot is a rigid body and has a planar motion with nonholonomic constraints.

The longitudinal velocity, lateral velocity, and yaw rate are denoted by v_x, v_y, and $\dot{\theta}$ in the body frame $\{b\}$ and are expressed in terms of the left/right wheel rotational speeds ω_l, ω_r as

$$\mathbf{v}(t) = G(\Lambda)w(t) = G(\Lambda)\begin{bmatrix} R_e\omega_l(t) \\ R_e\omega_r(t) \end{bmatrix} \tag{4}$$

where $\mathbf{v}(t) = [v_x(t),\, v_y(t),\, \dot{\theta}(t)]^\top$, the wheel rolling radius is denoted by R_e, and $G(\Lambda)$ includes the model parameter vector $\Lambda = [x_{IC},\, y_{IC,l},\, y_{IC,r},\, \alpha_l,\, \alpha_r]$ as follows

$$G(\Lambda) = \frac{1}{\tilde{y}}\begin{bmatrix} -y_{IC,r}\alpha_l & y_{IC,l}\alpha_r \\ -x_{IC}\alpha_l & x_{IC}\alpha_r \\ -\alpha_l & \alpha_r \end{bmatrix}, \quad \tilde{y} = y_{IC,l} - y_{IC,r}, \tag{5}$$

where IC, l is the instantaneous center of the front-left and rear-left tires of the robot and IC, r denotes the instantaneous center of the front-right and rear-right tires of the robot. In the schematic provided in Figure 8, due to nonholonomic constraints and since the longitudinal speed on the right side (i.e., rotational speed multiplied by the effective rolling radius R_e) is larger than the robot speed v_x, the instantaneous center IC, r is located on the right side (i.e., $y_{IC,r} < 0$ in the body frame).

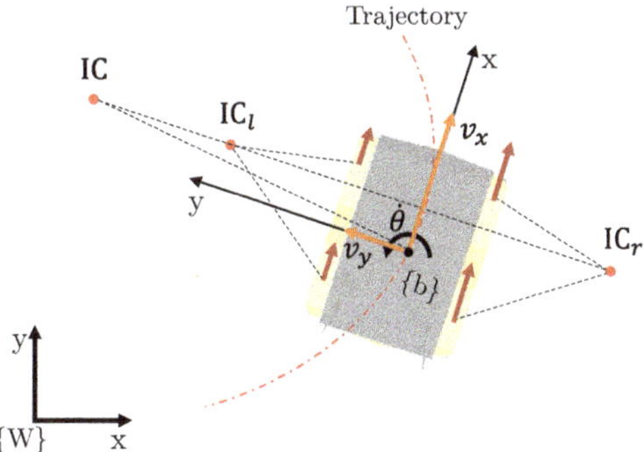

Figure 8. IC-based skid steer kinematics for the motion model.

The instantaneous center is expressed in $\{b\}$ as $(x_{IC,v}, y_{IC,v}) \in \mathbb{R}^2$, where $y_{IC,v} = \frac{v_x}{\dot{\theta}}$ [26]. The IC locations for the left and right wheels are expressed in $\{b\}$ as $(x_{IC,l}, y_{IC,l})$ and $(x_{IC,r}, y_{IC,r})$, respectively. It is assumed that the longitudinal position of ICs along the x-axis lie all on a parallel line to the Y-axis, i.e., $x_{IC} = x_{IC,v} = x_{IC,j} = \frac{v_y}{\dot{\theta}}, j \in \{l, r\}$ and have the same angular velocity. The lateral IC locations, which are bounded variables, are expressed as [21]:

$$y_{IC,j} = \frac{v_x - R_e \omega_j \alpha_j}{\dot{\theta}}, \quad \dot{\theta} = \frac{R_e(\omega_r - \omega_l)}{y_{IC,l} - y_{IC,r}} \tag{6}$$

where α_l and α_r are parameters accounting for model uncertainties (tire inflation and longitudinal slip ratios at each corner of the robot) and R_e is the tire rolling radius. The location of IC is bounded, i.e., $|x_{IC,v}| < \bar{x}$ and $|y_{IC,v}| < \bar{y}$ even reached in the proximity of straight trajectories where the numerator and denominator in (6) are of the same infinitesimal order which leads to finite values for x_{IC}, $y_{IC,j}$.

The boundedness of $y_{IC,v}$ need to be guaranteed for lateral stability and minimizing the robot's sideslip angle in harsh turning. Using the transformation between $\{b\}$ and the world frame, the robot states in $\{W\}$ are expressed as

$$\dot{\xi}(t) = {}^W R_b(t) \cdot \mathbf{v}(t) + \varrho, \quad {}^W R_b = \begin{bmatrix} \cos(\theta) & -\sin(\theta) & 0 \\ \sin(\theta) & \cos(\theta) & 0 \\ 0 & 0 & 1 \end{bmatrix}, \tag{7}$$

where $\theta(t)$ is the robot heading and $\varrho \in \mathbf{R}^3$ represents model uncertainties. Then, the parameter identification process consists of two steps: gathering representative data from on-board and infrastructure-mounted sensory data; and developing an optimization program to find the optimal parameter vector Λ^* through data set. The data collection consists of typically fast maneuvers on various surfaces in different trajectories based on the operational envelope of the mobile robot maintaining the lateral stability. The lateral stability is defined by a bounded sideslip angle $|\beta| < \bar{\beta}$ where $\beta \triangleq \tan^{-1}(\frac{v_y}{v_x})$ on various surface conditions. The wheel rotational speed measurement at each front-left, front-right, rear-left, and rear-right corners of the robot is used for the motion model by compensating the slip ratio component. The training data set (i.e., 12 different step-steer to the left and right, 18 random cornering, and 10 full/circular rotations in large and small path curvatures in indoor settings and on various surfaces) includes N_t independent segments with the training horizon d_t. The measured wheel speeds of each segment are used to predict the robot speeds in the body frame using (4) and determine the robot pose in $\{W\}$ using (7). The predicted pose $\hat{\zeta} \in \mathbb{R}^3$ and the ground truth at the end of each segment are included in the cost function

$$\Lambda^* = \min_{\Lambda} J(\Lambda), \quad J(\Lambda) = \sum_{i=1}^{N_h} ||\xi - \hat{\zeta}(\Lambda)_i||_2, \tag{8}$$

where $\hat{\zeta}(\Lambda)$ is the ground truth and $\zeta(k)$ is the predicted state based on the linearized slip-aware motion model in discrete times. Minimizing J results in the optimal parameter vector Λ^*. The trained model is evaluated over different data sets with the evaluation horizon d_e. In this context, the evaluation horizon represents the prediction horizon for specific applications. The evaluation horizon is the indication of the prediction horizon of the model in the application. Assessing variable evaluation horizons with respect to variable training horizons is reveals the impact of different prediction horizons in the application compared to the parameter identification process.

To analyse the impact of different training and evaluation horizons, the mean relative translation/rotation errors are provided in Figures 9 and 10. The analysis reveals that the best performance is achieved if the evaluation horizon is equal to (or less than) 0.5 m. The error increases for larger deviation but remains bounded and lower than 5%.

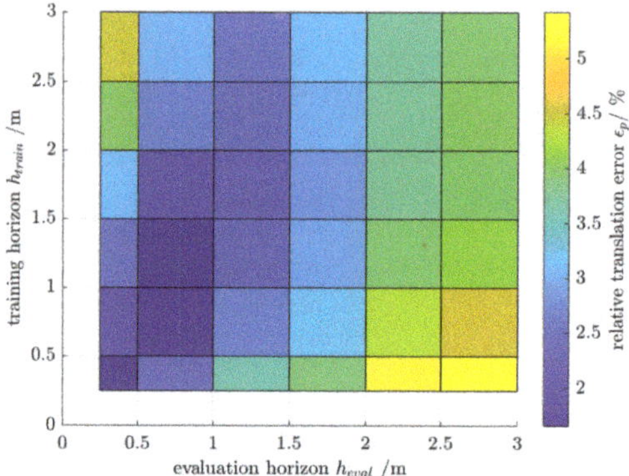

Figure 9. Relative translational error ϵ_p of the motion model parameter identification for varying training and evaluation horizons on the same ground classification (i.e., gravel or asphalt).

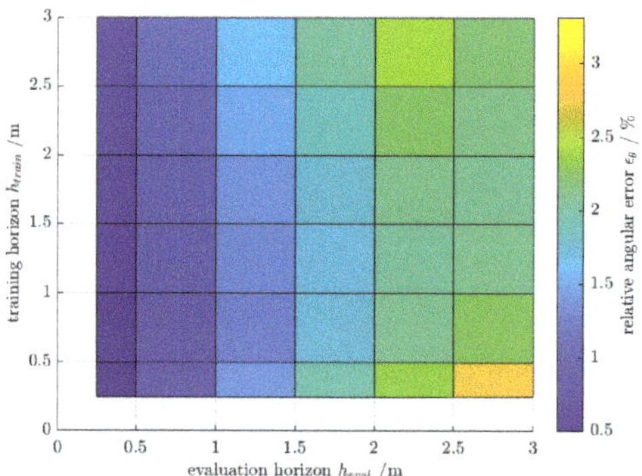

Figure 10. Relative angular error ϵ_θ of the motion model parameter identification for varying training and evaluation horizons on the same ground classification (i.e., gravel or asphalt).

3.2. Pose Prediction

The prediction model in (7), with elements from (4)–(6), is linearized around the operating point $(\xi^p(k), \mathbf{w}(k))$ at each time step k in discrete times, where $\xi^p(k) = [x(k), y(k), \theta(k)]^\top$ is the robot's pose by the ego motion thread. The linear affine prediction model can be written as:

$$\xi^p(k+1) = A(k)\xi^p(k) + B(k)\mathbf{w}(k) + \varrho(k), \tag{9}$$

whereas the zero-mean term ϱ is due to model uncertainties. The discrete-time realization is approximated by

$$A(k) := \phi^{A^c}_{t_{k+1}, t_k} \approx e^{A^c(t_k)T_s} \in \mathbb{R}^{3\times 3} \tag{10}$$

and

$$B(k) := \int_{t_k}^{t_{k+1}} \phi^{A^c(t_{k+1}), \tau} B^c(\tau) d\tau \approx \int_{t_k}^{t_{k+1}} e^{A^c(t_k)(t_{k+1}-\tau)} d\tau B^c(t_k) \tag{11}$$

whereas A^c, B^c are the continuous-time system and input matrices of the linearized prediction model, and $\phi^{A^c}_{t_i,t_j}$ for $t_i > t_j$ is the continuous-time state transition matrix expressed by the Peano-Baker series; the realization is assumed to not vary a lot in each interval $[t_k, t_{k+1}]$, which is valid for the proposed cooperative mobile robot localization model with the sampling time $T_s = 25$ ms. As a result, the bound of uncertainty due to the sampling time for discretization (in the slip-aware motion model) at the maximum speed of 1 m/s, at which the robot may experience wheel longitudinal slip, is 25 mm. Then, the expected state prediction from the ego motion thread is

$$\bar{\xi}^p(k+1) = A(k)\bar{\xi}^p(k) + B(k)\bar{\mathbf{w}}(k), \tag{12}$$

whereas $\bar{\xi}^p(k) = \mathbb{E}\{\xi^p(k)\}$ and $\bar{\mathbf{w}}(k) = \mathbb{E}\{\mathbf{w}(k)\}$; the joint covariance for $\mathbf{x} = [\xi^p(k), \mathbf{w}(k)]^\top$ is then given by

$$\text{cov}(\mathbf{x}) = \begin{bmatrix} Q_\xi(k) & 0 \\ 0 & Q_w(k) \end{bmatrix} = \mathbb{E}\left\{ \begin{bmatrix} \xi^p(k) - \bar{\xi}^p(k) \\ \mathbf{w}(k) - \bar{\mathbf{w}}(k) \end{bmatrix} \left[(\xi^p(k) - \bar{\xi}^p(k))^\top, (\mathbf{w}(k) - \bar{\mathbf{w}}(k))^\top \right] \right\}. \tag{13}$$

The predicted covariance is

$$Q_\xi(k+1) = \mathbb{E}\{[\xi^p(k+1) - \bar{\xi}^p(k+1)][\xi^p(k+1) - \bar{\xi}^p(k+1)]^\top\} \tag{14}$$

in which

$$\xi^p(k+1) - \bar{\xi}^p(k+1) = A(k)[\xi^p(k) - \bar{\xi}^p(k)] + \varrho(k). \tag{15}$$

Then, by using $\text{cov}(\mathbf{x})$, the predicted covariance from the slip-aware motion model yields:

$$Q_\xi(k+1) = A(k)Q_\xi(k+1)A^\top(k) + B(k)Q_w(k)B^\top(k). \tag{16}$$

3.3. Augmented Localization

The visual thread and the ego motion thread communicate within the ROS framework through WiFi for the specific mobile robot test platform. To ensure proper data synchronization, time stamps are used to associate the visual-based localization (i.e., state estimation of $\bar{\xi}^v(k)$) to the corresponding pose estimation $\bar{\xi}^p_k$ by the slip-aware model description. Delay in the communication, which is less than 20 ms for the tests conducted within 10 m of the stationary visual node (i.e., infrastructure-mounted stereo camera with the fisheye lens), is ignored in this section for the CI fusion. This is a valid assumption considering the sampling time $T_s = 25$ ms for the pose prediction in the slip-aware motion model, the fusion part's sampling time (i.e., 50 ms), and the maximum robot speed of 1 m/s at which the robot may experience wheel's longitudinal slip. Denoting the estimation error in the slip-aware motion model at time step k by $\tilde{\xi}^p(k) = \xi^p(k) - \hat{\xi}^p(k)$, and the visual thread by $\tilde{\xi}^v(k) = \xi^v(k) - \hat{\xi}^v(k)$, we utilize the covariance intersection method having the upper bound of the mean square estimation error and the *consistency* condition.

Remark 2. *The asymptotic stable state transition matrix of the error dynamics $\tilde{\xi}^p$ in the motion model* (9), *and the geometrical filters for the visual-based depth estimation guarantee that the mean square estimation error (MSE) for the pose prediction model and the visual localization are upper bounded, i.e., $\tilde{Q}_p(k) := \mathbb{E}\{\tilde{\xi}^p(k)\tilde{\xi}^{p\top}(k)\} \leq Q_p(k)$ and $\tilde{Q}_v(k) := \mathbb{E}\{\tilde{\xi}^v(k)\tilde{\xi}^{v\top}(k)\} \leq Q_v(k)$. As a result, the error covariance $\tilde{Q}_v(k)$ and $\tilde{Q}_p(k)$ of the estimated states from the two threads are consistent.*

The estimated states from the ego motion thread and the visual thread are then fused using CI which is a convex combination of the covariances of the estimated states and guarantees a consistent error covariance (i.e., $\tilde{Q}_f \leq Q_f$). The CI is a geometric interpretation of

$$\tilde{Q}_f = W_p \tilde{Q}_p W_p^\top + W_p \tilde{Q}_{pv} W_v^\top + W_v \tilde{Q}_{vp} W_p^\top + W_v \tilde{Q}_v W_v^\top, \tag{17}$$

in which for all choices of $\tilde{Q}_{pv}$, the covariance ellipses of the bound Q_f at level c,

$$\mathcal{E}_{Q_f}^c := \{z \in \mathbb{R} : z^\top Q_f^{-1} z < c\}, \tag{18}$$

lies within the intersection of covarinace ellipses of Q_p and Q_v, i.e., $\mathcal{E}_{Q_f}^c \subset \mathcal{E}_{Q_p}^c \cap \mathcal{E}_{Q_v}^c$.

The weights W_p, W_v will be obtained by minimizing a performance index on the bound Q_f, e.g., $\text{tr}(Q_f)$ or $\det(Q_f)$, and consequently the covariance $\tilde{Q}_f$. The CI update strategy finds Q_f which encloses the intersection area $\mathcal{E}_{Q_p}^c \cap \mathcal{E}_{Q_v}^c$ and is consistent, although no knowledge about Q_{pv} is available. The upper bounds of the covariance matrix elements for visual pose estimates is set to constant values derived from the error analysis (discussed in the next section) For the case where $\tilde{Q}_{pv} \neq 0$, the covariance Q_f can be given by

$$Q_f = [W_p,\ W_v] \underbrace{\begin{bmatrix} Q_p & \tilde{Q}_{pv} \\ \tilde{Q}_{vp}^\top & Q_v \end{bmatrix}}_{Q} \begin{bmatrix} W_p^\top \\ W_v^\top \end{bmatrix}, \tag{19}$$

in which the optimal W_p, W_v that minimize $\text{tr}(Q_f)$ is obtained from the following constrained optimization program

$$\min_W \text{tr}(Q_f)$$

$$\text{s.t.:} W_p, + W_v = \mathbf{I}, \tag{20}$$

where $\mathbf{I}$ is the identity matrix with the proper dimension. The trace minimization program in (20) yields $(Q_f)^{-1} = (Q_p^{-1}\tilde{Q}_{pv} - \mathbf{I})(Q_v - \tilde{Q}_{pv}^\top Q_p^{-1}\tilde{Q}_{pv})^{-1}(\tilde{Q}_{pv}^\top Q_p^{-1} - \mathbf{I}) + Q_p^{-1}$. As a result, the fusion of the estimated states from the ego motion and the visual threads is

$$\hat{\xi}^f(k) = Q_f(k)\left[W_p(Q_p(k))^{-1}\hat{\xi}^p(k)\right.$$

$$\left. + (1 - W_p)(Q_v(k))^{-1}\hat{\xi}^v(k)\right],$$

$$[Q_f(k)]^{-1} = W_p(Q_p(k))^{-1} + (1 - W_p)(Q_v(k))^{-1}, \tag{21}$$

where $W_p \in [0, 1]$ adjusts the assigned weights to $\hat{\xi}^p$ and $\hat{\xi}^v$ minimizing the performance index $\text{tr}(Q^f)$ of the updated covariance.

According to the consistency in Remark 2 and the property of CI, it holds that

$$\mathbb{E}\{(\hat{\xi}^f(k) - \bar{\xi}(k))(\hat{\xi}^f(k) - \bar{\xi}(k))^\top\} \leq Q_f(k). \tag{22}$$

The heading of the robot is fused once the robot is close to the camera, thus, measurements are more accurate and reliable. The slip-aware observer and fusion is described in Algorithm 1.

Algorithm 1: Augmented Slip-Aware Localization

Input : Stereo image (with fisheye lens distortion), robot's wheel speed, and initial estimate $\hat{\xi}^f(0)$

Output: Robot position and heading states $\hat{\xi}^f(k)$

while $k \geq 0$ **do**

    1.   `Undistortion & object Detect.`

    (i) Use (3) for u, v;

    (ii) $z\bar{u} = K_j\bar{x}$ with K_j in (1) for stereo images $\mathcal{I}_j(k), j \in \{l, r\}$;

    (iii) $\mathcal{B}_d(k) \leftarrow$ detected bounding box by Yolov4;

    2.   `ROI-based frustum for pointcloud (PC) processing (visual`
    `thread)`

    if $p^i(k) \in Int(\mathcal{B}_d(k)), p^i(k) \in \mathcal{I}_{l,r}(k)$ **then**

        (i) Extract ORB feat. $\{f_j^i \in \mathcal{F} | f_j^i \in Int(\mathcal{B}_d(k))\}$;

        (ii) Match features f_l^i, f_r^i to form the PC;

        (iii) Calculate the depth and estimate $\hat{\xi}^v(k)$ by 2D projection of PC;

        (iv) Plausibility check on states over horizon N_h

    else

        Recheck for occlusion in long distances $z^i \geq z_{th}$

    end

    3.   `Adaptive set allocation`

    For each wheel $q \in \bar{\mathcal{S}}, \bar{\mathcal{S}} := \{1, 2, 3, 4\}$

    if $|\dot{\omega}_q(k)| \geq \bar{\omega} \triangleq \frac{1}{n}\sum_{k-n+1}^{k} \dot{\omega}_q(k)$ **then**

        $\mathcal{S}^p(k) \leftarrow \bar{\mathcal{S}} \setminus \{q\}$;

    end

    4.   `Slip-aware pose estimation`

    if $\mathcal{S}^p(k) \neq$ **then**

        (i) Form $G(\Lambda)$ to estimate $\hat{\xi}^p$ on the discrete-time model of (7) (i.e., (9));

        (ii) Use the trained model for Λ^*; MSE for the pose prediction model is bounded, i.e., $\tilde{Q}_p(k) := \mathbb{E}\{\tilde{\xi}^p(k)\tilde{\xi}^{p\top}(k)\} \leq Q_p(k))$;

        (iii) Use CI on $\hat{\xi}^v(k)$ and $\hat{\xi}^p$

        Estimate $\hat{\xi}^f(k)$ by (21) with $Q_f(k) \leftarrow CI(Q_{p,v}^{-1}(k))$; with consistency

        $\mathbb{E}\{(\hat{\xi}^f(k) - \bar{\xi}(k))(\hat{\xi}^f(k) - \bar{\xi}(k))^T\} \leq Q_f(k)$

    else

        $\hat{\xi}^f(k) \leftarrow \hat{\xi}^f(k-1)$

    end

end

---

4. Experiments and Discussion

The proposed infrastructure-aided localization framework is experimentally evaluated in this section through tests with harsh turning, cornering with acceleration/deceleration, and accelerated straight maneuvers which all include longitudinal slip at each wheel. The reference measurement and system setup is first discussed, then the experimental evaluations are provided. The wheel slip during harsh cornering, with nonholonomic constraints, results is reduced pose estimation accuracy for the existing odometry-based motion models which rely on wheel rotational speed. This has been addressed in this paper by the proposed slip-aware motion model (considering instantaneous centers of rotation) and the a multi-modal data fusion with the visual thread (even with distortion challenges imposed by low-cost fisheye lens).

The ground truth trajectory is recorded with the optical motion capture system *Vicon Vantage V5*. The test setup is composed of the *Vicon* system, the autonomous mobile robot (*Jackal* AGV), and the stationary stereo camera T265. The T265 is fixed mounted on a tripod at a height of 2 m and capturing the whole area where the tests are conducted. The robot is operating under the normal path-tracking mode and starting in front of the tripod of T265, with the speed between 0.4 and 1 m/s, and mild and harsh cornering in tight and wide trajectories. In the proposed motion model, the wheel slip is indirectly quantified as a kinematic model parameter.

To detect the robot and initial setup of the stereo camera in the environment, passive markers are mounted on top of the robot and the stationary stereo camera, as shown in Figure 11, having sufficient distance for a rotation invariant geometry which is essential to ensure a unique pose and proper localization results using the *Vicon* system.

Figure 12 shows the visual point cloud of the robot detected under occlusion (by a human/user) in a long distance.

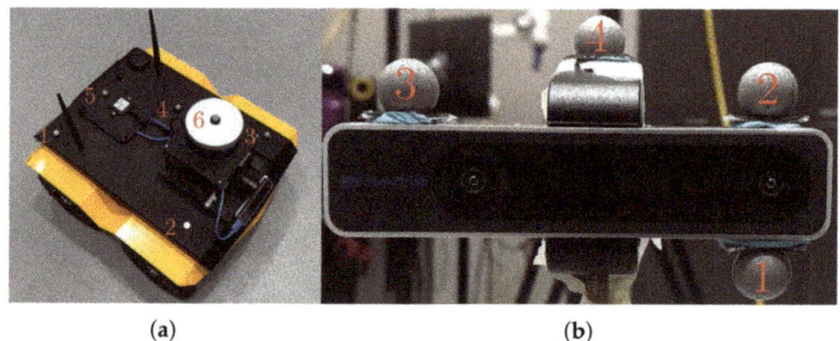

(a)   (b)

Figure 11. The experimental setup using *Vicon* (**a**) *Clearpath's Jackal* robot equipped with 16-line LiDARs (from RoboSense or Velodyne) for motion planning and controls in dynamic environments. (**b**) Infrastructure-mounted low-cost stereo vision for the augmented localization through dedicated short-range communication with the on-board state estimator.

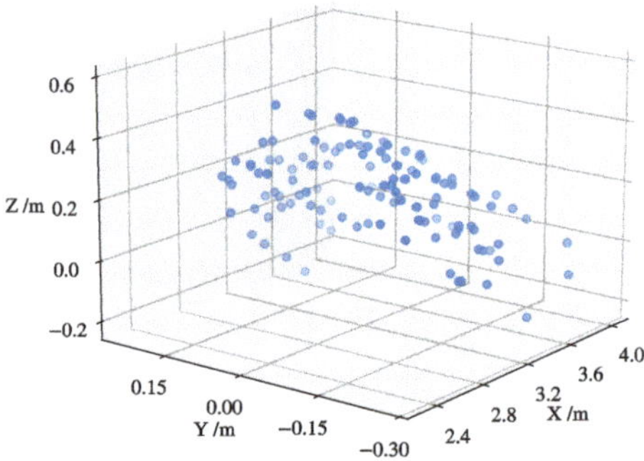

Figure 12. Robot point cloud with a statistical outlier filter for a detection with partial occlusion in a dynamic environment in a far (i.e., 7.8 m) range. This depicts the effect of far detection and partial occlusion (by an object/human) on the quality and sparsity of the point cloud used for clustering and pose estimation; with a predicted longitudinal dimension of 2 m in x-direction, the point cloud does not corresponds the robot dimension. The CI based fusion resolves partial occlusion/detection as will be illustrated for pose estimation later in this section.

The ROI-based point cloud processing, which generates point cloud within the 2D bounding box of the detected robot, reduced computation time up to 67% as has experimentally been tested with the robot in dynamic indoor environments with human presence. The processing time for the pose prediction based on the slip-aware motion model is almost <5 ms. There is no exhaustive recursive algorithm associate with the motion model part. The visual thread with the ROI-based processing takes up to 16 ms in various harsh turn and random cornering maneuvers. The fusion part with the trace optimization program on the visual and motion threads take up to 35 ms on the utilized embedded system in dynamic environments with human presence.

The position estimation error by the stereo visual thread is shown in Figure 13 for a maneuver with several tight cornering. The largest error of 21 cm is for the situation in which the robot is occluded (by a human/user in a shared working indoor environment) in a far (i.e., 7.8 m) distance. The slip-aware motion model helps CI to recover the robot pose guaranteeing consistency of the estimation error covariance, i.e., $\mathbb{E}\{(\tilde{\xi}^f(k)[\tilde{\xi}^f(k)]^\top\} \leq Q_f(k)$.

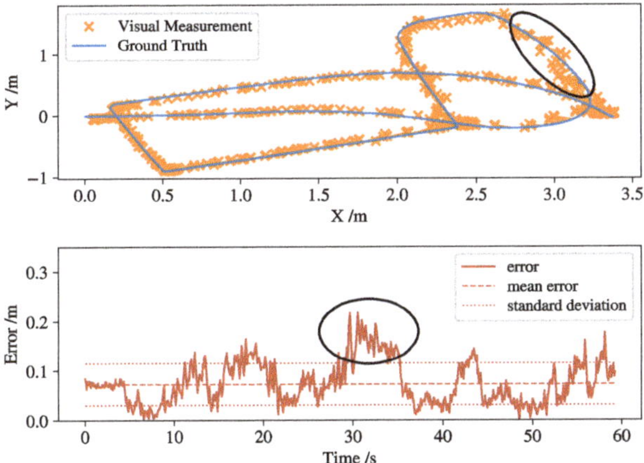

Figure 13. Position estimation results based on Euclidean center of the point cloud from the mobile robot. The T265 camera is located at position (0,0) facing the longitudinal x-direction. The largest error occurs at the maximum relative position (indicated with a black ellipse) between the robot and infrastructure-mounted stereo camera.

The heading fusion result is depicted in Figure 14, where the heading prediction by the ego motion thread (without visual thread updates) is shown in dotted lines; this heading has large estimation error due to the harsh cornering scenarios and inaccuracies in the position of instantaneous center for the slip-aware ego motion model. The prediction fused with pose update from the visual thread in Figure 14 confirms better performance even with occlusion in this perceptually degraded test. This is due to the fact that the heading estimator (by the visual thread) employs multiple geometrical and nonholonomic constraints for the robot motion.

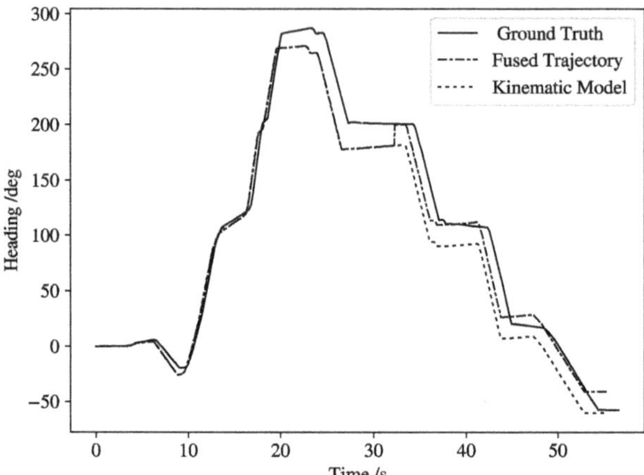

Figure 14. Orientation estimation by the infrastructure-aided localization framework.

The position fusion results are illustrated in Figure 15 which confirms improvements in the estimation error when CI is applied using the visual thread to address uncertainties in the slip-aware ego motion thread in such arduous scenarios. The position prediction is fused with visual thread data, intentionally at each 200 ms to evaluate the effectiveness in large sample time updates or possible packet drop. Once the heading estimates are corrected by CI, the localization data is accurate with the root mean square error (RMSE) $\leq 17\%$ for several tests even with intermittent CI updates. The triangular shapes show the effect of the fusion process in which the kinematic motion model has been corrected and fused with the visual thread data. The kinematic model, a dead reckoning system, suffers from fault propagation and has an higher uncertainty as well as biased position prediction. Once the position is corrected with the visual localization, the corrected position and new initial position for the dead reckoning system moves close to the ground truth. Increasing the frequency of the update by the CI fusion will smooth the final estimates.

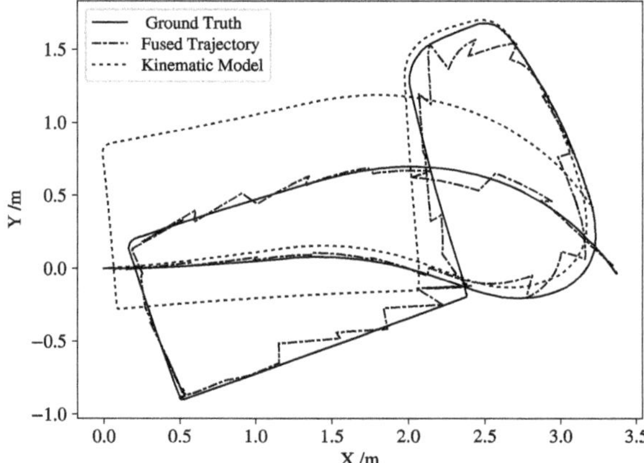

Figure 15. Position estimates by the infrastructure-aided localization; slip-aware motion model handles occlusion and uncertainties in the point cloud computation for the robot detection in far distances.

5. Conclusions

An augmented state estimation framework was proposed for localization of autonomous mobile robots in dynamic environments using infrastructure-mounted visual sensors and on-board data. The proposed system is composed of a visual tracking thread based on a stationary low-cost fisheye stereo camera mounted in the environment and a slip-aware ego motion thread that uses proprioceptive sensory data from a skid-steer mobile robot to enhance accuracy and reduce variance of the estimated states. The position and heading of the robot was estimated using the visual thread with a region of interest-based 3D point cloud processing which reduced the computation up to 67% in dynamic indoor environments with human presence. This significantly enhances the real-time processing capability of the infrastructure-mounted sensing unit for localization and tracking of multi robots in indoor settings. A slip-aware kinematic model was developed for the ego motion thread to predict the robot pose, then, covariance intersection with guaranteed consistency was used to update the pose prediction with visual estimates, addressing slippage and occlusion for wheel odometry based state estimators and visual based localization in dynamic environments. The experimental results confirmed RMSE $\leq 17\%$ and an average position accuracy of 7 cm for various tests even with intermittent (e.g., 0.2 s) CI updates. The real time capability of the state estimation framework was confirmed by the computation time 35 ms for ROI-based visual processing and the fusion (through trace minimization). The future avenues include: (i) Using a motion model in the visual thread to enhance the consistency of the pose estimation; (ii) Integrating the IMU data into the ego motion thread and developing a motion model connecting wheel speeds, longitudinal slips, and robot dynamics within an optimization problem constrained to the robot kinematic/dynamic constrains to enhance orientation estimation.

Author Contributions: Conceptualization, E.H. and D.F.; methodology, D.F. and E.H.; writing, D.F. and N.P.B. and E.H.; simulation, D.F. and N.P.B.; data analysis, D.F. and E.H.; visualization, D.F.; review and editing, D.F. and E.H. and N.P.B.; funding acquisition E.H.; project administration E.H.; supervision, E.H. and D.F.; D.F. is with the Institute for Regulation and Control Systems, Karlsruhe Institute of Technology (KIT), Germany, and conducted his Master's thesis at the NODE lab in Canada (under supervision of E.H.). All authors have read and agreed to the published version of the manuscript.

Funding: This work is supported by the Natural Science and Engineering Research Council of Canada, Discovery Grants RGPIN-05097-2020.

Institutional Review Board Statement: Not applicable.

Informed Consent Statement: Not applicable.

Data Availability Statement: Not applicable.

Acknowledgments: The authors thank University of Waterloo's RoboHub for their valuable technical support, and Christian Bürkert Foundation for their financial support.

Conflicts of Interest: The authors declare no conflict of interest.

References

1. Sun, D.; Geißer, F.; Nebel, B. Towards effective localization in dynamic environments. In Proceedings of the 2016 IEEE/RSJ International Conference on Intelligent Robots and Systems (IROS), Daejeon, Korea, 9–14 October 2016; pp. 4517–4523. [CrossRef]
2. Campos, C.; Elvira, R.; Rodríguez, J.J.G.; Montiel, J.M.; Tardós, J.D. Orb-slam3: An accurate open-source library for visual, visual–inertial, and multimap slam. *IEEE Trans. Robot.* **2021**, *37*, 1874–1890. [CrossRef]
3. Hess, W.; Kohler, D.; Rapp, H.; Andor, D. Real-time loop closure in 2D LIDAR SLAM. In Proceedings of the 2016 IEEE International Conference on Robotics and Automation (ICRA), Stockholm, Sweden, 16–21 May 2016; pp. 1271–1278.
4. Shan, T.; Englot, B. Lego-loam: Lightweight and ground-optimized lidar odometry and mapping on variable terrain. In Proceedings of the 2018 IEEE/RSJ International Conference on Intelligent Robots and Systems (IROS), Madrid, Spain, 1–5 October 2018; pp. 4758–4765.
5. Yang, P.; Freeman, R.A.; Lynch, K.M. Multi-agent coordination by decentralized estimation and control. *IEEE Trans. Autom. Control* **2008**, *53*, 2480–2496. [CrossRef]

6. He, X.; Hashemi, E.; Johansson, K.H. Event-Triggered Task-Switching Control Based on Distributed Estimation. *IFAC PapersOnLine* **2020**, *53*, 3198–3203. [CrossRef]

7. Chung, H.Y.; Hou, C.C.; Chen, Y.S. Indoor intelligent mobile robot localization using fuzzy compensation and Kalman filter to fuse the data of gyroscope and magnetometer. *IEEE Trans. Ind. Electron.* **2015**, *62*, 6436–6447. [CrossRef]

8. Qin, T.; Li, P.; Shen, S. Vins-mono: A robust and versatile monocular visual-inertial state estimator. *IEEE Trans. Robot.* **2018**, *34*, 1004–1020. [CrossRef]

9. Berntorp, K. Joint wheel-slip and vehicle-motion estimation based on inertial, GPS, and wheel-speed sensors. *IEEE Trans. Control Syst. Technol.* **2016**, *24*, 1020–1027. [CrossRef]

10. Zhou, S.; Liu, Z.; Suo, C.; Wang, H.; Zhao, H.; Liu, Y.H. Vision-based dynamic control of car-like mobile robots. In Proceedings of the 2019 International Conference on Robotics and Automation (ICRA), Montreal, QC, Canada, 20–24 May 2019; pp. 6631–6636.

11. Chae, H.W.; Choi, J.H.; Song, J.B. Robust and Autonomous Stereo Visual-Inertial Navigation for Non-Holonomic Mobile Robots. *IEEE Trans. Veh. Technol.* **2020**, *69*, 9613–9623. [CrossRef]

12. Tian, Y.; Sarkar, N. Control of a mobile robot subject to wheel slip. *J. Intell. Robot. Syst.* **2014**, *74*, 915–929. [CrossRef]

13. Kubelka, V.; Oswald, L.; Pomerleau, F.; Colas, F.; Svoboda, T.; Reinstein, M. Robust data fusion of multimodal sensory information for mobile robots. *J. Field Robot.* **2015**, *32*, 447–473. [CrossRef]

14. Delibasis, K.K.; Plagianakos, V.P.; Maglogiannis, I. Real time indoor robot localization using a stationary fisheye camera. In *Proceedings of the IFIP International Conference on Artificial Intelligence Applications and Innovations*; Springer: Berlin/Heidelberg, Germany, 2013; pp. 245–254.

15. Janković, N.V.; Ćirić, S.V.; Jovičić, N.S. System for indoor localization of mobile robots by using machine vision. In Proceedings of the 2015 23rd Telecommunications Forum Telfor (TELFOR), Belgrade, Serbia, 24–26 November 2015; pp. 619–622.

16. Mamduhi, M.H.; Hashemi, E.; Baras, J.S.; Johansson, K.H. Event-triggered Add-on Safety for Connected and Automated Vehicles Using Road-side Network Infrastructure. *IFAC-PapersOnLine* **2020**, *53*, 15154–15160. [CrossRef]

17. Zickler, S.; Laue, T.; Birbach, O.; Wongphati, M.; Veloso, M. SSL-vision: The shared vision system for the RoboCup Small Size League. In *Robot Soccer World Cup*; Springer: Berlin/Heidelberg, Germany, 2009; pp. 425–436.

18. Shim, J.H.; Cho, Y.I. A mobile robot localization via indoor fixed remote surveillance cameras. *Sensors* **2016**, *16*, 195. [CrossRef]

19. Ramer, C.; Sessner, J.; Scholz, M.; Zhang, X.; Franke, J. Fusing low-cost sensor data for localization and mapping of automated guided vehicle fleets in indoor applications. In Proceedings of the 2015 IEEE International Conference on Multisensor Fusion and Integration for Intelligent Systems (MFI), San Diego, CA, USA, 14–16 September 2015; pp. 65–70.

20. Ordonez, C.; Gupta, N.; Reese, B.; Seegmiller, N.; Kelly, A.; Collins, E.G., Jr. Learning of skid-steered kinematic and dynamic models for motion planning. *Robot. Auton. Syst.* **2017**, *95*, 207–221. [CrossRef]

21. Liu, F.; Li, X.; Yuan, S.; Lan, W. Slip-aware motion estimation for off-road mobile robots via multi-innovation unscented Kalman filter. *IEEE Access* **2020**, *8*, 43482–43496. [CrossRef]

22. Gosala, N.; Bühler, A.; Prajapat, M.; Ehmke, C.; Gupta, M.; Sivanesan, R.; Gawel, A.; Pfeiffer, M.; Bürki, M.; Sa, I.; et al. Redundant perception and state estimation for reliable autonomous racing. In Proceedings of the 2019 International Conference on Robotics and Automation (ICRA), Montreal, QC, Canada, 20–24 May 2019; pp. 6561–6567.

23. Wang, D.; Low, C.B. Modeling and analysis of skidding and slipping in wheeled mobile robots: Control design perspective. *IEEE Trans. Robot.* **2008**, *24*, 676–687. [CrossRef]

24. Rabiee, S.; Biswas, J. A friction-based kinematic model for skid-steer wheeled mobile robots. In Proceedings of the 2019 International Conference on Robotics and Automation (ICRA), Montreal, QC, Canada, 20–24 May 2019; pp. 8563–8569.

25. Huang, J.; Wen, C.; Wang, W.; Jiang, Z.P. Adaptive output feedback tracking control of a nonholonomic mobile robot. *Automatica* **2014**, *50*, 821–831. [CrossRef]

26. Mandow, A.; Martinez, J.L.; Morales, J.; Blanco, J.L.; Garcia-Cerezo, A.; Gonzalez, J. Experimental kinematics for wheeled skid-steer mobile robots. In Proceedings of the 2007 IEEE/RSJ International Conference on Intelligent Robots and Systems, San Diego, CA, USA, 29 October–2 November 2007; pp. 1222–1227.

27. Marín, L.; Vallés, M.; Soriano, Á.; Valera, Á.; Albertos, P. Multi sensor fusion framework for indoor-outdoor localization of limited resource mobile robots. *Sensors* **2013**, *13*, 14133–14160. [CrossRef]

28. Szeliski, R. *Computer Vision: Algorithms and Applications*; Springer Science & Business Media: Berlin/Heidelberg, Germany, 2010.

29. Kannala, J.; Brandt, S.S. A generic camera model and calibration method for conventional, wide-angle, and fish-eye lenses. *IEEE Trans. Pattern Anal. Mach. Intell.* **2006**, *28*, 1335–1340. [CrossRef]

30. Bochkovskiy, A.; Wang, C.Y.; Liao, H.Y.M. YOLOv4: Optimal Speed and Accuracy of Object Detection. *arXiv* **2020**, arXiv:2004.10934.

31. Rublee, E.; Rabaud, V.; Konolige, K.; Bradski, G. ORB: An efficient alternative to SIFT or SURF. In Proceedings of the 2011 International Conference on Computer Vision, Barcelona, Spain, 6–13 November 2011; pp. 2564–2571.

32. Gupta, S.; Kumar, M.; Garg, A. Improved object recognition results using SIFT and ORB feature detector. *Multimed. Tools Appl.* **2019**, *78*, 34157–34171. [CrossRef]

33. Holz, D.; Ichim, A.E.; Tombari, F.; Rusu, R.B.; Behnke, S. Registration with the point cloud library: A modular framework for aligning in 3-D. *IEEE Robot. Autom. Mag.* **2015**, *22*, 110–124. [CrossRef]

34. Munaro, M.; Rusu, R.B.; Menegatti, E. 3D robot perception with point cloud library. *Robot. Auton. Syst.* **2016**, *78*, 97–99. [CrossRef]

Article

Mutli-Robot Cooperative Object Transportation with Guaranteed Safety and Convergence in Planar Obstacle Cluttered Workspaces via Configuration Space Decomposition

Panagiotis Vlantis [1], Charalampos P. Bechlioulis [1,*] and Kostas J. Kyriakopoulos [2]

1   Department of Electrical and Computer Engineering, University of Patras, 26504 Rio, Greece
2   School of Mechanical Engineering, National Technical University of Athens, 15772 Athens, Greece
*   Correspondence: chmpechl@upatras.gr

Abstract: In this work, we consider the autonomous object transportation problem employing a team of mobile manipulators within a compact planar workspace with obstacles. As the object is allowed to translate and rotate and each robot is equipped with a manipulator consisting of one or more moving links, the overall system (object and mobile manipulators) should adapt its shape in a flexible way so that it fulfills the transportation task with safety. To this end, we built a sequence of configuration space cells, each of which defines an allowable set of configurations of the object, as well as explicit intervals for each manipulator's states. Furthermore, appropriately designed under- and over-approximations of the free configuration space are used in an innovative way to guide the configuration space's exploration without loss of completeness. In addition, we coupled methodologies based on Reference Governors and Prescribed Performance Control with harmonic maps, in order to design a distributed control law for implementing the transitions specified by the high-level planner, which possesses guaranteed invariance and global convergence properties, thus avoiding the requirement for synchronized motion as inherently dictated by the majority of the related works. Furthermore, the proposed low-level control law does not require continuous information exchange between the robots, which rely only on measurements of the object's configuration and their own states. Finally, a transportation scenario within a complex warehouse workspace demonstrates the proposed approach and verifies its efficiency.

Keywords: object transportation; cooperative control; mobile manipulator

Citation: Vlantis, P.; Bechlioulis, C.P.; Kyriakopoulos, K.J. Mutli-Robot Cooperative Object Transportation with Guaranteed Safety and Convergence in Planar Obstacle Cluttered Workspaces via Configuration Space Decomposition. *Robotics* **2022**, *11*, 148. https://doi.org/10.3390/robotics11060148

Academic Editor: António Paulo Moreira

Received: 10 October 2022
Accepted: 7 December 2022
Published: 9 December 2022

Publisher's Note: MDPI stays neutral with regard to jurisdictional claims in published maps and institutional affiliations.

1. Introduction

Until recently, the most common type of robot used on a global scale was single industrial-style arms operating on automated production lines and in well-defined and protected obstacle-free environments. Their main advantages were robust and reliable operation, high precision, and repeatability in their movement; consequently, they were employed in repetitive and high-load manual tasks in order to reduce production costs and increase the productivity. In recent years, however, the research activity shifted towards pushing robots out of a "sterile" workplace and enabling them to co-exist with humans and unknown obstacles in a fully functional industrial environment. Additionally, efforts have been put to create multi-robot systems that exhibit collective and cooperative behavior, since many practical problems are impossible to be solved with a single robot, either due to physical limitations and/or due to limited resources or information. These reasons highlight the importance of a cooperative behavior, where each member contributes to the achievement of a final goal.

The present work focuses on the object transportation problem from an initial point to an end point-goal within an arbitrary obstacle-cluttered indoor environment employing multiple collaborative mobile manipulator systems. The problem at hand applies to a variety of real-world applications in industry, such as moving objects on a production line or storing objects in a warehouse, particularly when the object to be transported is bulky or heavy. Towards this direction, a cooperative multi-robot system is designed, which involves a high-level planner that builds a sequence of intermediate configurations that are implemented by local controllers based on the reference governor and prescribed performance control techniques. In this way, the specifications of the problem are achieved, as each robot is tasked with fulfilling the object trajectory plan by acting completely individually.

1.1. Related Work

Since 1960, significant progress has been made in the field of robotics and new functionalities and capabilities have been added, with the direct consequence of ever expanding the range of applications in which a robot can take part. One of them is the transportation of an object from an initial point to an end-point within a workspace, which is the subject of this work. This problem was originally defined in the framework of the multi-cooperative robot system [1], with research oriented to the distribution of tasks and the cooperation of the individual robotic systems. A problem of this kind has innumerable practical applications; the most widespread are the transport of dangerous objects, automation in an industrial environment, the transport of very bulky objects, the rescue of injured people etc. Since then, various strategies have been proposed to solve the considered problem that involves various sub-tasks such as trajectory planning, obstacle avoidance, robustness, etc. The communication between the robots of the system has also been thoroughly studied. Yamada states that, regardless of system configuration (centralized or decentralized), communication needs may be omitted and replaced by specific behavioral mechanisms based on local information [2]. However, such strategies have the disadvantage of requiring a set of predefined behaviors to handle new challenges and obstacles. On the other hand, Munoz claims that communication can significantly improve the performance of multi-robot systems, positively influencing coordination, cooperation, and conflict resolution [3]. In particular, the object transportation problem has been addressed by the well-known object closure strategy, in which a team of robots enclose the object so that the position of the object can be controlled by reference to the position of each robot [4–8]. Another widespread technique is that of leader–follower, where a robot determines the movement of the object and defines the behavior of the others [9–13]. In addition, more recent strategies make use of swarm intelligence, a variation of which is followed in this work, in which homogeneous robots are used, which are based on decentralized and collective behaviors [14]. Finally, techniques based on machine learning and artificial intelligence have also appeared, which understandably require high computing power [15].

The problem of cooperation is distinguished by increased complexity and requires the combination of many individual research processes in order to achieve it [16]. The most critical issue to be solved, without even the slightest discount, is that of safety, as the robots' operating space is now accessible by humans and other valuable equipment. It is therefore necessary to develop reliable safety systems [17], either on the design via robots with elastic and low-inertia moving parts or alongside algorithmic active solutions to avoid collisions. Additionally, collaboration with physical contact involves the development of forces between the robots and the commonly grasped, which affect both the safety as well as the smooth operation of the robots. Until recently, industrial robots were in a protected environment without obstacles. Therefore, only position control was effectively used for their movement. However, when robots are introduced in unstructured and uncertain environments, there are immediate effects on the safety and stability of the overall control system. For example, a potential force applied to the robot's body, depending on its magnitude and direction, will create a position error that the controller will try to compensate for with unknown results. In general, the greater the stiffness of the environment is, the greater forces and moments will be developed. Possible results of such an unavoidable scenario include, among others, the breakdown of the robot and even injury of a human or damage within the environment [18].

Cooperative manipulation particularly has been well-studied in the literature, especially the centralized scheme. In [19], a hybrid position/force control was presented. In [20], the overall closed-chain system is treated as an augmented object, with its inertial properties expressed via a single inertia matrix. The authors in [21] propose a centralized motion planning methodology based on dipolar navigation functions for nonholonomic mobile manipulators. The concept of object impedance control is also presented in [22]. Nevertheless, despite its efficacy, centralized control is less robust, since all robots depend on a central system and its complexity rises sharply and heavy inter-robot communication is required, as the number of team-robots increases. On the other hand, decentralized control usually depends on either explicit communication or off-line knowledge of the desired object trajectory, e.g., [23,24]. Furthermore, position–force hybrid control schemes, where the position of the object is controlled towards a given direction in the workspace and the internal forces on the object are controlled close to the origin are presented in [25–27]. Moreover, in other leader–follower schemes, e.g., [28,29], the leader has to transmit on-line the desired object trajectory to the follower.

Alternatively, implicit communication has been adopted in various decentralized studies on mobile manipulators. Kosuge et al. [30], presented a leader-follower scheme for holonomic manipulators in free space, where the leader implements a reference trajectory through an impedance scheme, while the follower estimates it through the motion of the object. However, the estimation error is kept small only for fixed velocity profiles. Regarding non-holonomic mobile robots, the follower's passive caster behavior was adopted in [31]. Although, the stability of the follower's contact is established, it is not mentioned how the object's trajectory can be controlled. Alternatively, the authors in [32] designed a motion coordination controller with no explicit communication for a group of physically connected robots using only interaction force measurements. In a similar direction but following a pushing-only strategy, refs. [33,34] employed a visual occlusion notion to guide the robot swarm to the goal position without exchanging any information. Finally, in [35] the robots coordinate their actions by sensing the motion of the object itself.

1.2. Contribution

In this work, we generalize our previous effort [36] by presenting a methodology for coordinating the transportation of an object that is rigidly grasped by a team of mobile manipulators, which operate within a compact planar workspace with obstacles of arbitrary shape. Owing to the object rotation and the manipulators' motion, our scheme takes into consideration the varying configuration of the robotic system, as opposed to [36], in order to build a plan that can safely drive the robotic system to the goal configuration. More specifically, we devise a high-level planner which is tasked with building a sequence of adjacent configuration space cells of the overall system (i.e., robots and object) that connect the system's initial and desired configurations, each of which defines an allowable set of configurations for the object, as well as explicit intervals for each manipulator's states. The main contributions of this work are summarized as follows:

- **Completeness:** We innovatively introduce appropriately designed under- and over-approximations of the free configuration space in order to guide the configuration space's exploration by selecting the cells that need further subdivision, thus establishing the completeness of our approach (i.e., if there exists a feasible path to go to the goal configuration, our algorithm will discover it, otherwise it halts when the problem cannot be solved when the initial and the goal configurations belong to disconnected parts of the workspace).
- **Safety and Convergence:** We combine methodologies based on Reference Governors and Prescribed Performance Control with our recently proposed harmonic maps approach [37] in order to design a distributed control law that implements specified cell transitions with guaranteed invariance and convergence properties.
- **Lean Communication:** Contrary to majority of the related literature, the proposed low-level control law does not require continuous information exchange between the

robots (e.g., via a local network), thus rendering the expected latency negligible, since it relies exclusively on measurements of the object's current configuration and the state of the corresponding robot in order to compute the respective control inputs. Regarding potential delays in the local measurements since our approach is a feedback control approach certain levels of robustness against measurement delays are expected.

To the best of our knowledge, there is no other approach that addresses the coupled path and motion planning problems with guaranteed safety. For instance, probabilistic methods can be employed for obtaining a trajectory of the states of the augmented robotic system but the low-level controllers employed for realizing it would require exact coordination/synchronization of the independent units for tracking, otherwise the robots would risk to collide with the obstacles and among them, because of the transient behavior of its individual robot's controller. On the contrary, our approach adopts a decentralized control scheme that executes the transitions with provable guarantees of safety and convergence.

1.3. Outline

The outline of this work is given as follows. First, we present some preliminary notation and definitions in Section 2. Next, the problem at hand is formulated in Section 3. In Section 4, we present the control scheme that drives the robotic system to the specified goal configuration while ensuring collision avoidance with the workspace boundary, and we elaborate on the closed-loop system's properties in Section 5. Finally, we provide simulation results verifying the efficacy of our approach in Section 6.

2. Preliminaries

Throughout this chapter, we shall use $\mathbb{R}$ to denote the set of real numbers and $\mathbb{N}$ to denote the set of natural numbers starting from zero. Moreover, we shall use $\mathfrak{I}_N \triangleq \{1, 2, \ldots, N\}$ (resp. $\mathfrak{I}_N^\star \triangleq \{0, 1, 2, \ldots, N\}$) to denote the set consisting of all natural numbers up to N, starting from 1 (resp. 0). Additionally, given sets A and B, we use ∂A, $\text{int}(A)$, $\text{cl}(A)$ to denote the boundary, interior, closure respectively, and $A \setminus B$ to denote the complement of B with respect to A.

Given a coordinate frame $\mathcal{F}_O$ in $\mathbb{R}^2$ and two points $P_A, P_B \in \mathbb{R}^2$, we will use ${}^{\{O\}}_{P_A} P_B$ to denote the position of point P_B relative to point P_A, expressed with respect to $\mathcal{F}_O$. Given frames $\mathcal{F}_A, \mathcal{F}_B, \mathcal{F}_C$, we will use ${}^{\{A\}}_{\{B\}} P_{\{C\}} \in \mathbb{R}^2$ to denote the position of the origin of frame $\mathcal{F}_C$ relative to the origin of frame $\mathcal{F}_B$, expressed with respect to $\mathcal{F}_A$. Accordingly, given frames $\mathcal{F}_A, \mathcal{F}_B$, we will use ${}^{\{A\}}_{\{B\}} R \in \mathbb{R}^{2 \times 2}$ to denote the rotation matrix corresponding to the relative orientation of $\mathcal{F}_B$ with respect to $\mathcal{F}_A$.

Given a rotation angle θ, let $R(\theta)$ be the rotation matrix defined as

$$R(\theta) \triangleq \begin{bmatrix} \cos\theta & -\sin\theta \\ \sin\theta & \cos\theta \end{bmatrix}.$$

For two given coordinate frames $\mathcal{F}_A, \mathcal{F}_B$, we define ${}^{\{A\}}_{\{B\}} \mathcal{T}$ as the homogeneous transformation from frame $\mathcal{F}_B$ to $\mathcal{F}_A$, defined as

$$\,^{\{A\}}_{\{B\}} \mathcal{T} \triangleq \begin{bmatrix} {}^{\{A\}}_{\{B\}} R & {}^{\{A\}}_{\{A\}} P_{\{B\}} \\ 0 & 1 \end{bmatrix}.$$

We recall that the following equation holds for any given point P:

$$\begin{bmatrix} {}^{\{A\}}_{\{A\}} P \\ 1 \end{bmatrix} = {}^{\{A\}}_{\{B\}} \mathcal{T} \cdot \begin{bmatrix} {}^{\{B\}}_{\{B\}} P \\ 1 \end{bmatrix}$$

where ${}^{\{A\}}_{\{A\}} P$ is the position of P with respect to frame $\mathcal{F}_A$ and ${}^{\{B\}}_{\{B\}} P$ is the position of P with respect to frame $\mathcal{F}_B$. For brevity's shake, we shall abuse notation slightly and write ${}^{\{A\}}_{\{A\}} P = {}^{\{A\}}_{\{B\}} \mathcal{T} \cdot {}^{\{B\}}_{\{B\}} P$ instead of the above when convenient.

3. Problem Formulation

We consider a compact workspace $\mathcal{W} \subseteq \mathbb{R}^2$ enclosed by a static outer boundary $\partial \mathcal{W}_0$ and N_o inner static boundaries $\partial \mathcal{W}_i$, $i \in \mathfrak{I}_{N_o}$, with $N_o \in \mathbb{N}$. More specifically, we assume that $\mathcal{W}$ can be written as follows:

$$\mathcal{W} \triangleq \overline{\mathcal{O}}_0 \setminus \bigcup_{i \in \mathfrak{I}_{N_o}} \mathcal{O}_i. \tag{1}$$

where $\mathcal{O}_0$ denotes the area that lies outside of $\partial \mathcal{W}_0$ with $\overline{\mathcal{O}}_0 \triangleq \mathbb{R}^2 \setminus \mathcal{O}_0$, and $\mathcal{O}_i$ denotes the area enclosed by $\partial \mathcal{W}_i$, for all $i \in \mathfrak{I}_{N_o}$ (see Figure 1). We shall also use $\overline{\mathcal{W}}$ to denote the complement of $\mathcal{W}$ with respect to $\mathbb{R}^2$, i.e., $\overline{\mathcal{W}} \triangleq \mathbb{R}^2 \setminus \mathcal{W}$, which is assumed to be closed. In addition, the workspace outer boundary $\partial \mathcal{W}_0 \triangleq \partial \mathcal{O}_0$ and its inner boundaries $\partial \mathcal{W}_i \triangleq \partial \mathcal{O}_i$, $i \in \mathfrak{I}_{N_o}$ are considered to be disjoint Jordan curves. Without loss of generality, we assume that $\mathcal{W}$ is embedded with the arbitrarily positioned and oriented inertial frame $\mathcal{F}_\mathcal{W}$.

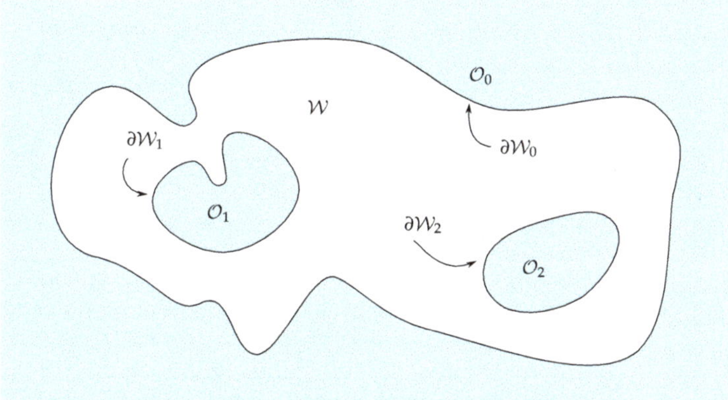

Figure 1. Typical workspace.

We now consider an object $\mathcal{L} \subset \mathbb{R}^2$ whose body is a compact, closed, polygonal 2-manifold, able to translate and rotate freely within $\mathcal{W}$ as long as it is not in contact with the workspace boundary. Let $\mathcal{F}_\mathcal{L}$ be a fixed coordinate frame arbitrarily embedded in $\mathcal{L}$. We shall use $p_\mathcal{L}$ and $\theta_\mathcal{L}$ to denote the current position and orientation of $\mathcal{L}$ with respect to $\mathcal{F}_\mathcal{W}$, i.e.,:

$$p_\mathcal{L} \triangleq {}^{\{\mathcal{W}\}}_{\{\mathcal{W}\}} P_{\{\mathcal{L}\}} \quad R(\theta_\mathcal{L}) \triangleq {}^{\{\mathcal{W}\}}_{\{\mathcal{L}\}} R.$$

Object $\mathcal{L}$ is considered a rigid body and let $M_\mathcal{L}$, $P_{\mathcal{L},\text{com}}$, $I_\mathcal{L}$ denote the object's mass, its center of mass, and its moment of inertia about $P_{\mathcal{L},\text{com}}$, respectively, expressed with respect to frame $\mathcal{F}_\mathcal{L}$. Assuming that $P_{\mathcal{L},\text{com}}$ coincides with the origin of $\mathcal{F}_\mathcal{L}$, the dynamics of $\mathcal{L}$ is given by:

$$M_\mathcal{L} \cdot \ddot{p}_\mathcal{L} = \tau_{\mathcal{L},p}$$

$$I_\mathcal{L} \cdot \ddot{\theta}_\mathcal{L} = \tau_{\mathcal{L},\theta}$$

where $\tau_{\mathcal{L},p} \in \mathbb{R}^2$ and $\tau_{\mathcal{L},\theta} \in \mathbb{R}$ are the force and torque applied externally to the object. Lastly, we define $\mathcal{L}(p,\theta)$ as the footprint of $\mathcal{L}$, i.e., the space of $\mathcal{W}$ that the body of $\mathcal{L}$ occupies when $p_\mathcal{L} = p$ and $\theta_\mathcal{L} = \theta$.

In order to transport object $\mathcal{L}$ from an initial configuration to a desired one, a team of $N_\mathcal{R} \geq 2$ cooperating mobile manipulators is employed. More specifically, each robot $\mathcal{R}_i$, $i \in \mathfrak{I}_{N_\mathcal{R}}$ consists of a holonomic base platform $\mathcal{B}_i$ and a manipulator $\mathcal{A}_i$ which is attached to the base and is equipped with an end-effector $\mathcal{A}_{i,E}$ that rigidly grasps object $\mathcal{L}$ at a specified point, and is thus able to exert a wrench onto it. The kinematics and dynamics of each mobile manipulator $\mathcal{R}_i$, $i \in \mathfrak{I}_{N_\mathcal{R}}$ is described in detail in Section 3.1 and Section 3.2, respectively. It is also assumed that the bodies of $\mathcal{B}_i$ and $\mathcal{A}_i$ can be described by compact closed and connected 2-manifolds, for all $i \in \mathfrak{I}_{N_\mathcal{R}}$.

Thus, given an initial configuration $q_{\mathcal{L},\text{init}} = [p_{\mathcal{L},\text{init}}^T, \theta_{\mathcal{L},\text{init}}]^T$ and a desired configuration $q_{\mathcal{L},\text{goal}} = [p_{\mathcal{L},\text{goal}}^T, \theta_{\mathcal{L},\text{goal}}]^T$ for the object $\mathcal{L}$, our goal is to design a control scheme for the mobile manipulators $\mathcal{R}_i$, $i \in \mathfrak{I}_{N_\mathcal{R}}$ which can drive the object to its destination, if a path between the two configurations exists, while ensuring that neither the object not the robots will collide with the workspace boundary $\partial \mathcal{W}$. In addition, if the given problem is infeasible (i.e., no collision-free path connecting the given configurations exists) our control scheme should be able to conclude so in finite time.

3.1. Mobile Manipulator Kinematics

For each $i \in \mathfrak{I}_{N_\mathcal{R}}$, let $\mathcal{F}_{\mathcal{B}_i}$ be a body-fixed frame arbitrarily embedded in $\mathcal{B}_i$. Without loss of generality, we assume that the origin of $\mathcal{F}_{\mathcal{B}_i}$ coincides with the center of rotation of the base platform $\mathcal{B}_i$. For brevity's shake, let p_i and θ_i denote the current position and orientation of $\mathcal{F}_{\mathcal{B}_i}$ with respect to $\mathcal{F}_\mathcal{W}$, i.e.,:

$$p_i \triangleq {}^{\{W\}}_{\{W\}}P_{\{\mathcal{B}_i\}} \quad R(\theta_i) \triangleq {}^{\{W\}}_{\{\mathcal{B}_i\}}R.$$

Furthermore, we will use $\mathcal{B}_i(p, \theta)$ to denote the footprint of the base platform of robot $\mathcal{R}_i$ when it is centered at p with orientation θ.

Regarding the manipulator $\mathcal{A}_i$ affixed to robot $\mathcal{R}_i$, we assume that it consists of one or more links $\mathcal{A}_{i,j}$, $j \in N_{\mathcal{A}_i}$ which are connected such that they form an open chain. Furthermore, the first link $\mathcal{A}_{i,1}$ is rigidly affixed to the base platform $\mathcal{B}_i$, whereas the end-effector is rigidly affixed to the last link $\mathcal{A}_{i,N_{\mathcal{A}_i}}$, for all $i \in \mathfrak{I}_{N_\mathcal{R}}$. The indexing of the remaining links of each manipulator is such that the body of link $\mathcal{A}_{i,j+1}$ is able to either rotate or slide about the joint it shares with link $\mathcal{A}_{i,j}$. For each manipulator $\mathcal{A}_i$, we shall use $q_{i,j}$ and $\mathfrak{D}_{q_{i,j}}$ to denote the state and domain, respectively, of the j-th degree of freedom corresponding to the joint between links $\mathcal{A}_{i,j}$ and $\mathcal{A}_{i,j+1}$, for all $j \in \mathfrak{I}_{N_{\mathcal{A}_i}-1}$ and $i \in \mathfrak{I}_{N_\mathcal{R}}i$. We remark that each domain $\mathfrak{D}_{q_{i,j}}$ is a subset of either $\mathbb{R}$ or $\mathbb{S}^1$ depending on whether the joint is prismatic or revolute, respectively. The augmented state vector z_i of robot $\mathcal{R}_i$ as follows

$$z_i \triangleq \begin{bmatrix} p_i^T, & \theta_i, & q_i^T \end{bmatrix}^T$$

where q_i is the stacked vector of joint states of manipulator $\mathcal{A}_i$, for all $i \in \mathfrak{I}_{N_\mathcal{R}}$. Similarly for each $i \in \mathfrak{I}_{N_\mathcal{R}}$ and $j \in \mathfrak{I}_{N_{\mathcal{A}_i}}$, let $\mathcal{F}_{\mathcal{A}_{i,j}}$ be a body-fixed frame arbitrarily embedded in $\mathcal{A}_{i,j}$. Additionally, we affix an arbitrary coordinate frame $\mathcal{F}_{E_i}$ at the point of contact between the end-effector of manipulator $\mathcal{A}_i$ and the object $\mathcal{L}$. For the shake of simplicity and without harming generality, we assume herein that a) the origin of frame $\mathcal{F}_{\mathcal{A}_{i,j+1}}$ lies on the axis of rotation or sliding of the j-th joint, and b) the origin of frame $\mathcal{F}_{E_i}$ coincides with the corresponding contact point (see Figure 2).

Regarding each robot's forward kinematics, we shall use $\mathcal{T}_{\mathcal{B}_i}(p, \theta)$ to denote the rigid transformation from $\mathcal{F}_{\mathcal{B}_i}$ to $\mathcal{F}_\mathcal{W}$ when the robot's is placed at p with orientation θ, i.e. ${}^{\{W\}}_{\{W\}}P_{\{\mathcal{B}_i\}} = p$ and ${}^{\{W\}}_{\{\mathcal{B}_i\}}R = R(\theta)$. Additionally, let $\mathcal{T}_{\mathcal{A}_i}(q_i)$ be the forward kinematics of manipulator $\mathcal{A}_i$, i.e., $\mathcal{T}_{\mathcal{A}_i}(q_i) \triangleq {}^{\{\mathcal{A}_{i,1}\}}_{\{E_i\}}\mathcal{T}$. Since the manipulator of each robot is rigidly attached to its base, there exists a fixed homogeneous transformation, denoted by $\mathcal{T}_{\mathcal{B}_i,\mathcal{A}_i}$ between the base $\mathcal{B}_i$ and the manipulator's first link $\mathcal{A}_{i,1}$, i.e.: $\mathcal{T}_{\mathcal{B}_i,\mathcal{A}_i} \triangleq {}^{\{\mathcal{B}_i\}}_{\{\mathcal{A}_{i,1}\}}\mathcal{T}$. The forward kinematics $\mathcal{T}_{\mathcal{R}_i}(p, \theta, q)$ of robot $\mathcal{R}_i$ is given by:

$$\mathcal{T}_{\mathcal{R}_i}(p, \theta, q) \triangleq \mathcal{T}_{\mathcal{B}_i}(p, \theta) \cdot \mathcal{T}_{\mathcal{B}_i,\mathcal{A}_i} \cdot \mathcal{T}_{\mathcal{A}_i}(q).$$

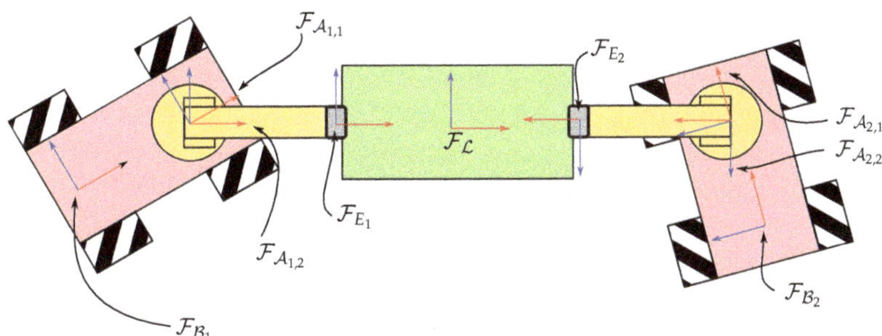

Figure 2. Example of robotic system consisting of two mobile manipulators carrying a rectangular object. Each platform is equipped with a 2-link manipulator, which is able to rotate about the joint with the base.

Lastly, for each $i \in \mathfrak{I}_{N_{\mathcal{R}}}$, we shall use $\mathcal{J}_{\mathcal{R}_i}$ and $\mathcal{J}_{\mathcal{A}_i}$ to denote the Jacobian matrices of robot $\mathcal{R}_i$ and its manipulator $\mathcal{A}_i$, i.e.,:

$$\begin{bmatrix} {}^{\{\mathcal{W}\}}\dot{P}_{\{E_i\}} \\ {}^{\{\mathcal{W}\}}\Omega_{\{E_i\}} \end{bmatrix} = \mathcal{J}_{\mathcal{R}_i}(z_i) \cdot z_i$$

and

$$\begin{bmatrix} {}^{\{\mathcal{B}_i\}}\dot{P}_{\{E_i\}} \\ {}^{\{\mathcal{B}_i\}}\Omega_{\{E_i\}} \end{bmatrix} = \mathcal{J}_{\mathcal{A}_i}(q_i) \cdot \dot{q}_i$$

where $\Omega_{\{E_i\}}$ is the angular velocity of end-effector $\mathcal{A}_{i,E}$.

3.2. Mobile Manipulator Dynamics

The dynamics of each robot $\mathcal{R}_i$, $i \in \mathfrak{I}_{N_{\mathcal{R}}}$ is assumed to obey the standard Euler–Lagrange model, i.e.,:

$$M_{\mathcal{R}_i}(z_i) \cdot \ddot{z}_i + C_{\mathcal{R}_i}(z_i, \dot{z}_i) \cdot \dot{z}_i + G_{\mathcal{R}_i}(z_i) = \tau_{m,i} - \left(\mathcal{J}_{\mathcal{R}_i}(z_i) \right)^T \cdot \tau_{e,i} \tag{2}$$

where $M_{\mathcal{R}_i}$, $C_{\mathcal{R}_i}$, $G_{\mathcal{R}_i} \in \mathbb{R}^{(3+N_{\mathcal{A}_i}) \times (3+N_{\mathcal{A}_i})}$ are the corresponding inertia, Coriolis and gravity matrices, $\tau_{m,i} \in \mathbb{R}^{(3+N_{\mathcal{A}_i})}$ is the wrench applied by the robot's actuators to the robot, and $\tau_{e,i} \in \mathbb{R}^3$ is the wrench applied by the robot to the object $\mathcal{L}$ via its end-effector.

4. Control Design

To address the aforementioned problem, we design a hybrid control scheme which consists of:

(a)   a high-level controller that given an initial configuration $q_{\mathcal{L},\text{init}}$ and a final configuration $q_{\mathcal{L},\text{goal}}$ configuration, can compute a sequence of reachable intermediate goals for the robotic system, if a solution to the above problem exists, or determine its infeasibility otherwise (completeness), and

(b)   a low-level controller which utilizes appropriate workspace transformations in order to drive the object and the mobile manipulators from each goal to the next while avoiding collisions with the workspace boundary (safety and convergence).

More specifically, the high-level controller, presented in Section 4.1, constructs a partitioning of the system's configuration space into cells by adaptively subdividing the domain of the robotic system's degrees of freedom until a sequence of connected cells containing $q_{\mathcal{L},\text{init}}$ and $q_{\mathcal{L},\text{goal}}$ is found (if one exists). Then, for each cell, intermediate goals for the object's position $p_{\mathcal{L}}$ and orientation $\theta_{\mathcal{L}}$ are computed, as described in Section 4.2, and a suitable low-level control law is employed for driving the system to the corresponding goal configuration while ensuring forward invariance of the current configuration space cell.

4.1. Configuration Space Decomposition

In this subsection, we present the hierarchical cell decomposition scheme that shall be employed for designing a sequence of high-level, feasible instructions that define a "path" leading to the desired configuration. Before doing so, we shall first take a closer look at the configuration space $\mathfrak{C}$ of the aforementioned robotic system. Throughout this subsection, we shall model this system as one virtual robot $\mathfrak{R}$ consisting of $N_{\mathfrak{R}} = 1 + \sum_{i \in \mathfrak{I}_{N_{\mathcal{R}}}} N_{\mathcal{A}_i}$ connected components, which correspond to the object $\mathcal{L}$, the base platform $\mathcal{B}_i$ and the links $\mathcal{A}_{i,j}$ of each mobile manipulator $\mathcal{R}_i$, for all $j \in \mathfrak{I}_{N_{\mathcal{A}_i}}$ and $i \in \mathfrak{I}_{N_{\mathcal{R}}}$.

One can readily see that the components of $\mathfrak{R}$ form an undirected tree $\mathfrak{T}(\mathfrak{n}, \mathfrak{e})$, where $\mathfrak{n}$ is the set of components and $\mathfrak{e} \subset \mathfrak{n} \times \mathfrak{n}$ is the set of connections between the nodes. We shall use $\mathfrak{R}_i$ to denote the i-th component of $\mathfrak{R}$. A connection $(i, j) \in \mathfrak{e}$ implies that the j-th component is able to move (rotate, translate, slice) relative to the i-th component about a pivot point $\mathfrak{P}_{i,j}$. Furthermore, given $i \in \mathfrak{I}_{N_{\mathfrak{R}}}$, we will use $\mathfrak{n}_c^i$ to denote the children of component $\mathfrak{R}_i$, i.e., the set of components $\mathfrak{R}_j$ such that $(i, j) \in \mathfrak{e}$, for all $j \in \mathfrak{I}_{N_{\mathfrak{R}}}$. Moreover $\mathfrak{n}_p^i$ will be used to denote the parent $\mathfrak{R}_j$ of component $\mathfrak{R}_i$, i.e., the sole component such that $(j, i) \in \mathfrak{e}$, if one exists. Accordingly, we define $\mathfrak{n}_d^i$ and $\mathfrak{n}_a^i$ as the set of descendants and ancestors, respectively, of component $\mathfrak{R}_i$. Without loss of generality, we can choose the indexing of the components such that the first component of $\mathfrak{R}$ is the root of $\mathfrak{T}$, corresponding to the object $\mathcal{L}$. For simplicity's shake, we will use $\mathcal{F}_{\mathfrak{R}_i}$, $i \in \mathfrak{I}_{N_{\mathfrak{R}}}$ to denote the coordinate frames embedded in each component of $\mathfrak{R}$ and we shall refer to their origins as the reference point of the corresponding component, respectively. Furthermore, let $p_{\mathfrak{R}} \triangleq [x_{\mathfrak{R}}, y_{\mathfrak{R}}]^T \in \mathbb{R}^2$ and $\theta_{\mathfrak{R}} \in \mathfrak{D}_\theta \subseteq \mathbb{S}^1$ denote the relative position of the robotic system's reference point and the relative orientation of its coordinate frame $\mathcal{F}_{\mathfrak{R}_0}$ with respect to the workspace's coordinate frame $\mathcal{F}_{\mathcal{W}}$, respectively.

Regarding the coupling between components, we will refer to the joint between two connected components as prismatic (resp. revolute) if the child is able to slide (resp. rotate) about the corresponding pivot point. We will use $\mathfrak{q}_i$ and $\mathfrak{D}_{\mathfrak{q}_i}$, $i \in \mathfrak{I}_{N_{\mathfrak{q}}}$, to denote the degree of freedom and its domain, respectively, corresponding to the joint between the i-th component and its parent, where $N_{\mathfrak{q}} \triangleq \sum_{j \in \mathfrak{I}_{N_{\mathcal{R}}}} (N_{\mathcal{A}_j} - 1)$, Without loss of generality, since each component other than the root has exactly one parent, we assume that each pivot $\mathfrak{P}_{i,j}$ coincides with the origin of frame $\mathcal{F}_{\mathfrak{R}_j}$. Furthermore, by treating the orientation $\theta_{\mathcal{L}}$ of the object as a virtual joint state, the state z of the virtual robotic system $\mathfrak{R}$ is defined as follows

$$z \triangleq [p_{\mathfrak{R}}^T, \theta_{\mathfrak{R}}, \mathfrak{q}^T]^T = [p_{\mathfrak{R}}^T, \mathfrak{q}^T]^T$$

where $\mathfrak{q} \triangleq [\mathfrak{q}_i]_{i \in \mathfrak{I}_{N_{\mathfrak{q}}}^\star}$, is the stacked vector of virtual joint parameters with $\mathfrak{q}_0 \triangleq \theta_{\mathfrak{R}}$ and $\mathfrak{q}_i \triangleq \mathfrak{q}_i$ for all $i \in \mathfrak{I}_{N_{\mathfrak{q}}}$.

Let us now consider the footprint of the robotic system while it moves within the workspace. We notice that, for each $i \in \mathfrak{I}_{N_{\mathfrak{R}}}$, the footprint of the individual component $\mathfrak{R}_i$, i.e., the area occupied by it at a given configuration, is defined by the position of its pivot point and the current value of its (virtual) joint parameter. We shall use $\mathfrak{R}_i(p, \mathfrak{q})$ to denote the footprint when the pivot point is placed at p and the joint parameter value is $\mathfrak{q}$. We also remark that, although each component may move freely with respect to its pivot point, any motion of theirs propagates directly to their children, thus potentially inducing a translation and/or rotation onto every one of its descendants $\mathfrak{n}_d^i$. Thus, the footprint of component $\mathfrak{R}_i$ can also be defined in terms of the current position $p_{\mathfrak{R}}$ of the robotic system and the (virtual) joint parameters of every component $\mathfrak{R}_j$ belonging to $\mathfrak{n}_a^i$. By remarking that the footprint $\mathfrak{R}(z)$ of the robotic system at a given configuration z is simply the union of the footprints of its individual components, i.e.,:

$$\mathfrak{R}(p_{\mathfrak{R}}, \mathfrak{q}) \triangleq \bigcup_{i \in \mathfrak{I}_{N_{\mathfrak{R}}}} \mathfrak{R}_i \left(p_{\mathfrak{R}}, [\mathfrak{q}_j]_{j \in \mathfrak{n}_a^i \cup i}^T \right) \tag{3}$$

we are now ready to formally define the set of admissible configurations to our problem. For brevity, $\mathfrak{R}(\mathfrak{q})$ will be used instead of $\mathfrak{R}(0,\mathfrak{q})$ where is deemed preferable. By noticing that the configuration space $\mathfrak{C}$ of this robotic system is a manifold diffeomorphic to $\mathbb{R}^2 \times \mathbb{S}^1 \times \mathfrak{D}_{\mathfrak{q}_1} \times \ldots \mathfrak{D}_{\mathfrak{q}_{N_\mathfrak{q}}}$. and recalling that neither the object $\mathcal{L}$ nor any of the robots $\mathcal{R}_i, i \in \mathfrak{I}_{N_\mathcal{R}}$ are allowed to collide with the workspace boundary $\partial \mathcal{W}$, the set $\mathfrak{C}_f$ of collision free configurations of $\mathfrak{R}$ is given by:

$$\mathfrak{C}_f = \{z \mid \mathcal{W} \cap \mathfrak{R}(z) = \varnothing \text{ and } z \in \mathfrak{C}\}. \tag{4}$$

Finally, let $\mathfrak{C}_o \triangleq \mathfrak{C} \setminus \mathfrak{C}_f$.

Now, in order to design a continuous "path" inside $\mathfrak{C}_f$ connecting the two given configurations $\mathfrak{q}_{\mathcal{L},\text{init}}$ and $\mathfrak{q}_{\mathcal{L},\text{goal}}$, we extend the methodology presented in [38]. More specifically, by designing a suitable cover of the free configuration space via recursive subdivision of the domain of $\mathfrak{q}$, our goal is to obtain a hierarchical partitioning of $\mathfrak{C}_f$. For each implicitly defined cell, we compute over suitable over- and under-approximations, whose shape is much simpler than the shape of the corresponding exact cell, which are used for both guiding the configuration space's exploration, as well as designing a high-level plan will drive the robotic system to its goal. To do so, we first consider the domain $\mathfrak{D}_{\mathfrak{q}_i}$ of the joint state $\mathfrak{q}_i$, for $i \in \mathfrak{I}_{N_\mathfrak{q}}^\star$. Furthermore, we shall refer to a set of the form $\mathcal{S}_{[\mathfrak{q}_{i1},\mathfrak{q}_{i2}]}^{\mathfrak{q}_i}$ as a simple slice of the parameter $\mathfrak{q}_i$, where $\mathfrak{q}_{i1}, \mathfrak{q}_{i2} \in \mathfrak{D}_{\mathfrak{q}_i}$. Furthermore, a set $\mathfrak{S}_i = \{\mathcal{S}_j^{\mathfrak{q}_i} \mid j \in \mathfrak{I}_{N_{\mathfrak{S}_i}}\}$ consisting of $N_{\mathfrak{S}_i}$ simple slices of $\mathfrak{q}_i$ shall be called a cover of $\mathfrak{D}_{\mathfrak{q}_i}$ if

$$\mathfrak{D}_{\mathfrak{q}_i} = \bigcup_{j \in \mathfrak{I}_{N_{\mathfrak{S}_i}}} \mathcal{S}_j^{\mathfrak{q}_i}$$

for all $k, \ell \in \mathfrak{I}_{N_{\mathfrak{S}_i}}$ with $k \neq \ell$. A compound slice $\widehat{\mathcal{S}}$ is defined as a set of simple slices of the form $\widehat{\mathcal{S}} = \{\mathcal{S}^{\mathfrak{q}_i} \mid i \in \mathfrak{I}_{N_\mathfrak{q}}^\star\}$. Respectively, a set $\widehat{\mathfrak{S}} = \{\mathfrak{S}_i \mid i \in \mathfrak{I}_{N_\mathfrak{q}}^\star\}$ is called a cover of the free configuration space $\mathfrak{C}_f$ if each $\mathfrak{S}_i$ is a cover of $\mathfrak{D}_{\mathfrak{q}_i}$. We note that a cover $\widehat{\mathfrak{S}}$ induces a partitioning of $\mathfrak{C}_f$ into regions

$$\mathfrak{C}_{\widehat{\mathcal{S}}} = \left\{ [p_\mathfrak{R}^T, \mathfrak{q}^T]^T \mid p_\mathfrak{R} \in \mathcal{W} \text{ and } \mathfrak{q} \in \widehat{\mathcal{S}} \right\} \cap \mathfrak{C}_f, \quad \widehat{\mathcal{S}} \in \widehat{\mathfrak{S}} \tag{5}$$

each of which consists of $N_{\widehat{\mathcal{S}}} \geq 0$ individually connected but pairwise disjoint subsets $\hat{\mathcal{C}}_{\widehat{\mathcal{S}},i}, i \in \mathfrak{I}_{N_{\mathfrak{C}_{\widehat{\mathcal{S}}}}}$. Taking a closer look at the connectedness of these cells, one can readily see that two configuration space cells $\hat{\mathcal{C}}_{\widehat{\mathcal{S}}_i}$ and $\hat{\mathcal{C}}_{\widehat{\mathcal{S}}_j}$ are connected iff $\widehat{\mathcal{S}}_i, \widehat{\mathcal{S}}_j$ are adjacent and the projections of $\hat{\mathcal{C}}_{\widehat{\mathcal{S}}_i}, \hat{\mathcal{C}}_{\widehat{\mathcal{S}}_j}$ onto the plane intersect. We recall that two distinct simple slices $\mathcal{S}_i^{\mathfrak{q}_k}$ and $\mathcal{S}_j^{\mathfrak{q}_k}$ are called adjacent if their intersection $\mathcal{S}_i^{\mathfrak{q}_k} \cap \mathcal{S}_j^{\mathfrak{q}_k}$ is not empty, whereas two compound slices $\widehat{\mathcal{S}}_i = \{\mathcal{S}_i^{\mathfrak{q}_k} \mid k \in \mathfrak{I}_{N_\mathfrak{q}}^\star\}$, and $\widehat{\mathcal{S}}_j = \{\mathcal{S}_j^{\mathfrak{q}_k} \mid k \in \mathfrak{I}_{N_\mathfrak{q}}^\star\}$, are called adjacent if $\mathcal{S}_i^{\mathfrak{q}_k}, \mathcal{S}_j^{\mathfrak{q}_k}$ are adjacent, for all $k \in \mathfrak{I}_{N_\mathfrak{q}}^\star$.

Similarly to the method employed in [38], in order to avoid explicitly computing the shape of a given configuration space cell, we shall define suitable over- and under-approximations of it, which, in addition, shall be used for guiding the configuration space exploration in a similar manner. In order to build these approximations of the set of free configurations corresponding to the compound slice $\widehat{\mathcal{S}} = \{\mathcal{S}^{\mathfrak{q}_i} \mid i \in \mathfrak{I}_{N_\mathfrak{q}}^\star\}$, we first compute an over-approximation $\overline{\mathfrak{R}}(\widehat{\mathcal{S}})$ and an under-approximation $\underline{\mathfrak{R}}(\widehat{\mathcal{S}})$ of the robotic system's footprint as follows:

$$\overline{\mathfrak{R}}(\widehat{\mathcal{S}}) = \bigcup_{\mathfrak{q} \in \mathcal{I}_{\widehat{\mathcal{S}}}} \mathfrak{R}(\mathfrak{q})$$

$$\underline{\mathfrak{R}}(\widehat{\mathcal{S}}) = \bigcap_{\mathfrak{q} \in \mathcal{I}_{\widehat{\mathcal{S}}}} \mathfrak{R}(\mathfrak{q}) \tag{6}$$

where

$$\mathcal{I}_{\widehat{\mathcal{S}}} = \mathcal{S}^{q_0} \times \mathcal{S}^{q_1} \times \cdots \times \mathcal{S}^{q}.$$

(7)

We remark that, although seemingly daunting at first sight, the computation of $\overline{\mathfrak{R}}\left(\widehat{\mathcal{S}}\right)$ and $\underline{\mathfrak{R}}\left(\widehat{\mathcal{S}}\right)$ can be significantly simplified by recalling that the footprint of each component does not necessarily depends on every component of q but only on those of its ancestors by virtue of the robotic system's tree-like structure. An example of such over- and under-approximation for a robotic system consisting of two connected components can be seen in Figure 3.

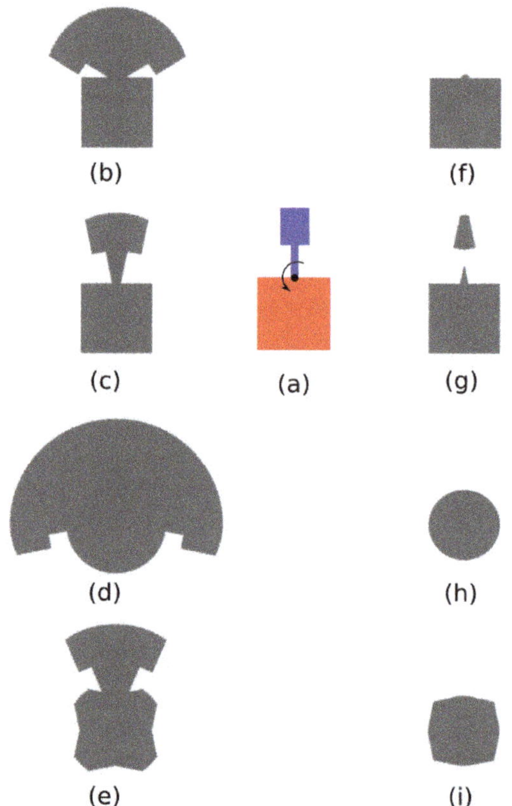

Figure 3. Over- and under-approximations of a robotic system (**a**) consisting of two components corresponding to slices: $\{0\} \times [2\pi - 1, 1]$ (**b,f**), $\{0\} \times [2\pi - 0.2, 0.2]$ (**c,g**), $[3\pi/2, pi/2] \times [2\pi - 0.2, 0.2]$ (**d,h**), $[2\pi - 0.2, 0.2] \times [2\pi - 0.2, 0.2]$ (**e,i**).

Following these definitions, the over-approximation $\overline{\mathfrak{C}}_{\widehat{\mathcal{S}}}$ and the under-approximation $\underline{\mathfrak{C}}_{\widehat{\mathcal{S}}}$ of a given partition $\mathfrak{C}_{\widehat{\mathcal{S}}}$, can be computed as follows:

$$\overline{\mathfrak{C}}_{\widehat{\mathcal{S}}} \triangleq \left\{ [p^T, q^T]^T \mid p \in \overline{\mathcal{W}}_{\widehat{\mathcal{S}}} \text{ and } q \in \mathcal{I}_{\widehat{\mathcal{S}}} \right\}$$
$$\underline{\mathfrak{C}}_{\widehat{\mathcal{S}}} \triangleq \left\{ [p^T, q^T]^T \mid p \in \underline{\mathcal{W}}_{\widehat{\mathcal{S}}} \text{ and } q \in \mathcal{I}_{\widehat{\mathcal{S}}} \right\}$$

(8)

where

$$\overline{W}_{\widehat{S}} \triangleq W \ominus \underline{\mathfrak{R}}\left(\widehat{S}\right)$$
$$\underline{W}_{\widehat{S}} \triangleq W \ominus \overline{\mathfrak{R}}\left(\widehat{S}\right)$$

(9)

with $A \ominus B$ denoting the Minkowski difference of sets A and B, and $\mathcal{I}_{\widehat{S}} = \mathcal{S}^{q_0} \times \mathcal{S}^{q_1} \times \cdots \times \mathcal{S}^q$. Obviously, each of $\overline{\mathfrak{C}}_{\widehat{S}}$ and $\underline{\mathfrak{C}}_{\widehat{S}}$ consists of individually connected but pairwise disjoint cells $\overline{C}_{\widehat{S},i}$, $i \in \mathfrak{I}_{N_{\overline{\mathfrak{C}}_{\widehat{S}}}}$ and $\underline{C}_{\widehat{S},i}$, $i \in \mathfrak{I}_{N_{\underline{\mathfrak{C}}_{\widehat{S}}}}$, respectively, which enclose or are enclosed by the cells of $\mathfrak{C}_{\widehat{S}}$.

At this point, we remark that the approximation of $\mathfrak{C}_f$ improves as one subdivides the configuration space into more and finer slices. Thus, choosing a sufficiently fine partitioning of $\mathfrak{C}$, a sequence of adjacent under-approximation cells connecting $q_{\mathcal{L},\text{init}}$ and $q_{\mathcal{L},\text{goal}}$ will appear, as long as one exists in the first place. Instead of choosing such a fine partitioning arbitrary, we design an adaptive subdivision scheme which makes also use of the space's over-approximations for choosing which slice to subdivide at each iteration. More specifically, we design an algorithm which given compound cover $\widehat{S}$, it tries to find a sequence $\underline{\Pi}$ of adjacent under-approximation cells connecting the initial and goal configurations. If no such path can be found, then our algorithm tries to connect the two given configurations with a sequence $\overline{\Pi}$ made of adjacent over-approximation cells instead. If such a path exists, then a slice corresponding to a cell of $\overline{\Pi}$ is selected according to a suitable heuristic and becomes subdivided, producing a new partitioning of $\mathfrak{C}_f$. Otherwise, if no such path can be found, then this obviously indicates that the problem at hand is infeasible (i.e., the two given configurations exist in disjoint components of the robotic system's configuration space) and our algorithm terminates. In short, one can readily verify that the following statements hold:

1. If there exists a path of adjacent under-approximation cells for a given cover $\widehat{S}$ containing $q_{\mathcal{L},\text{init}}$ and $q_{\mathcal{L},\text{goal}}$, then a solution to our problem exists.

2. If there exists a path of adjacent over-approximation cells for a given cover $\widehat{S}$ containing $q_{\mathcal{L},\text{init}}$ and $q_{\mathcal{L},\text{goal}}$, then whether our problem has a solution is unknown and further expansion of $\widehat{S}$ is in order.

3. If there is no path of adjacent over-approximation cells for a given cover $\widehat{S}$ containing $q_{\mathcal{L},\text{init}}$ and $q_{\mathcal{L},\text{goal}}$, then our problem is infeasible.

The proposed algorithm can be seen in Algorithm 1. More specifically, we begin the configuration space exploration with a rough partitioning of $\mathfrak{C}$ induced by a compound slice covering the entire domain of virtual joint parameters q. Then, we search for cells $\underline{C}_{\text{init}}$ and $\underline{C}_{\text{goal}}$ containing the robot's initial and final configurations, respectively, by subdividing $\widehat{S}$. If no such pair of cells exists, our initial problem is obviously infeasible and the algorithm terminates. Otherwise, we try to connect $\underline{C}_{\text{init}}$ and $\underline{C}_{\text{goal}}$ using the available under-approximation cells corresponding to $\widehat{S}$. If this attempt fails, then we try instead to find a path of over-approximation cells connecting $\overline{C}_{\text{init}}$ and $\overline{C}_{\text{goal}}$. If such a path cannot be found, this also implies that no solution exists and the algorithm terminates. Otherwise, a heuristic is utilized for selecting a compound slice in $\widehat{S}$ to be expanded and the process starts anew. The heuristic used, which can be seen in Algorithm 2, selects which slice of $\widehat{S}$ to expand as follows. Given a path $\overline{\Pi}$ of over-approximation cells, it essentially tries to construct a path made of the under-approximation cells that belong in the same compound slices as the elements of $\overline{\Pi}$. Failing to connect under-approximation cells belonging in two adjacent compound slices $\widehat{S}_i$ and $\widehat{S}_j$ indicates that the connectedness of the over- and under-approximation cells in this slices is not the same, which means that these slices need to be further expanded. Thus, the largest simple slice of these compound slices becomes subdivided and the function returns. Finally, we remark that the functions CONNECTUACELLS and CONNECTOACELLS employ standard graph search algorithms for constructing the corresponding paths based on a heuristic that penalizes cells with smaller slices (i.e., cells corresponding to larger slices are preferred).

Algorithm 1 Configuration space exploration algorithm.

---

function CONNECTCONFIGS($q_{\mathcal{L},\text{init}}$, $q_{\mathcal{L},\text{goal}}$)

    $\widehat{\mathfrak{S}} \leftarrow \left\{ \mathfrak{D}_{q_i} \mid i \in \mathfrak{I}^{\star}_{N_q} \right\}$

    loop

        $\widehat{\mathfrak{S}}, \underline{\mathcal{C}}_{\text{init}}, \overline{\mathcal{C}}_{\text{init}} \leftarrow$ FINDENCLOSINGCELLS($q_{\mathcal{L},\text{init}}$, $\widehat{\mathfrak{S}}$)

        $\widehat{\mathfrak{S}}, \underline{\mathcal{C}}_{\text{goal}}, \overline{\mathcal{C}}_{\text{goal}} \leftarrow$ FINDENCLOSINGCELLS($q_{\mathcal{L},\text{goal}}$, $\widehat{\mathfrak{S}}$)

        if $\overline{\mathcal{C}}_{\text{init}}$ is Nil **or** $\overline{\mathcal{C}}_{\text{goal}}$ is Nil **then**

            return Nil

        end if

        $\underline{\Pi} \leftarrow$ CONNECTUACELLS($\underline{\mathcal{C}}_{\text{init}}$, $\underline{\mathcal{C}}_{\text{goal}}$)

        if $\underline{\Pi}$ is Nil **then**

            $\overline{\Pi} \leftarrow$ CONNECTOACELLS($\overline{\mathcal{C}}_{\text{init}}$, $\overline{\mathcal{C}}_{\text{goal}}$)

            if $\overline{\Pi}$ is empty **then**

                return Nil

            else

                $\widehat{\mathfrak{S}} \leftarrow$ REFINE($\widehat{\mathfrak{S}}$, $\overline{\Pi}$, $\{\underline{\mathcal{C}}_{\text{init}}\}$, $\{\underline{\mathcal{C}}_{\text{goal}}\}$)

            end if

        else

            return $\underline{\Pi}$

        end if

    end loop

end function

Algorithm 2 Heuristic choosing next simple slice for subdivision.

---

function REFINE($\widehat{\mathfrak{S}}$, $\overline{\Pi}$, *src*, *dst*)

    if $len(\overline{\Pi}) = 1$ **then**

        return SUBDIVIDE($\widehat{\mathfrak{S}}$, $\overline{\Pi}[0]$)

    else

        cells $\leftarrow \{\}$

        connected $\leftarrow \{\}$

        if $len(\overline{\Pi}) == 2$ **then**

            cells $\leftarrow$ *dst*

        else

            $\overline{\mathcal{C}}_{\widehat{\mathcal{S}}_j} \leftarrow \overline{\Pi}[1]$

            for all $\underline{\mathcal{C}}_{\widehat{\mathcal{S}}_j}$ in $\underline{\mathfrak{C}}_{\widehat{\mathcal{S}}_j}$ **do**

                cells $\leftarrow$ *cells* $\cup \{\underline{\mathcal{C}}_{\widehat{\mathcal{S}}_j}\}$

            end for

        end if

        for all $\underline{\mathcal{C}}_{\widehat{\mathcal{S}}_i}$ in *src* **do**

            for all $\underline{\mathcal{C}}_{\widehat{\mathcal{S}}_j}$ in *cells* **do**

                if $\underline{\mathcal{C}}_{\widehat{\mathcal{S}}_i} \cap \underline{\mathcal{C}}_{\widehat{\mathcal{S}}_j} \neq \varnothing$ **then**

                    connected $\leftarrow$ *connected* $\cup \{\underline{\mathcal{C}}_{\widehat{\mathcal{S}}_j}\}$

                end if

            end for

        end for

        if $len(connected) > 0$ **then**

            return REFINE($\widehat{\mathfrak{S}}$, $\overline{\Pi}[1:]$, *connected*, *dst*)

        else

            return SUBDIVIDELONGEST($\widehat{\mathfrak{S}}$, $\widehat{\mathcal{S}}_i$, $\widehat{\mathcal{S}}_j$)

        end if

    end if

end function

4.2. Distributed Control Law

Given now a path $\underline{\Pi}$ of cells obtained by the high-level planner described in the previous sub-section, we shall now design a distributed control scheme for the mobile manipulators that ensures safe transitions from one cell to the next until the goal configuration $q_{\mathcal{L},\text{goal}}$ is reached. Let $\mathcal{C}_{\widehat{S}}$ be a cell in $\underline{\Pi}$ and let $\underline{\mathcal{C}}_{\widehat{S}}$ denote its the projection on the plane. We recall that $\underline{\mathcal{C}}_{\widehat{S}}$ is an under-approximation of the actual free configuration space, constructed by extruding $\underline{\mathcal{W}}_{\widehat{S}}$, which implies that, as long as $q \in \widehat{S}$, then $p_{\mathcal{L}}$ can safely occupy any position of $\underline{\mathcal{C}}_{\widehat{S}}$. We also note that $\underline{\mathcal{C}}_{\widehat{S}}$ is a non-empty, compact region of $\mathbb{R}^2$ with arbitrary connectedness and shape. Exploiting this fact, we can decouple the low-level control laws for: (a) the object's position $p_{\mathcal{L}}$, (b) the object's orientation $\theta_{\mathcal{L}}$, and (c) the joints q_i of each manipulator $\mathcal{R}_i$, $i \in \mathfrak{I}_{N_{\mathcal{R}}}$, as explained in the following.

For each intermediate cell of $\underline{\Pi}$, we can obtain goal sets corresponding to $p_{\mathcal{L}}$, $\theta_{\mathcal{L}}$ and q separately, by computing its intersection with the next one (which is non-empty by construction of $\underline{\Pi}$), Let us consider a pair of consecutive cells $\mathcal{C}_{\widehat{S}_i}$ and $\mathcal{C}_{\widehat{S}_j}$ in $\underline{\Pi}$. Regarding the object's position, in order to safely traverse from $\mathcal{C}_{\widehat{S}_i}$ to $\mathcal{C}_{\widehat{S}_j}$, it is sufficient that $p_{\mathcal{L}}$ reaches the set

$$\mathscr{G}_{p_{\mathcal{L}}}\left(\underline{\mathcal{C}}_{\widehat{S}_i}, \underline{\mathcal{C}}_{\widehat{S}_j}\right) \triangleq \underline{\mathcal{C}}_{\widehat{S}_i} \cap \underline{\mathcal{C}}_{\widehat{S}_j}. \tag{10}$$

We also note that $\mathscr{G}_{p_{\mathcal{L}}}\left(\underline{\mathcal{C}}_{\widehat{S}_i}, \underline{\mathcal{C}}_{\widehat{S}_j}\right)$ is generally made of one or more disjoint subsets of arbitrary connectedness and that, as long as the object's position reaches either of these, the system can cross to the next cell. Respectively, a goal set corresponding to the object's orientation can be obtained by computing the intersection of the corresponding simple slices of $\widehat{S}_i$ and $\widehat{S}_j$, i.e.,:

$$\mathscr{G}_{\theta_{\mathcal{L}}}\left(\mathcal{C}_{\widehat{S}_i}, \mathcal{C}_{\widehat{S}_j}\right) \triangleq S_i^{q_0} \cap S_j^{q_0}. \tag{11}$$

Goal sets for the joints of each mobile manipulator can be computed in a similar manner. Particularly, let $\mathscr{P}_{A_k}(\mathcal{C}_{\widehat{S}})$ denote the projection $\mathcal{C}_{\widehat{S}}$ along the dimensions corresponding to the degrees of freedom of $\mathcal{A}_k$. Obviously, $\mathscr{P}_{A_k}(\mathcal{C}_{\widehat{S}})$ is equal to the product of the simple slices of $\widehat{S}$ corresponding to q_k. Then, the corresponding goal set of q_k is given by

$$\mathscr{G}_{A_k}\left(\mathcal{C}_{\widehat{S}_i}, \mathcal{C}_{\widehat{S}_j}\right) \triangleq \mathscr{P}_{A_k}\left(\mathcal{C}_{\widehat{S}_i}\right) \cap \mathscr{P}_{A_k}\left(\mathcal{C}_{\widehat{S}_j}\right), \quad \forall k \in \mathfrak{I}_{N_{\mathcal{R}}}. \tag{12}$$

Thus, for successfully driving the robotic system from $\mathcal{C}_{\widehat{S}_i}$ to $\mathcal{C}_{\widehat{S}_j}$, we need to design decoupled control laws for the mobile manipulators which:

- ensure invariance of the current cell, i.e., $p_{\mathcal{L}} \in \mathscr{P}_{p_{\mathcal{L}}}\left(\mathcal{C}_{\widehat{S}_i}\right), \theta_{\mathcal{L}} \in S_i^{q_0}$ and $q_k \in \mathscr{P}_{A_k}\left(\mathcal{C}_{\widehat{S}_i}\right)$, $\forall k \in \mathfrak{I}_{N_{\mathcal{R}}}$, until the transition is complete, and

- ensure converge to the system's states to the corresponding goals sets $\mathscr{G}_{p_{\mathcal{L}}}\left(\mathcal{C}_{\widehat{S}_i}, \mathcal{C}_{\widehat{S}_j}\right)$, $\mathscr{G}_{\theta_{\mathcal{L}}}\left(\mathcal{C}_{\widehat{S}_i}, \mathcal{C}_{\widehat{S}_j}\right)$ and $\mathscr{G}_{A_k}\left(\mathcal{C}_{\widehat{S}_i}, \mathcal{C}_{\widehat{S}_j}\right)$, $k \in \mathfrak{I}_{N_{\mathcal{R}}}$.

Finally, the transition is considered complete after all states have reached the corresponding goal sets. We remark that, regarding the last cell of $\underline{\Pi}$, the goal sets corresponding to the object's position and orientation can taken equal to $\{p_{\mathcal{L},\text{goal}}\}$ and $\{\theta_{\mathcal{L},\text{goal}}\}$, respectively, while the joints of the manipulators need only to remain within the bounds imposed by the last cell.

Before we proceed with formulating the corresponding control laws, we must first formally state the following assumptions about our system.

Assumption 1. *Each robot $\mathcal{R}_k$, $k \in \mathfrak{I}_{N_{\mathcal{R}}}$ has exact knowledge of the object's and its own dynamic model, i.e., $M_{\mathcal{L}}$, $I_{\mathcal{L}}$, $P_{\mathcal{L},\text{com}}$ and $M_{\mathcal{R}_k}$, $C_{\mathcal{R}_k}$, $D_{\mathcal{R}_k}$, $G_{\mathcal{R}_k}$ are known.*

Assumption 2. *Each robot $\mathcal{R}_k$, $k \in \mathfrak{I}_{N_{\mathcal{R}}}$ has full knowledge of its own state z_k and the current configuration $q_{\mathcal{L}}$ of the object $\mathcal{L}$.*

Assumption 3. *The plan generated by the high-level planner is available to all robots. Furthermore, each robot $\mathcal{R}_k$, $k \in \mathfrak{I}_{N_\mathcal{R}}$ is able to communicate with the others only for announcing that it is ready to transition to the next cell, i.e., that $p_\mathcal{L} \in \mathscr{G}_{p_\mathcal{L}}\left(\mathcal{C}_{\widehat{\mathcal{S}}_i}, \mathcal{C}_{\widehat{\mathcal{S}}_j}\right)$, $\theta_\mathcal{L} \in \mathscr{G}_{\theta_\mathcal{L}}\left(\mathcal{C}_{\widehat{\mathcal{S}}_i}, \mathcal{C}_{\widehat{\mathcal{S}}_j}\right)$, and $q_k \in \mathscr{G}_{A_k}\left(\mathcal{C}_{\widehat{\mathcal{S}}_i}, \mathcal{C}_{\widehat{\mathcal{S}}_j}\right)$ (lean communication).*

Assumption 4. *Each mobile manipulator $\mathcal{R}_k$, $k \in \mathfrak{I}_{N_\mathcal{R}}$ is sufficiently redundant, i.e., it can independently apply a desired wrench to its end-effector while keeping q_k in $\mathscr{P}_{A_k}\left(\mathcal{C}_{\widehat{\mathcal{S}}_i}\right)$. Additionally, the lower diagonal $N_{A_k} - 1 \times N_{A_k} - 1$ block of $\left(M_{\mathcal{R}_k}\right)^{-1}\left(I - \left(\mathcal{J}_{\mathcal{R}_k}(z_k)\right)^{\dagger} \cdot \mathcal{J}_{\mathcal{R}_k}(z_k)\right)$ is non-singular.*

4.2.1. Object's Position

First, we shall design a suitable vector field for safely driving the object's position $p_\mathcal{L}$ to $\mathscr{G}_{p_\mathcal{L}}\left(\mathcal{C}_{\widehat{\mathcal{S}}_i}, \mathcal{C}_{\widehat{\mathcal{S}}_j}\right)$. To do so, we construct a transformation T_i of $\mathscr{F}_{p_\mathcal{L}}\left(\mathcal{C}_{\widehat{\mathcal{S}}_i}, \mathcal{C}_{\widehat{\mathcal{S}}_j}\right) \triangleq \mathscr{P}_{p_\mathcal{L}}\left(\mathcal{C}_{\widehat{\mathcal{S}}_i}\right) \setminus \mathscr{G}_{p_\mathcal{L}}\left(\mathcal{C}_{\widehat{\mathcal{S}}_i}, \mathcal{C}_{\widehat{\mathcal{S}}_j}\right)$ to the unit disk and collapse the selected component of $\mathscr{G}_{p_\mathcal{L}}\left(\mathcal{C}_{\widehat{\mathcal{S}}_i}, \mathcal{C}_{\widehat{\mathcal{S}}_j}\right)$ to a point, using the procedure described in [37]. By recalling that T_i is a diffeomorphism that collapses all inner obstacles of to isolated points, one can readily verify that the chance of a line connecting the image $q_\mathcal{L}^{[i]} \triangleq T_i(p_\mathcal{L})$ of the object's current position to the image $q_{\mathcal{L},d}^{[i]}$ of the current cell's goal is zero [39,40]. Therefore, the following velocity control law would safely drive the object's position to the goal set for almost all initial configurations:

$$v_{p_\mathcal{L}}^{[i]}(p_\mathcal{L}) \triangleq \left(J_{T_i}(p_\mathcal{L})\right)^{-1} \cdot \left(q_{\mathcal{L},d}^{[i]} - q_\mathcal{L}^{[i]}\right) \tag{13}$$

where J_{T_i} is the Jacobian matrix of T_i. In order to design a law for the desired force to be applied to the object $\mathcal{L}$ by the robots, we employ a novel methodology presented in [41], which allows us to extend the vector field from Equation (13) to second-order dynamics. The corresponding control law for the desired force applied to the object is formed by a term proportional to the error with respect to the reference governor's state (to keep it small) plus a damping term to avoid oscillations, as follows:

$$\tau_{\mathcal{L},p}^{des} = -M_\mathcal{L} \cdot \left(K_{p_\mathcal{L}}^{[i]} \cdot \left(p_\mathcal{L} - p_{\mathcal{L},G}^{[i]}\right) + \zeta_{p_\mathcal{L}}^{[i]} \cdot \dot{p}_\mathcal{L}\right)$$

$$\dot{p}_{\mathcal{L},G}^{[i]} = K_{p_\mathcal{L},G}^{[i]} \cdot \frac{v_{p_\mathcal{L}}^{[i]}\left(p_{\mathcal{L},G}^{[i]}\right)}{\left\|v_{p_\mathcal{L}}^{[i]}\left(p_{\mathcal{L},G}^{[i]}\right)\right\|} \cdot \tag{14}$$

$$\min\left(\left\|v_{p_\mathcal{L}}^{[i]}\left(p_{\mathcal{L},G}^{[i]}\right)\right\|, \sqrt{\Delta E_{p_\mathcal{L}}^{[i]}\left(p_\mathcal{L}, p_{\mathcal{L},G}^{[i]}\right)/K_{p_\mathcal{L}}^{[i]}}\right)$$

where

$$\Delta E_{p_\mathcal{L}}^{[i]}(p_\mathcal{L}, p_{\mathcal{L},G}^{[i]}) \triangleq K_{p_\mathcal{L}}^{[i]} \cdot d\left(p_{\mathcal{L},G}^{[i]}, \partial\mathscr{F}_{p_\mathcal{L}}\left(\mathcal{C}_{\widehat{\mathcal{S}}_i}, \mathcal{C}_{\widehat{\mathcal{S}}_j}\right)\right) - E_{p_\mathcal{L}}^{[i]}(p_\mathcal{L}, p_{\mathcal{L},G}^{[i]}) \tag{15}$$

$$E_{p_\mathcal{L}}^{[i]}(p_\mathcal{L}, p_{\mathcal{L},G}^{[i]}) \triangleq \frac{1}{2} \cdot \left(\|\dot{p}_\mathcal{L}\|^2 + K_{p_\mathcal{L}}^{[i]} \cdot \|p_\mathcal{L} - p_{\mathcal{L},G}^{[i]}\|^2\right) \tag{16}$$

$p_{\mathcal{L},G}^{[i]}$ is the (virtual) state of the reference governor, $d(x, \mathcal{X})$ is the distance of x from the set $\mathcal{X}$, $K_{p_\mathcal{L}}^{[i]}$, $K_{p_\mathcal{L},G}^{[i]}$ are fixed, positive gains and $\zeta_{p_\mathcal{L}}^{[i]}$ is a virtual damping.

4.2.2. Object's Orientation

To drive the orientation $\theta_\mathcal{L}$ of the object to the specified goal set $\mathscr{G}_{\theta_\mathcal{L}}\left(\mathcal{C}_{\widehat{\mathcal{S}}_i}, \mathcal{C}_{\widehat{\mathcal{S}}_j}\right)$ while ensuring that it remains within $\mathcal{S}_i^{qo} = \left[\theta_l^{[i]}, \theta_l^{[i]}\right]$, we design the desired torque $\tau_{\mathcal{L},\theta}$ applied to the object based on the Prescribe Performance Control (PPC) methodology. We assume that $\mathcal{G}_{\theta_{\mathcal{L}}}\left(\mathcal{C}_{\hat{S}_i}, \mathcal{C}_{\hat{S}_j}\right)$ is of the form $\left[\theta_{l,G}^{[i]}, \theta_{u,G}^{[i]}\right]$, which can be ensured by designing the partitioning scheme of the configuration space planner described in Section 4.1 such that the compound slices which form a valid cover are overlapping. We now define the following two performance functions:

$$\begin{aligned}
\underline{\rho}_{\theta_{\mathcal{L}}}^{[i]}(t) &\triangleq \theta_{l,G}^{[i]} + \left(\theta_l^{[i]} - \theta_{l,G}^{[i]}\right) \cdot e^{-\lambda_{\theta_{\mathcal{L}}} \cdot t} \\
\overline{\rho}_{\theta_{\mathcal{L}}}^{[i]}(t) &\triangleq \theta_{u,G}^{[i]} + \left(\theta_l^{[i]} - \theta_{u,G}^{[i]}\right) \cdot e^{-\lambda_{\theta_{\mathcal{L}}} \cdot t}
\end{aligned} \tag{17}$$

where t denotes the time and $\lambda_{\theta_{\mathcal{L}}}$ is a positive constant. The corresponding control law is given by

$$\tau_{\mathcal{L},\theta}^{des} = I_{\mathcal{L}} \cdot \left(-K_{\theta_{\mathcal{L}},2}^{[i]} \cdot \left(\dot{\theta}_{\mathcal{L}} - v_{\theta_{\mathcal{L}}}^{[i]}\right) + \dot{v}_{\theta_{\mathcal{L}}}^{[i]} - a_{\theta_{\mathcal{L}}}^{[i]} \cdot \ln\left(\frac{\theta_{\mathcal{L}} - \underline{\rho}_{\theta_{\mathcal{L}}}^{[i]}}{\overline{\rho}_{\theta_{\mathcal{L}}}^{[i]} - \theta_{\mathcal{L}}}\right)\right) \tag{18}$$

$$v_{\theta_{\mathcal{L}}}^{[i]} \triangleq \frac{b_{\theta_{\mathcal{L}}}^{[i]} - K_{\theta_{\mathcal{L}},1}^{[i]} \cdot \ln\left(\frac{\theta_{\mathcal{L}} - \underline{\rho}_{\theta_{\mathcal{L}}}^{[i]}}{\overline{\rho}_{\theta_{\mathcal{L}}}^{[i]} - \theta_{\mathcal{L}}}\right)}{a_{\theta_{\mathcal{L}}}^{[i]}} \tag{19}$$

$$\begin{aligned}
a_{\theta_{\mathcal{L}}}^{[i]} &\triangleq \frac{1}{\theta_{\mathcal{L}} - \underline{\rho}_{\theta_{\mathcal{L}}}^{[i]}} + \frac{1}{\overline{\rho}_{\theta_{\mathcal{L}}}^{[i]} - \theta_{\mathcal{L}}} \\
b_{\theta_{\mathcal{L}}}^{[i]} &\triangleq \frac{\dot{\underline{\rho}}_{\theta_{\mathcal{L}}}^{[i]}}{\theta_{\mathcal{L}} - \underline{\rho}_{\theta_{\mathcal{L}}}^{[i]}} + \frac{\dot{\overline{\rho}}_{\theta_{\mathcal{L}}}^{[i]}}{\overline{\rho}_{\theta_{\mathcal{L}}}^{[i]} - \theta_{\mathcal{L}}}
\end{aligned} \tag{20}$$

with $K_{\theta_{\mathcal{L}},1}^{[i]}$ and $K_{\theta_{\mathcal{L}},2}^{[i]}$ being positive gains. Notice that the logarithmic term attains a high positive or negative value, when the orientation of the object approaches the upper or lower performance function defined in (17), thus confining it strictly within them. Hence, the orientation never escapes the set of viable orientations of the cell and moreover converges to the set of orientations requested by the planner.

4.2.3. Manipulators

Considering now the control scheme for the mobile manipulators, we remark that, by virtue of Assumption 1 and Assumption 2 and assuming a common initialization policy for the virtual states of the reference governors corresponding to the object's position and orientation, respectively, each robot is able to compute the desired total force $\tau_{\mathcal{L},p}$ and torque $\tau_{\mathcal{L},\theta}$ that should be applied to the object. Thus, the wrench $\tau_{e,k}$ that each robot $\mathcal{R}_k$, $k \in \mathbb{J}_{N_{\mathcal{R}}}$ should apply to the object via its end-effector is given by

$$\tau_{e,k} = \frac{1}{N_{\mathcal{R}}} \cdot \left[\tau_{\mathcal{L},\theta} - \left(\tau_{\mathcal{L},p}\right)^T \cdot {}_{\{E_k\}}P_{\mathcal{L},com}^{\perp}\right] \tag{21}$$

where ${}_{\{E_k\}}P_{\mathcal{L},com}^{\perp} = R\left(\frac{\pi}{2}\right) \cdot R(\theta_{\mathcal{L}}) \cdot {}_{\{E_k\}}P_{\mathcal{L},com}$ with ${}_{\{E_k\}}P_{\mathcal{L},com}$ being the position of the object's center of mass relative to the contact point of manipulator $\mathcal{A}_k$. Furthermore, each robot must also ensure that $q_k \in \mathcal{P}_{\mathcal{A}_k}\left(\mathcal{C}_{\hat{S}_i}\right)$ while driving q_k to $\mathcal{G}_{\mathcal{A}_k}\left(\mathcal{C}_{\hat{S}_i}, \mathcal{C}_{\hat{S}_j}\right)$. To do so, we shall exploit the redundancy of each robot to design a force in the null-space of $\mathcal{J}_{\mathcal{R}_k}$ which can ensure that the aforementioned specifications are met without affecting the force applied to the object. We now recall the dynamics of mobile manipulator $\mathcal{R}_k$:

$$M_{\mathcal{R}_k}(z_k) \cdot \ddot{z}_k + C_{\mathcal{R}_k}(z_k, \dot{z}_k) \cdot \dot{z}_k + G_{\mathcal{R}_k}(z_k) = \tau_{m,k} - \left(\mathcal{J}_{\mathcal{R}_k}(z_k)\right)^T \cdot \tau_{e,k} \tag{22}$$

Assuming known dynamic parameters and state, we can design

$$\tau_{m,k} = C_{\mathcal{R}_k}(z_k, \dot{z}_k) \cdot \dot{z}_k + G_{\mathcal{R}_k}(z_k) + \tau_{m,k,1} + \left(I - \left(\mathcal{J}_{\mathcal{R}_k}(z_k) \right)^{\dagger} \cdot \mathcal{J}_{\mathcal{R}_k}(z_k) \right) \cdot \tau_{m,k,2} \qquad (23)$$

where $\tau_{m,k,1}$ and $\tau_{m,k,2}$ are new virtual inputs to be defined later and $\left(\mathcal{J}_{\mathcal{R}_k} \right)^{\dagger}$ denotes the pseudo-inverse of $\mathcal{J}_{\mathcal{R}_k}$. Substituting the above in Equation (22) yields:

$$M_{\mathcal{R}_k}(z_k) \cdot \ddot{z}_k = \tau_{m,k,1} + \left(I - \left(\mathcal{J}_{\mathcal{R}_k}(z_k) \right)^{\dagger} \cdot \mathcal{J}_{\mathcal{R}_k}(z_k) \right) \cdot \tau_{m,k,2} - \left(\mathcal{J}_{\mathcal{R}_k}(z_k) \right)^{T} \cdot \tau_{e,k} \qquad (24)$$

We now consider the above dynamical model in the robot's task-space:

$$M'_{\mathcal{R}_k} \cdot \begin{bmatrix} \ddot{p}_{E_k} \\ \ddot{\theta}_{E_k} \end{bmatrix} + C'_{\mathcal{R}_k} \cdot \begin{bmatrix} \dot{p}_{E_k} \\ \dot{\theta}_{E_k} \end{bmatrix} = \left(\mathcal{J}_{\mathcal{R}_k}^{T} \right)^{\dagger} \cdot \tau_{m,k,1} - \tau_{e,k} \qquad (25)$$

where $\ddot{p}_{E_k}$ and $\ddot{\theta}_{E_k}$ are the position and orientation of the corresponding end-effector's contact point and

$$M'_{\mathcal{R}_k} = \left(\mathcal{J}_{\mathcal{R}_k}^{T} \right)^{\dagger} \cdot M_{\mathcal{R}_k} \cdot \left(\mathcal{J}_{\mathcal{R}_k} \right)^{\dagger}$$
$$C'_{\mathcal{R}_k} = -\left(\mathcal{J}_{\mathcal{R}_k}^{T} \right)^{\dagger} \cdot M_{\mathcal{R}_k} \cdot \left(\mathcal{J}_{\mathcal{R}_k} \right)^{\dagger} \cdot \dot{\mathcal{J}}_{\mathcal{R}_k} \cdot \left(\mathcal{J}_{\mathcal{R}_k} \right)^{\dagger}. \qquad (26)$$

Let $\begin{bmatrix} p_{E_k}^{T} & \theta_{E_k} \end{bmatrix}^{T} = \mathcal{T}_{\mathcal{L},E_k}(p_{\mathcal{L}}, \theta_{\mathcal{L}})$ be the rigid transformation between the positions and orientations of the corresponding points. It holds that

$$\begin{bmatrix} \dot{p}_{E_k} \\ \dot{\theta}_{E_k} \end{bmatrix} = J_{\mathcal{L},E_k}(p_{\mathcal{L}}, \theta_{\mathcal{L}}) \cdot \begin{bmatrix} \dot{p}_{\mathcal{L}} \\ \dot{\theta}_{\mathcal{L}} \end{bmatrix} \qquad (27)$$

$$\begin{bmatrix} \ddot{p}_{E_k} \\ \ddot{\theta}_{E_k} \end{bmatrix} = J_{\mathcal{L},E_k}(p_{\mathcal{L}}, \theta_{\mathcal{L}}) \cdot \begin{bmatrix} \ddot{p}_{\mathcal{L}} \\ \ddot{\theta}_{\mathcal{L}} \end{bmatrix} + \dot{J}_{\mathcal{L},E_k}(p_{\mathcal{L}}, \theta_{\mathcal{L}}) \cdot \begin{bmatrix} \dot{p}_{\mathcal{L}} \\ \dot{\theta}_{\mathcal{L}} \end{bmatrix} \qquad (28)$$

with $J_{\mathcal{L},E_k}$ denoting the Jacobian matrix of this rigid transformation. Therefore, Equation (25) can be re-written with respect to to the object's state as follows

$$M''_{\mathcal{R}_k} \cdot \begin{bmatrix} \ddot{p}_{\mathcal{L}} \\ \ddot{\theta}_{\mathcal{L}} \end{bmatrix} + C''_{\mathcal{R}_k} \cdot \begin{bmatrix} \dot{p}_{\mathcal{L}} \\ \dot{\theta}_{\mathcal{L}} \end{bmatrix} = \left(\mathcal{J}_{\mathcal{R}_k}^{T} \right)^{\dagger} \cdot \tau_{m,k,1} - \tau_{e,k} \qquad (29)$$

with

$$M''_{\mathcal{R}_k} = M'_{\mathcal{R}} \cdot J_{\mathcal{L},E_k}$$
$$C''_{\mathcal{R}_k} = C'_{\mathcal{R}_k} \cdot J_{\mathcal{L},E_k} + M'_{\mathcal{R}_k} \cdot \dot{J}_{\mathcal{L},E_k}. \qquad (30)$$

We notice that achieving our indented behavior, i.e., the object obeying the dynamics imposed by Equations (14) and (18) while distributing the load equally between the robots, is equivalent to

$$\tau_{e,k} = \frac{1}{N_{\mathcal{R}}} \left[\begin{matrix} M_{\mathcal{L}} \cdot \ddot{p}_{\mathcal{L}} \\ I_{\mathcal{L}} \cdot \ddot{\theta}_{\mathcal{L}} - M_{\mathcal{L}} \cdot (\ddot{p}_{\mathcal{L}})^{T} \cdot {}_{\{E_k\}} P^{\perp}_{\mathcal{L},com} \end{matrix} \right] = \frac{1}{N_{\mathcal{R}}} M_{\mathcal{L},E_k} \cdot \begin{bmatrix} \ddot{p}_{\mathcal{L}} \\ \ddot{\theta}_{\mathcal{L}} \end{bmatrix} \qquad (31)$$

where $M_{\mathcal{L},E_k}$ is the fragment of the object's inertia, as perceived by the manipulator $\mathcal{R}_k$. Substituting the above into Equation (25) yields

$$\left(M''_{\mathcal{R}_k} + \frac{1}{N_{\mathcal{R}}} \cdot M_{\mathcal{L},E_k} \right) \cdot \begin{bmatrix} \ddot{p}_{\mathcal{L}} \\ \ddot{\theta}_{\mathcal{L}} \end{bmatrix} + C''_{\mathcal{R}_k} \cdot \begin{bmatrix} \dot{p}_{\mathcal{L}} \\ \dot{\theta}_{\mathcal{L}} \end{bmatrix} = \left(\mathcal{J}_{\mathcal{R}_k}^{T} \right)^{\dagger} \cdot \tau_{m,k,1}. \qquad (32)$$

As such, we can see that selecting

$$\tau_{m,k,1} \triangleq \mathcal{J}_{\mathcal{R}_k} \cdot \left(\left(M''_{\mathcal{R}_k} + \frac{1}{N_{\mathcal{R}}} \cdot M_{\mathcal{L},E_k} \right) \cdot \begin{bmatrix} \tau_{\mathcal{L},p}^{des} / M_{\mathcal{L}} \\ \tau_{\mathcal{L},\theta}^{des} / I_{\mathcal{L}} \end{bmatrix} + C''_{\mathcal{R}_k} \cdot \begin{bmatrix} \dot{p}_{\mathcal{L}} \\ \dot{\theta}_{\mathcal{L}} \end{bmatrix} \right) \tag{33}$$

will achieve the desired behavior, assuming all $N_{\mathcal{R}}$ robots execute the same control law.

Considering again Equation (24), we shall now design $\tau_{m,k,2}$ appropriately in order to satisfy the manipulator joint limit specifications. We recall that the projection of $\tau_{m,k,2}$ with respect to $I - \left(\mathcal{J}_{\mathcal{R}_k}(z_k) \right)^{\dagger} \cdot \mathcal{J}_{\mathcal{R}_k}(z_k)$ has no effect on the wrench applied to the attached object. Now, let $L_{A,k}$, $L_{B,k}$, $L_{C,k}$, $L_{D,k}$ be matrices such that $L_{A,k} \in \mathbb{R}^{3\times3}$ and

$$L_k \triangleq M_{\mathcal{R}_k}^{-1} \cdot \left(I - \left(\mathcal{J}_{\mathcal{R}_k}(z_k) \right)^{\dagger} \cdot \mathcal{J}_{\mathcal{R}_k}(z_k) \right) = \begin{bmatrix} L_{A,k} & L_{B,k} \\ L_{C,k} & L_{D,k} \end{bmatrix}. \tag{34}$$

By recalling that $L_{D,k}$ is assumed to be invertible according to Assumption 4, we employ the Prescribed Performance Control method along with back-stepping to design $\tau_{m,k,2}$ as follows:

$$\tau_{m,k,2} \triangleq \begin{bmatrix} 0 \\ 0 \\ 0 \\ (L_{D,k})^{-1} \cdot v_{k,B} \end{bmatrix} \tag{35}$$

where

$$v_{k,B} \triangleq \begin{bmatrix} -(\dot{q}_{k,1} - v_{k,b,1}) + \dot{v}_{k,b,1} - a_{q_{k,1}} \cdot \ln\left(\frac{q_{k,1} - \underline{\rho}_{q_{k,1}}}{\overline{\rho}_{q_{k,1}} - q_{k,1}} \right) \\ -(\dot{q}_{k,1} - v_{k,b,1}) + \dot{v}_{k,b,2} - a_{q_{k,2}} \cdot \ln\left(\frac{q_{k,2} - \underline{\rho}_{q_{k,2}}}{\overline{\rho}_{q_{k,2}} - q_{k,2}} \right) \\ \vdots \\ -\left(\dot{q}_{k,N_{A_k}-1} - v_{k,b,N_{A_k}-1} \right) + \dot{v}_{k,b,N_{A_k}-1} - a_{q_{k,N_{A_k}}-1} \cdot \ln\left(\frac{q_{k,N_{A_k}-1} - \underline{\rho}_{q_{k,N_{A_k}-1}}}{\overline{\rho}_{q_{k,N_{A_k}-1}} - q_{k,N_{A_k}-1}} \right) \end{bmatrix} \tag{36}$$

$$v_{k,b,\ell} \triangleq \frac{b_{q_{k,\ell}} - \ln\left(\frac{q_{k,\ell} - \underline{\rho}_{q_{k,\ell}}}{\overline{\rho}_{q_{k,\ell}} - q_{k,\ell}} \right)}{a_{q_{k,\ell}}} \tag{37}$$

$$a_{q_{k,l}} \triangleq \frac{1}{q_{k,l} - \underline{\rho}_{q_{k,l}}} + \frac{1}{\overline{\rho}_{q_{k,l}} - q_{k,l}} \tag{38}$$

$$b_{q_{k,l}} \triangleq \frac{\dot{\underline{\rho}}_{q_{k,l}}}{q_{k,l} - \underline{\rho}_{q_{k,l}}} + \frac{\dot{\overline{\rho}}_{q_{k,l}}}{\overline{\rho}_{q_{k,l}} - q_{k,l}}$$

is the reference velocity control law, $\underline{\rho}_{q_{k,\ell}}$ and $\overline{\rho}_{q_{k,\ell}}$ are performance functions which smoothly "shrink" $\mathscr{P}_{A_k}\left(\mathcal{C}_{\widehat{S}_i} \right)$ to $\mathscr{G}_{A_k}\left(\mathcal{C}_{\widehat{S}_i}, \mathcal{C}_{\widehat{S}_j} \right)$, given by

$$\underline{\rho}_{q_{k,\ell}}(t) \triangleq \overline{q}_{G,k,\ell} + \left(\overline{q}_{k,\ell} - \overline{q}_{G,k,\ell} \right) \cdot e^{-\lambda_q \cdot t}$$
$$\overline{\rho}_{q_{k,\ell}}(t) \triangleq \underline{q}_{G,k,\ell} + \left(\underline{q}_{k,\ell} - \underline{q}_{G,k,\ell} \right) \cdot e^{-\lambda_q \cdot t} \tag{39}$$

with $\overline{q}_{k,\ell}$, $\underline{q}_{k,\ell}$ and $\overline{q}_{G,k,\ell}$, $\underline{q}_{G,k,\ell}$ being the lower and upper bounds of the joint parameters of A_k corresponding to $\mathscr{P}_{A_k}\left(\mathcal{C}_{\widehat{S}_i} \right)$ and $\mathscr{G}_{A_k}\left(\mathcal{C}_{\widehat{S}_i}, \mathcal{C}_{\widehat{S}_j} \right)$, respectively. Similar to the orientation control design, the input control signal $\tau_{m,k,2}$ was designed to constrain the evolution of the manipulators state within the corresponding upper and lower performance functions to enforce the necessary safety and convergence properties.

5. Stability Analysis

In this section, we provide an analysis of the robotic system's stability properties under the proposed control scheme.

Proposition 1. *(Safety). Given two adjacent under-approximation cells $\underline{\mathcal{C}}_{\widehat{S}_i}$ and $\underline{\mathcal{C}}_{\widehat{S}_j}$, the object's configuration will asymptotically converge to $\mathscr{G}_{p_{\mathcal{L}}}\left(\underline{\mathcal{C}}_{\widehat{S}_i}, \underline{\mathcal{C}}_{\widehat{S}_j}\right) \times \mathscr{G}_{\theta_{\mathcal{L}}}\left(\underline{\mathcal{C}}_{\widehat{S}_i}, \underline{\mathcal{C}}_{\widehat{S}_j}\right)$ for almost all initial configurations under control laws in Equations (14) and (18). Furthermore, the set $\mathscr{F}_{p_{\mathcal{L}}}\left(\underline{\mathcal{C}}_{\widehat{S}_i}, \underline{\mathcal{C}}_{\widehat{S}_j}\right) \times S_i^{q_0}$ is invariant.*

Proof. We begin this proof by first recalling that, as long as object's orientation and robot joints remain within the bounds imposed by $\underline{\mathcal{C}}_{\widehat{S}_i}$, control of the object's position and orientation can be safely decoupled. Regarding the object's position, one can readily verify that since T_i is a diffeomorphism in $\mathscr{F}_{p_{\mathcal{L}}}\left(\underline{\mathcal{C}}_{\widehat{S}_i}, \underline{\mathcal{C}}_{\widehat{S}_j}\right)$ (see [37]), the reference velocity control law i is Lipschitz, has exactly one critical point which is located at the transformed goal configuration and is inward pointing at the outer boundary of $\mathscr{F}_{p_{\mathcal{L}}}\left(\underline{\mathcal{C}}_{\widehat{S}_i}, \underline{\mathcal{C}}_{\widehat{S}_j}\right)$. Then by invocation of Theorem 2 in [41], the control law in Equation (14) ensures invariance of cell and convergence to the goal set of $p_{\mathcal{L}}$ for almost all initial configurations.

Regarding the object's orientation, we define the following coordinate transformation

$$
\begin{aligned}
z_{1,i} &= \ln\left(\frac{\theta_{\mathcal{L}} - \underline{\rho}_{\theta_{\mathcal{L}}}^{[i]}}{\overline{\rho}_{\theta_{\mathcal{L}}}^{[i]} - \theta_{\mathcal{L}}}\right) \\
z_{2,i} &= \dot{\theta}_{\mathcal{L}} - v_{\theta_{\mathcal{L}}}^{[i]}
\end{aligned}
\tag{40}
$$

and consider the following Lyapunov candidate:

$$
V = \frac{1}{2} \cdot z_{1,i}^2 + \frac{1}{2} \cdot z_{2,i}^2.
\tag{41}
$$

The time derivatives of $z_{1,i}$ and $z_{2,i}$ are given by

$$
\begin{aligned}
\dot{z}_{1,i} &= a_{\theta_{\mathcal{L}}}^{[i]} \cdot z_{2,i} + a_{\theta_{\mathcal{L}}}^{[i]} \cdot v_{\theta_{\mathcal{L}}}^{[i]} - b_{\theta_{\mathcal{L}}}^{[i]} \\
\dot{z}_{2,i} &= \frac{\tau_{\mathcal{L},\theta}}{I_{\mathcal{L}}} - \dot{v}_{\theta_{\mathcal{L}}}^{[i]}
\end{aligned}
\tag{42}
$$

Thus, computing the derivative of V with respect to time yields

$$
\begin{aligned}
\dot{V} &= z_{1,i} \cdot \dot{z}_{1,i} + z_{2,i} \cdot \dot{z}_{2,i} \\
&= a_{\theta_{\mathcal{L}}}^{[i]} \cdot z_{1,i} \cdot z_{2,i} + z_{1,i} \cdot \left(a_{\theta_{\mathcal{L}}}^{[i]} \cdot v_{\theta_{\mathcal{L}}}^{[i]} - b_{\theta_{\mathcal{L}}}^{[i]}\right) + z_{2,i} \cdot \left(\frac{\tau_{\mathcal{L},\theta}}{I_{\mathcal{L}}} - \dot{v}_{\theta_{\mathcal{L}}}^{[i]}\right)
\end{aligned}
\tag{43}
$$

Noting that $a_{\theta_{\mathcal{L}}}^{[i]} \cdot v_{\theta_{\mathcal{L}}}^{[i]} - b_{\theta_{\mathcal{L}}}^{[i]} = -K_{\theta_{\mathcal{L}},1}^{[i]} z_{1,i}^2$ and substituting the control law for $\tau_{\mathcal{L},\theta}$ to the above, we obtain

$$
\dot{V} = -\left(K_{\theta_{\mathcal{L}},1}^{[i]} \cdot z_{1,i}^2 + K_{\theta_{\mathcal{L}},2}^{[i]} \cdot z_{2,i}^2\right).
\tag{44}
$$

Since $\dot{V}$ is negative definite, assuming that the initial value of $\theta_{\mathcal{L}}$ lies within the specified bounds, the proposed control law ensures that $S_i^{q_0}$ remains invariant and that $\theta_{\mathcal{L}}$ will asymptotically converge to $\left(\theta_{l,G}^{[i]} + \theta_{u,G}^{[i]}\right)/2$. $\square$

Proposition 2. *(Safety). Given two adjacent under-approximation cells $\underline{\mathcal{C}}_{\widehat{S}_i}$ and $\underline{\mathcal{C}}_{\widehat{S}_{j'}}$, under the control law Equation (23), the joint states q_k of mobile manipulator $\mathcal{R}_k$ will converge to $\mathscr{G}_{A_k}\left(\underline{\mathcal{C}}_{\widehat{S}_i}, \underline{\mathcal{C}}_{\widehat{S}_j}\right)$. Furthermore, the set $\mathscr{P}_{A_k}\left(\underline{\mathcal{C}}_{\widehat{S}_i}\right)$ is invariant.*

Proof. We consider once again Equation (24). Since $M_{\mathcal{R}_k}$ is an inertia matrix, we know that its inverse exists, thus multiplying both sides with $\left(M_{\mathcal{R}_k}\right)^{-1}$ and substituting Equation (35) yields:

$$
\begin{aligned}
\ddot{z}_k &= \left(M_{\mathcal{R}_k}(z_k)\right)^{-1} \cdot \left(I - \left(\mathcal{J}_{\mathcal{R}_k}(z_k)\right)^{\dagger} \cdot \mathcal{J}_{\mathcal{R}_k}(z_k)\right) \cdot \tau_{m,k,2} + \\
&\quad \left(M_{\mathcal{R}_k}(z_k)\right)^{-1} \cdot \tau_{m,k,1} - \left(M_{\mathcal{R}_k}(z_k)\right)^{-1} \cdot \left(\mathcal{J}_{\mathcal{R}_k}(z_k)\right)^{T} \cdot \tau_{e,k} \\
&= \begin{bmatrix} L_{B,k} \cdot (L_{D,k})^{-1} \\ I \end{bmatrix} \cdot v_{k,B} + \left(M_{\mathcal{R}_k}(z_k)\right)^{-1} \cdot \tau_{m,k,1} - \left(M_{\mathcal{R}_k}(z_k)\right)^{-1} \cdot \left(\mathcal{J}_{\mathcal{R}_k}(z_k)\right)^{T} \cdot \tau_{e,k}
\end{aligned}
\tag{45}
$$

We note that the term $\tau_{m,k,2}$, which is designed to ensure satisfaction of joint parameter specifications, has no effect on the stability properties involving the object's position $p_{\mathcal{L}}$ and orientation $\theta_{\mathcal{L}}$ by virtue of $I - \left(\mathcal{J}_{\mathcal{R}_k}(z_k)\right)^{\dagger} \cdot \mathcal{J}_{\mathcal{R}_k}(z_k)$. Therefore, the last two r.h.s. terms of Equation (45) are bounded by design and vanish as the object approaches the specified configuration corresponding to the current cell. As such, the dynamics of the joint parameters can be written as:

$$
\ddot{q}_k = v_{k,B} + w_B
\tag{46}
$$

where the term w_B corresponding to $\tau_{m,k,1}$ and $\tau_{e,k}$ and can be viewed as a bounded and vanishing disturbance. We now define the following coordinate transformation for each joint value $q_{k,\ell}$, $\ell \in \mathfrak{I}_{N_{A_k}-1}$:

$$
z_{q,1,\ell} = \ln\left(\frac{q_{k,\ell} - \underline{\rho}_{q_{k,\ell}}}{\overline{\rho}_{q_{k,\ell}} - q_{k,\ell}}\right)
\tag{47}
$$

$$
z_{q,2,\ell} = \dot{q}_{k,\ell} - v_{k,b,\ell}
$$

and consider the Lyapunov candidate

$$
V_\ell = \frac{1}{2} \cdot z_{q,1,\ell}^2 + \frac{1}{2} \cdot z_{q,2,\ell}^2.
\tag{48}
$$

Following the same procedure as above, we derive that

$$
\begin{aligned}
\dot{V}_\ell &= -\left(z_{q,1,\ell}^2 + z_{q,2,\ell}^2\right) + z_{q,2,\ell} \cdot w_{B,\ell} \\
&\leq -z_{q,1,\ell}^2 - z_{q,2,\ell}^2 + |z_{q,2,\ell}| \cdot |w_{B,\ell}| \\
&\leq -z_{q,1,\ell}^2 - |z_{q,2,\ell}| \cdot |w_{B,\ell}| + (w_{B,\ell})^2
\end{aligned}
\tag{49}
$$

which implies that $z_{q,1,\ell}, z_{q,2,\ell}$ and the control law are globally uniformly bounded (Lemma 2.28 [42]), and, thus, concludes the proof. $\square$

Theorem 1. *(Convergence). The robotic system under the distributed control law in Equation (23) will successfully drive the object $\mathcal{L}$ to its goal configuration $q_{\mathcal{L},\text{init}}$, from almost all initial configurations.*

Proof. First, we note that, by virtue of Equation (4) and the design of Equation (23), the total force and torque applied to the object's center of mass by the robotic system is equal to the desired ones specified by Equation (14) and Equation (18), since the remaining terms either cancel the robot's dynamics or are projected along the kernel space of $\mathcal{J}_{\mathcal{R}_k}$, respectively. As such, according to Proposition 1, the object is guaranteed traverse from one cell to another till it arrives to the desired configuration $q_{\mathcal{L},\text{goal}}$, starting from almost any initial configuration $q_{\mathcal{L},\text{init}}$, as long as the robots do not collide with the workspace boundary. However, according to Proposition 2, the configurations of the mobile manipulators remain within the bounds specified by the under-approximation cell $\underline{\mathcal{C}}_{\hat{S}_i}$, which, by design of the high-level planner, implies that the robotic system's footprint cannot intersect the workspace's boundary. $\square$

6. Simulation Results

To demonstrate the efficacy of the proposed control scheme, we consider a robotic system consisting of two mobile manipulators holding a rectangular object within the workspace depicted in Figure 4. The robotic system was initialized at $q_{\mathcal{L},\text{init}} = [0.9, 2, 1.57]^T$ and $q_{1,1} = q_{2,1} = 0$ whereas the desired configuration of the object was set to $q_{\mathcal{L},\text{goal}} = [5, 8, 4.663]^T$. The intervals for the object's orientation and robot joints generated by the high-level planner can be seen in Table 1, whereas the control parameters selected during this simulation are given in Table 2. Notice that our planner extracted a viable sequence of configuration cells despite the fact that the feasible configuration space becomes very narrow particularly when the robotic systems has to transverse a corner thus verifying the completeness of our approach. Figure 5 shows the trajectory executed by the robotic system under the proposed control law, whereas plots of the object's position, orientation and corresponding rates can be seen in Figure 6, Figure 7, Figure 8, and Figure 9, respectively. It should be noted that the transition between successive cells is executed by the proposed low level control algorithms without harming either the safety or the convergence properties. Accordingly, Figures 10 and 11 show the evolution of each manipulator's state, as well as the computed lower and upper bounds corresponding to each cell. The total force and torque applied to the object is also displayed in Figure 12 and Figure 13, respectively. As one can verify from the aforementioned figures, the robotic system successfully reaches the goal configuration while satisfying the specifications corresponding to $\theta_{\mathcal{L}}, q_{1,1}, q_{2,1}$. A video of the aforementioned transportation task can be found in the following url: https://youtu.be/AQ_8z3tysRo (accessed on 6 December 2022).

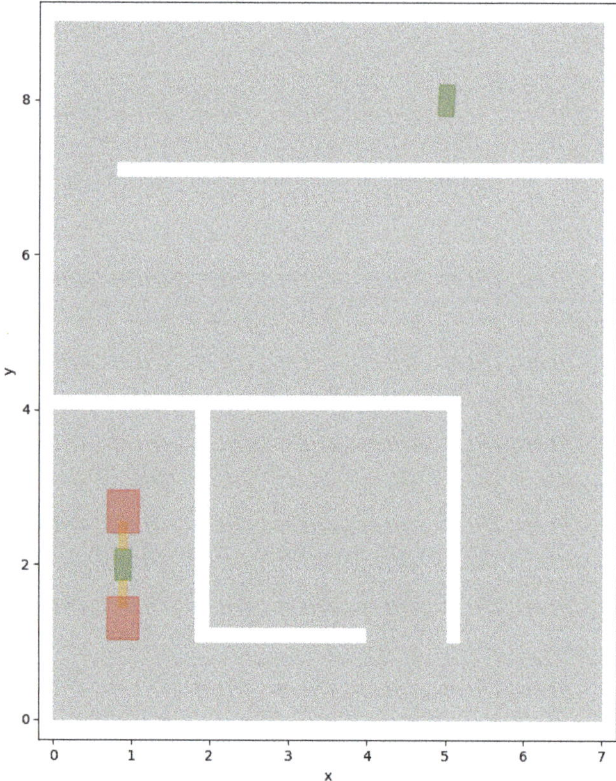

Figure 4. Initial and goal configuration of the robotic system and object, respectively.

Notice that the simulated workspace consists of both narrow and wide areas, which results in both situations that the robotic system needs to undergo major reconfiguration and situations where the system can navigate without the need to alter its configuration, demonstrating the adaptive nature of the algorithm. Adding more than two robots in this example would render the overall space around the object overcrowded by the robots carrying it, thus limiting its flexibility and not demonstrating the searching capabilities of the proposed algorithm with respect to the shape of the robotic system that leads to feasible paths. In other words, when multiple robots, grasping the object uniformly at its boundary, are adopted, the expected motion of each robot with respect to the object would be very constrained in order to avoid collisions with neighboring robots, i.e., the robotic system would travel as a rigid formation. Consequently, we selected to demonstrate the most reconfigurable case in order to show how the proposed algorithm seeks and finds viable configurations in the workspace that avoid inter-robot collisions and collisions with the environment and fulfill the transportation task.

Table 1. Lower and upper bounds of the intervals corresponding to each cell, as generated by the planner.

Cell ID	θ_l	θ_u	$\overline{q}_{1,1}$	$\underline{q}_{1,1}$	$\overline{q}_{2,1}$	$\underline{q}_{2,1}$
1	1.470	1.867	−0.100	0.567	−0.100	0.567
2	1.667	2.065	−0.100	0.567	−0.567	0.100
3	1.863	2.456	−0.100	0.567	−1.035	0.367
4	1.863	2.456	−0.100	0.333	−1.870	−0.835
5	2.256	2.652	−0.100	0.333	−1.503	−0.835
6	2.452	2.849	−0.100	0.333	−1.035	−0.368
7	2.649	2.947	−0.100	0.217	−0.567	0.100
8	2.747	3.045	−0.050	0.108	−0.050	0.108
9	2.895	3.191	−0.050	0.108	−0.050	0.108
10	3.091	3.388	−0.050	0.158	−0.158	0.050
11	3.287	3.584	0.067	0.284	−0.284	−0.067
12	3.484	3.781	0.184	0.518	−0.518	−0.184
13	3.681	3.977	0.418	0.985	−0.985	−0.418
14	3.877	4.172	0.651	0.984	−0.984	−0.651
15	4.072	4.370	0.418	0.985	−0.985	−0.418
16	4.270	4.576	0.184	0.518	−0.518	−0.184
17	4.476	4.762	0.067	0.284	−0.284	−0.067
18	4.564	4.762	0.067	0.284	−0.284	−0.067
19	4.564	4.762	−0.050	0.108	−0.108	0.050
20	4.564	4.762	−0.050	0.108	−0.108	0.050
21	4.564	4.762	−0.985	0.050	−0.108	0.050
22	4.564	4.762	−1.453	−0.885	−0.108	0.050
23	4.564	4.762	−1.687	−1.353	−0.108	0.050
24	4.564	4.762	−1.687	−1.528	−0.108	−0.050
25	4.564	4.762	−1.687	−1.528	−0.168	−0.008
26	4.564	4.762	−1.687	−1.528	−0.284	−0.067
27	4.564	4.762	−1.687	−1.528	−0.518	−0.184
28	4.564	4.762	−1.687	−1.528	−0.985	−0.418
29	4.564	4.762	−1.687	−1.528	−1.453	−0.885
30	4.564	4.762	−1.687	−1.528	−1.687	−1.353

Table 2. Simulation parameters.

Parameter	Value
$M_{\mathcal{L}}$	1 kg
$I_{\mathcal{L}}$	1 kg m^2
$K_{p_{\mathcal{L}}}$	50
$\zeta_{p_{\mathcal{L}}}$	5
$K_{p_{\mathcal{L}},G}$	5
$\lambda_{\theta_{\mathcal{L}}}$	2
$K_{\theta_{\mathcal{L}},1}$	1
$K_{\theta_{\mathcal{L}},2}$	1
λ_q	2

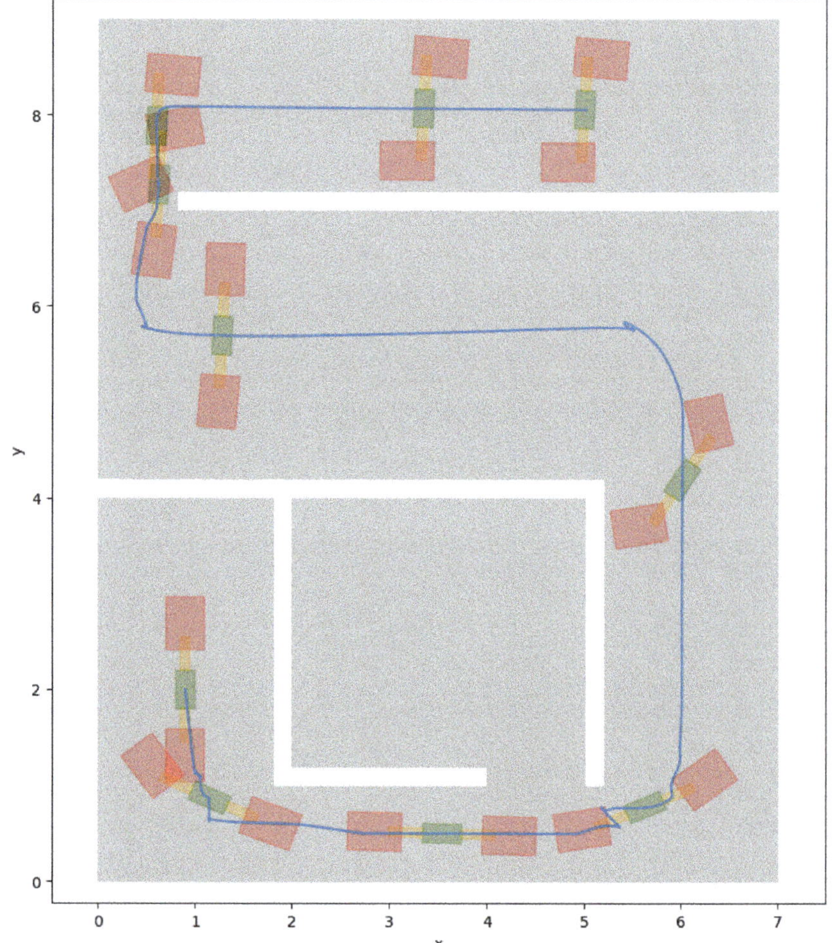

Figure 5. Path executed by the robotic system during the simulations (blue line), as well as the footprint of the robotic system at various instants.

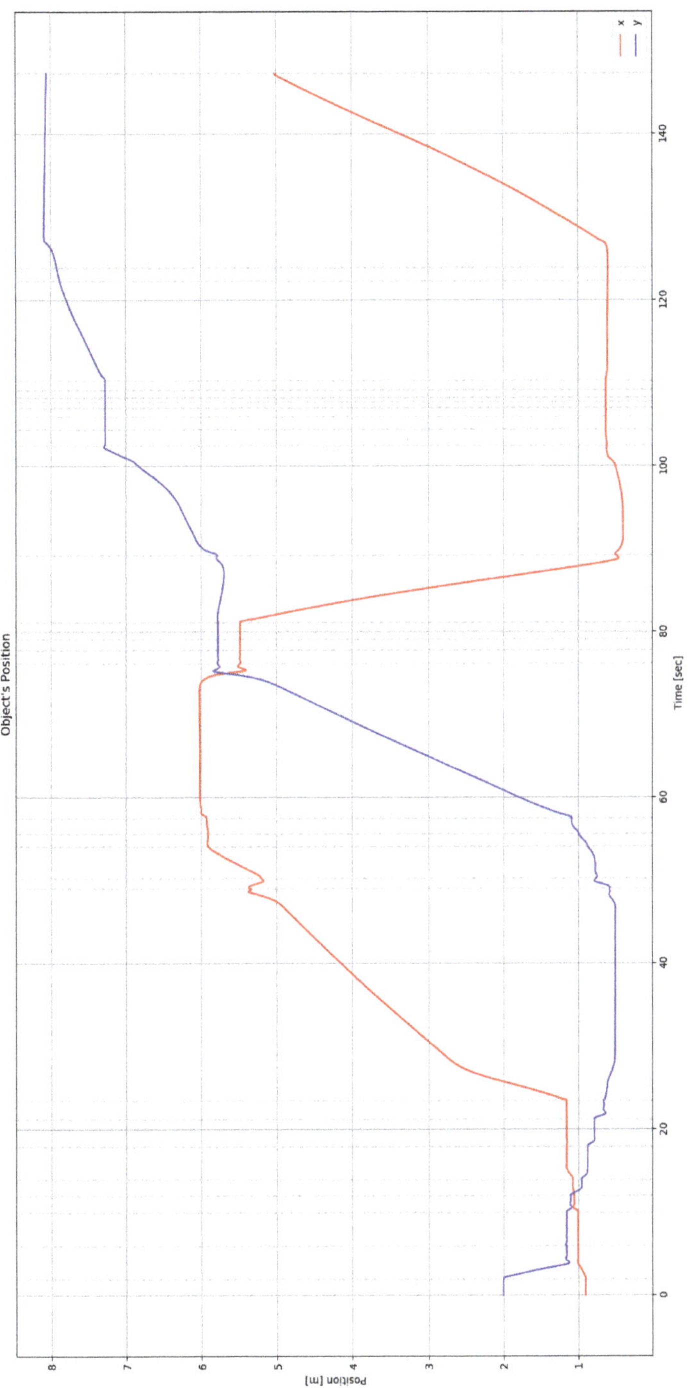

Figure 6. Evolution of the object's position $p_{\mathcal{L}}$ over time. The vertical dashed lines indicate transitions between consecutive cells.

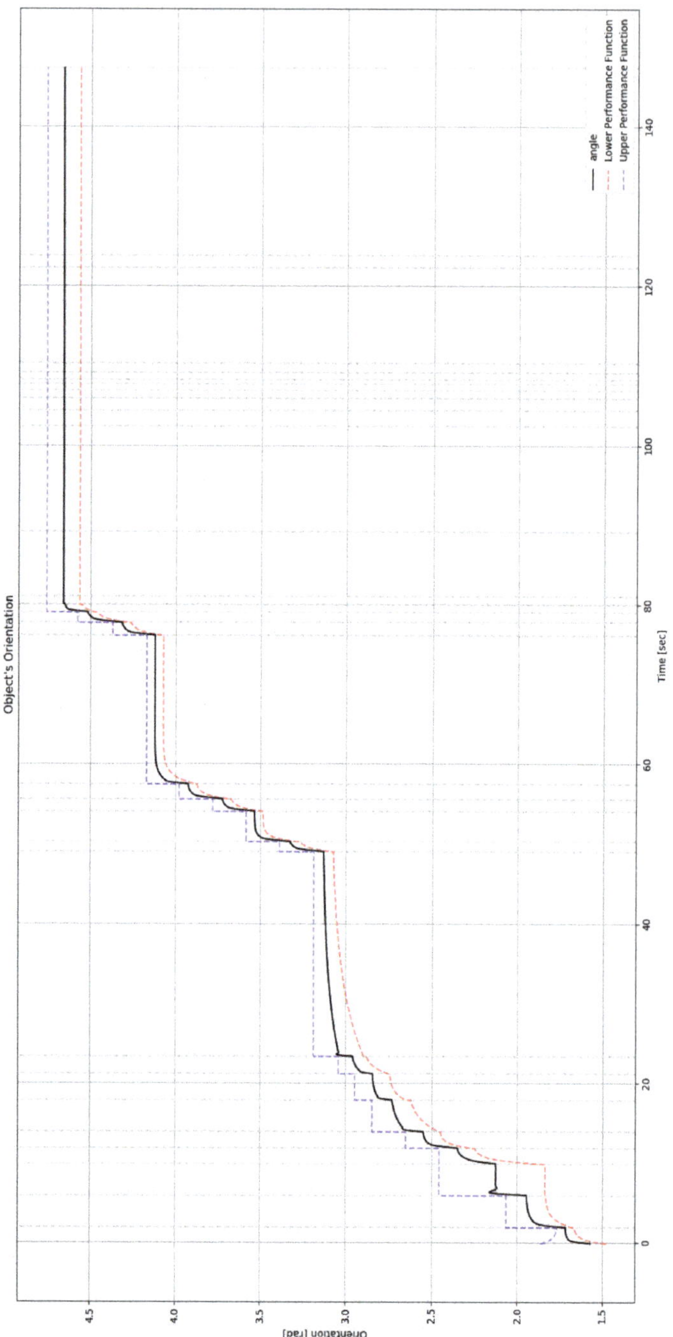

Figure 7. Evolution of the object's orientation $\theta_{\mathcal{L}}$ over time (solid line), as well as the corresponding performance functions $\underline{\rho}_{\theta_{\mathcal{L}}}$ and $\overline{\rho}_{\theta_{\mathcal{L}}}$.

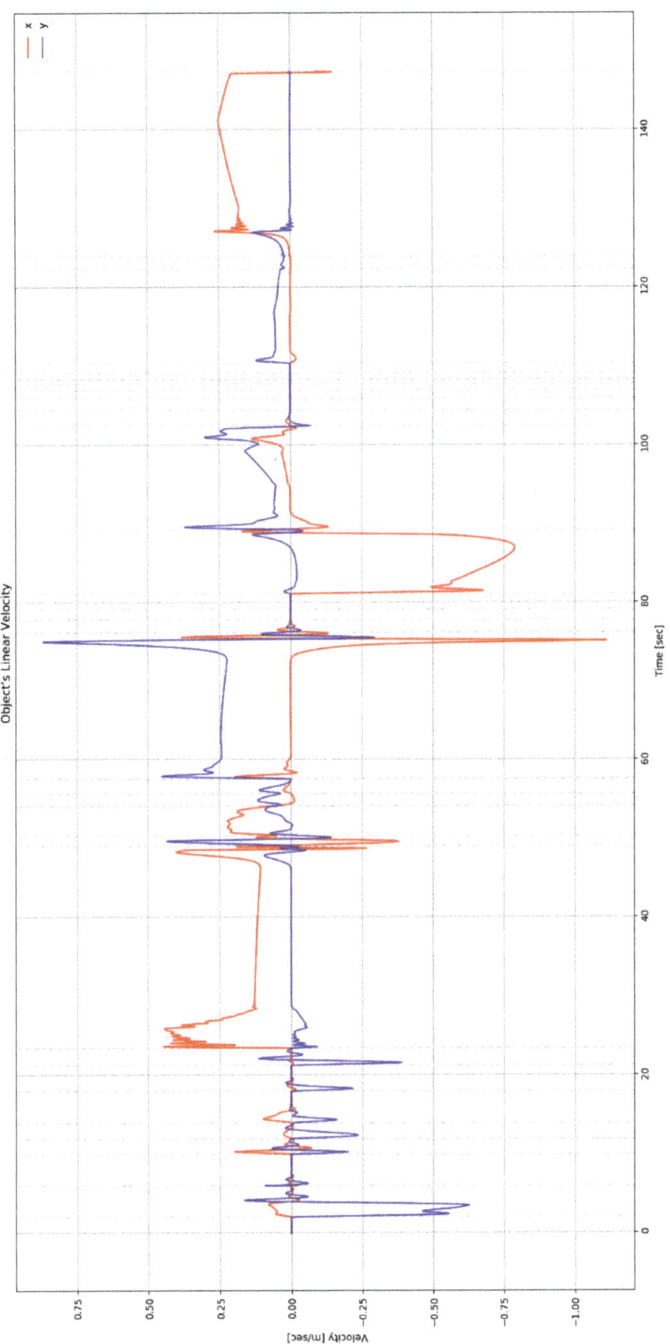

Figure 8. Object's linear velocity $\dot{p}_{\mathcal{L}}$.

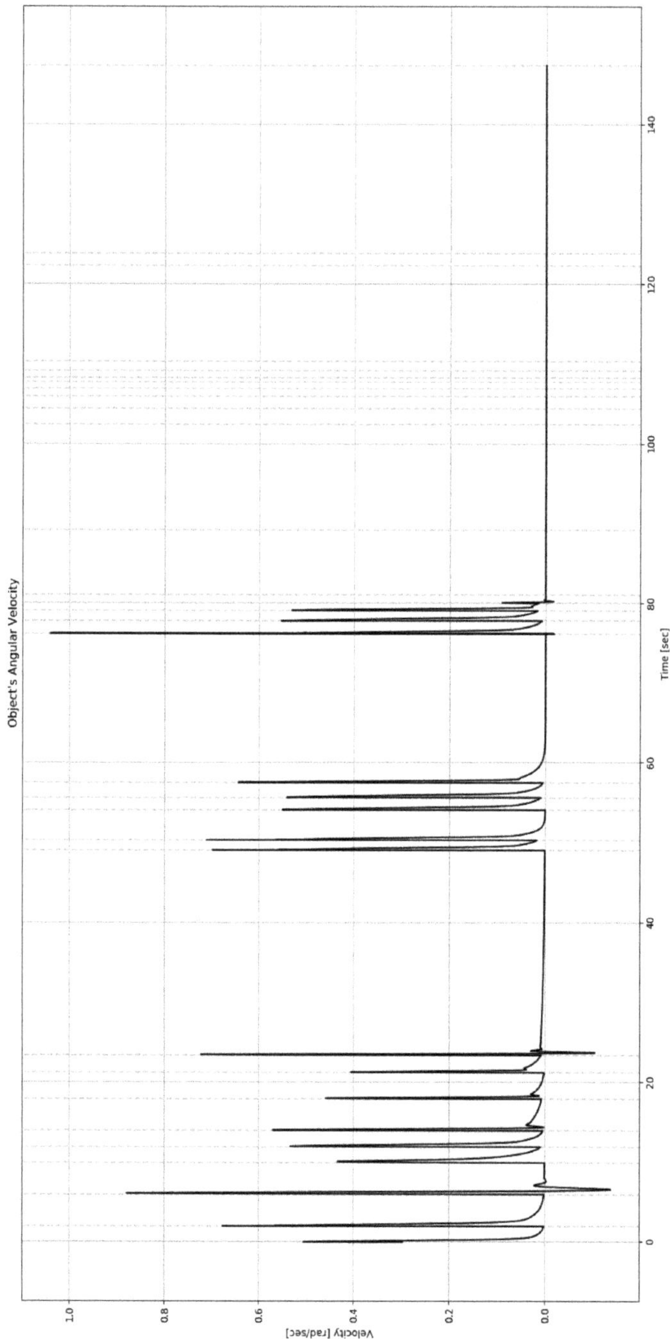

Figure 9. Object's angular velocity $\dot{\theta}_{\mathcal{L}}$.

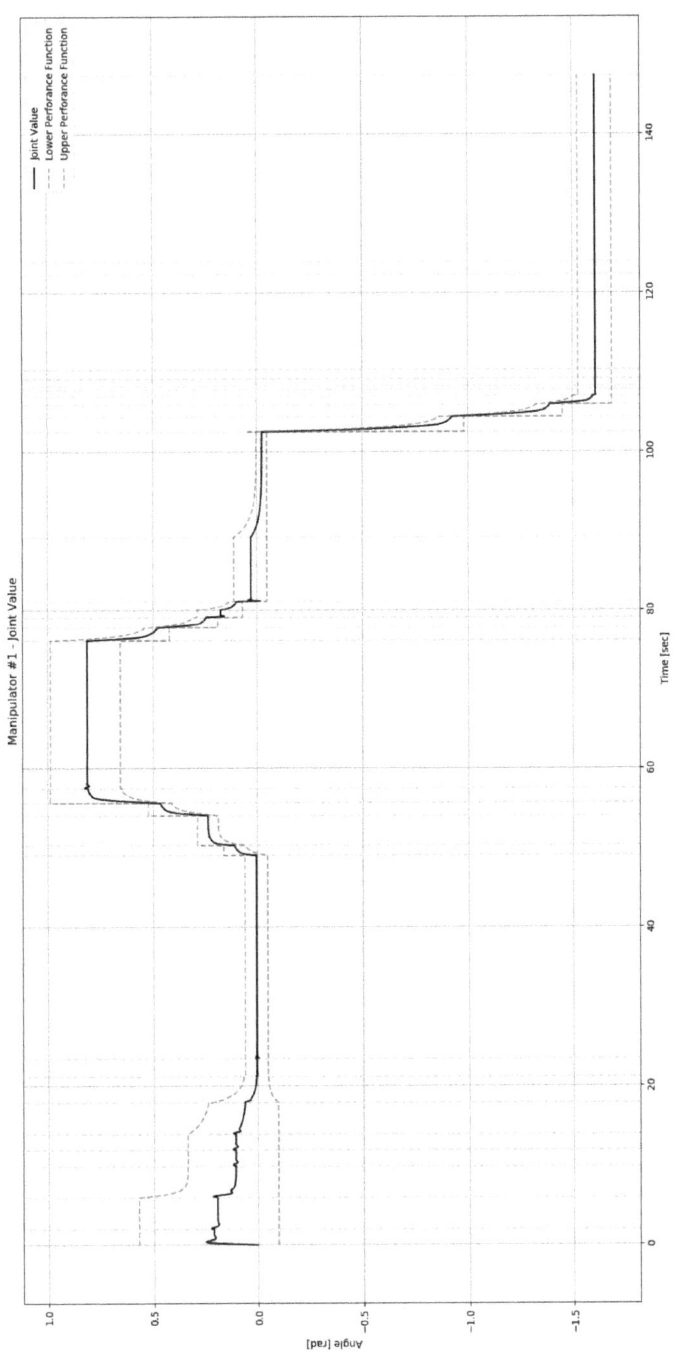

Figure 10. Evolution of joint value $q_{1,1}$ with corresponding lower and upper bounds.

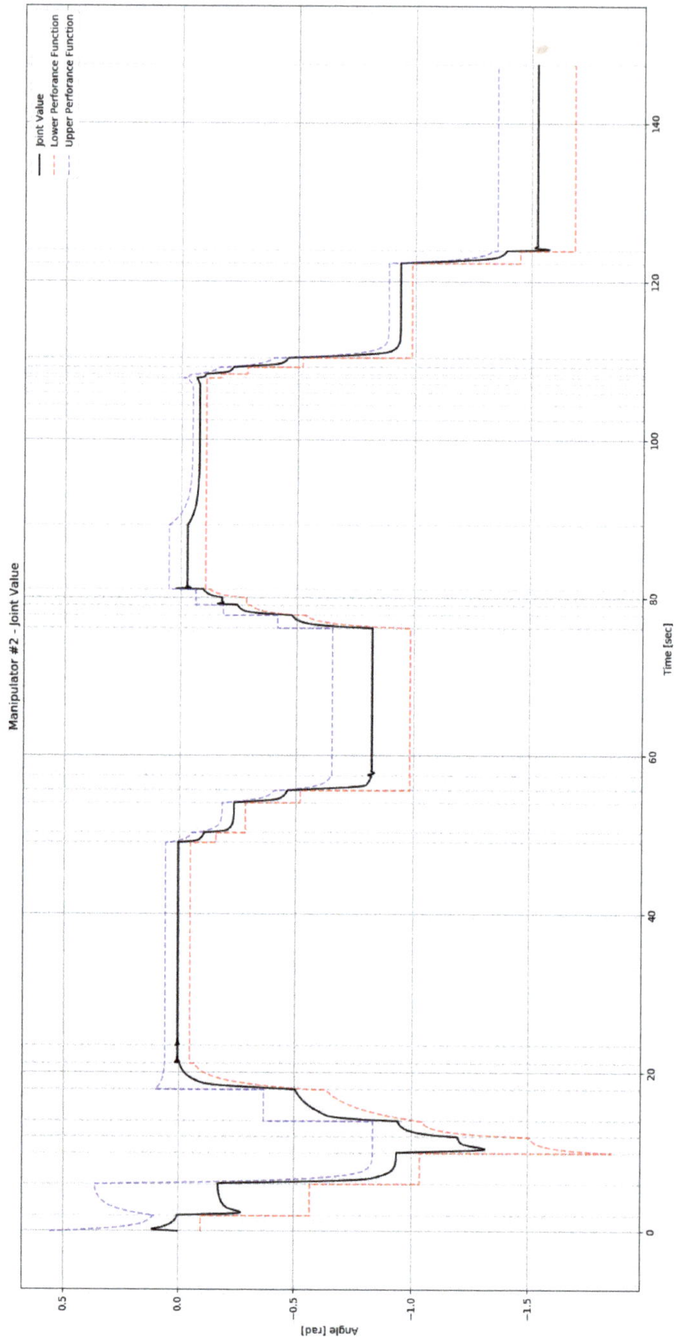

Figure 11. Evolution of joint value $q_{2,1}$ with corresponding lower and upper bounds.

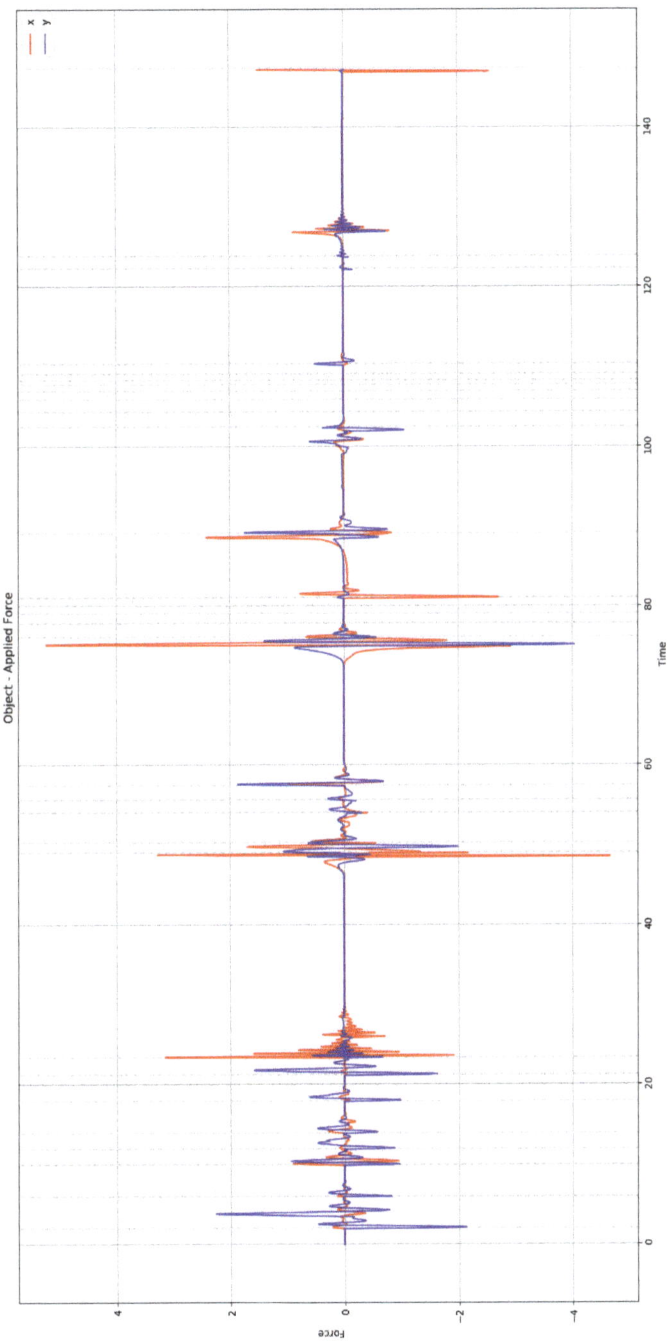

Figure 12. Total force $\tau_{\mathcal{L},p}^{des}$ applied to the object by the robots.

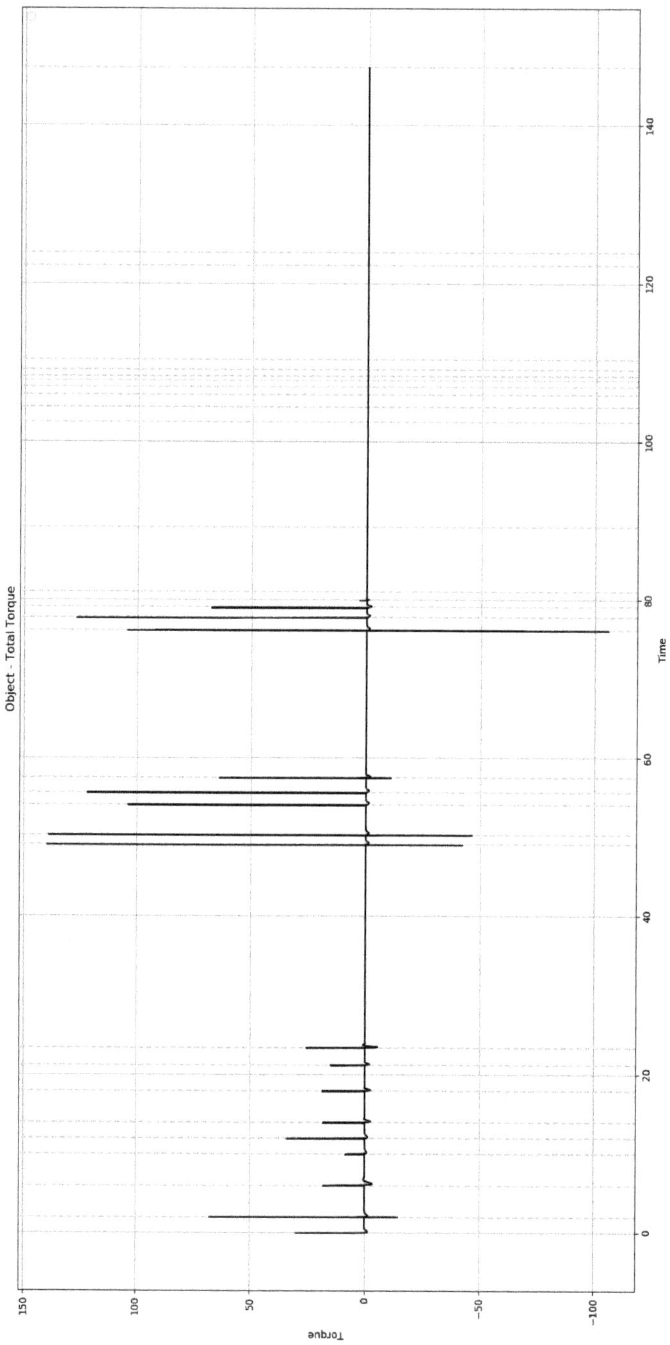

Figure 13. Total torque $\tau_{\mathcal{L},\theta}^{des}$ applied to the object by the robots.

7. Conclusions and Future Directions

In this work, we presented a hybrid control scheme for addressing the cooperative transportation problem for a team of mobile manipulators carrying an object within a planar workspace. Particularly, a high-level planner was designed for computing a sequence of feasible cells by adaptively subdividing the system's configuration space using a hierarchical cell decomposition scheme. In addition, a distributed low-level control law was employed for realizing the given plan with guaranteed collision avoidance and convergence properties. Finally, simulation results validating the proposed scheme's efficacy were provided. Future research efforts will be devoted towards extending the proposed framework for human–robot collaborative transportation tasks within obstacle cluttered workspaces, where the robots are in charge of taking over the load while avoiding collisions and the human performs only high-level planning. More work is also needed towards devising optimal performance criteria to quantify the achieved response and guide appropriately the selection of the control parameters as well as to evaluate the robustness level against actuation limitations, disturbances, measurement delays, and noise.

Author Contributions: Conceptualization, P.V. and C.P.B. and K.J.K.; methodology, P.V. and C.P.B.; software, P.V.; validation, P.V.; formal analysis, P.V. and C.P.B.; investigation, P.V. and C.P.B. and K.J.K.; resources, K.J.K.; data curation, P.V.; writing—original draft preparation, P.V.; writing—review and editing, C.P.B. and K.J.K.; visualization, P.V.; supervision, C.P.B. and K.J.K.; project administration, C.P.B. and K.J.K.; funding acquisition, K.J.K. All authors have read and agreed to the published version of the manuscript.

Funding: This work was carried out with an SSF scholarshib funded by the Act "Enhancement of human research potential through doctoral research" from the resources of the "Human Resources Development, Education and Lifelong Learning" Program, 2014-2020 with the co-financing of the European Social Fund (ESF) and the Greek State.

Conflicts of Interest: The authors declare no conflict of interest.

References

1. Gerkey, B.P.; Matarić, M.J. Pusher-watcher: An approach to fault-tolerant tightly-coupled robot coordination. In Proceedings of the IEEE International Conference on Robotics and Automation, Washington, DC, USA, 11–15 May 2002; Volume 1, pp. 464–469.
2. Yamada, S.; Saito, J. Adaptive action selection without explicit communication for multi-robot box-pushing. In Proceedings of the IEEE International Conference on Intelligent Robots and Systems, Phoenix, AZ, USA, 15–19 March 1999; Volume 3, pp. 1444–1449.
3. Munoz Melendez, A.; Drogoul, A. *Analyzing Multi-Robot Box-Pushing*; Technical Report; Universidad de Colima, Colima, Mexico, 2004.
4. Wang, Z.D.; Kumar, V. Object closure and manipulation by multiple cooperating mobile robots. In Proceedings of the IEEE International Conference on Robotics and Automation, Washington, DC, USA, 11–15 May 2002; Volume 1, pp. 394–399.
5. Fink, J.; Michael, N.; Kumar, V. Composition of vector fields for multi-robot manipulation via caging. In Proceedings of the International Conference on Robotics Science and Systems, Orlando, FL, USA, 7–10 July 2008; Volume 3, pp. 25–32.
6. Eoh, G.; Lee, S.; Lee, T.H.; Lee, B. Distributed Object Transportation Using Virtual Object. *J. Ind. Intell. Inf.* **2014**, *2*, 20–25. [CrossRef]
7. Dai, Y.; Kim, Y.; Wee, S.; Lee, D.; Lee, S. Symmetric caging formation for convex polygonal object transportation by multiple mobile robots based on fuzzy sliding mode control. *ISA Trans.* **2016**, *60*, 321–332. [CrossRef] [PubMed]
8. Wan, W.; Shi, B.; Wang, Z.; Fukui, R. Multirobot Object Transport via Robust Caging. *IEEE Trans. Syst. Man, Cybern. Syst.* **2020**, *50*, 270–280. [CrossRef]
9. Ferrante, E.; Brambilla, M.; Birattari, M.; Dorigo, M. Socially-mediated negotiation for obstacle avoidance in collective transport. In *Distributed Autonomous Robotic Systems*; Springer Tracts in Advanced Robotics; Springer: Berlin/Heidelberg, Germany, 2012; Volume 83, pp. 571–583.
10. Habibi, G.; Kingston, Z.; Xie, W.; Jellins, M.; McLurkin, J. Distributed centroid estimation and motion controllers for collective transport by multi-robot systems. In Proceedings of the IEEE International Conference on Robotics and Automation, Seattle, WA, USA, 26–30 May 2015; pp. 1282–1288.
11. Machado, T.; Malheiro, T.; Monteiro, S.; Erlhagen, W.; Bicho, E. Multi-constrained joint transportation tasks by teams of autonomous mobile robots using a dynamical systems approach. In Proceedings of the IEEE International Conference on Robotics and Automation, Stockholm, Sweden, 16–21 May 2016, pp. 3111–3117.
12. Wang, Z.; Schwager, M. Kinematic multi-robot manipulation with no communication using force feedback. In Proceedings of the PIEEE International Conference on Robotics and Automation, Stockholm, Sweden, 16–21 May 2016, pp. 427–432.

13. Wang, Z.; Yang, G.; Su, X.; Schwager, M. OuijaBots: Omnidirectional Robots for Cooperative Object Transport with Rotation Control Using No Communication. In *Distributed Autonomous Robotic Systems*; Springer Proceedings in Advanced Robotics; Springer: Cham, Switzerland, 2018; Volume 6, pp. 117–131.

14. Hekmatfar, T.; Masehian, E.; Mousavi, S.J. Cooperative object transportation by multiple mobile manipulators through a hierarchical planning architecture. In Proceedings of the 2014 2nd RSI/ISM International Conference on Robotics and Mechatronics, ICRoM 2014, Tehran, Iran, 15–17 October 2014; pp. 503–508.

15. Wang, Y.; De Silva, C.W. Multi-robot box-pushing: Single-agent Q-learning vs. team Q-learning. In Proceedings of the IEEE International Conference on Intelligent Robots and Systems, Beijing, China, 9–13 October 2006; pp. 3694–3699.

16. Goldman, C.V.; Zilberstein, S. Decentralized control of cooperative systems: Categorization and complexity analysis. *J. Artif. Intell. Res.* **2004**, *22*, 143–174. [CrossRef]

17. Haddadin, S.; De Luca, A.; Albu-Schäffer, A. Robot collisions: A survey on detection, isolation, and identification. *IEEE Trans. Robot.* **2017**, *33*, 1292–1312. [CrossRef]

18. Haddadin, S.; Khoury, A.; Rokahr, T.; Parusel, S.; Burgkart, R.; Bicchi, A.; Albu-Schäffer, A. On making robots understand safety: Embedding injury knowledge into control. *Int. J. Robot. Res.* **2012**, *31*, 1578–1602. [CrossRef]

19. Uchiyama, M.; Dauchez, P. A symmetric hybrid position/force control scheme for the coordination of two robots. In Proceedings of the IEEE International Conference on Robotics and Automation, Philadelphia, PA, USA, 24–29 April 1988; pp. 350–356.

20. Khatib, O. Object Manipulation in a Multi-effector Robot System. In *Proceedings of the 4th International Symposium on Robotics Research*; MIT Press: Cambridge, MA, USA, 1988; Volume 4, pp. 137–144.

21. Tanner, H.; Loizou, S.; Kyriakopoulos, K. Nonholonomic navigation and control of cooperating mobile manipulators. *IEEE Trans. Robot. Autom.* **2003**, *19*, 53–64. [CrossRef]

22. Erhart, S.; Hirche, S. Model and Analysis of the Interaction Dynamics in Cooperative Manipulation Tasks. *IEEE Trans. Robot.* **2016**, *32*, 672–683. [CrossRef]

23. Khatib, O.; Yokoi, K.; Chang, K.; Ruspini, D.; Holmberg, R.; Casal, A. Vehicle/arm coordination and multiple mobile manipulator decentralized cooperation. In Proceedings of the IEEE International Conference on Intelligent Robots and Systems, Osaka, Japan, 4–8 November 1996; Volume 2, pp. 546–553.

24. Verginis, C.K.; Mastellaro, M.; Dimarogonas, D.V. Robust cooperative manipulation without force/torque measurements: Control design and experiments. *IEEE Trans. Control Syst. Technol.* **2020**, *28*, 713–729. [CrossRef]

25. Zhang, Y.; Zou, M.; Xiao, H.; Wen, J.; Wang, Y. Cooperative-manipulation scheme of routh-hurwitz type for simultaneous repetitive motion planning of two-manipulator robotic systems. In Proceedings of the 28th Chinese Control and Decision Conference, CCDC 2016, Yinchuan, China, 28–30 May 2016; pp. 4409–4414.

26. Petitti, A.; Franchi, A.; Di Paola, D.; Rizzo, A. Decentralized motion control for cooperative manipulation with a team of networked mobile manipulators. In Proceedings of the IEEE International Conference on Robotics and Automation, Stockholm, Sweden, 16–21 May 2016, pp. 441–446.

27. Noohi, E.; Zefran, M. Modeling the interaction force during a haptically-coupled cooperative manipulation. In Proceedings of the 25th IEEE International Symposium on Robot and Human Interactive Communication, RO-MAN 2016, New York, NY, USA, 26–31 August 2016; pp. 119–124.

28. Pajak, I. Real-Time Trajectory Generation Methods for Cooperating Mobile Manipulators Subject to State and Control Constraints. *J. Intell. Robot. Syst. Theory Appl.* **2019**, *93*, 649–668. [CrossRef]

29. He, Y.; Wu, M.; Liu, S. An optimisation-based distributed cooperative control for multi-robot manipulation with obstacle avoidance. *IFAC-PapersOnLine* **2020**, *53*, 9859–9864. [CrossRef]

30. Kosuge, K.; Oosumi, T. Decentralized control of multiple robots handling an object. In Proceedings of the IEEE International Conference on Intelligent Robots and Systems, Osaka, Japan, 4–8 November 1996; Volume 1, pp. 318–323.

31. Stilwell, D.J.; Bay, J.S. Toward the development of a material transport system using swarms of ant-like robots. In Proceedings of the IEEE International Conference on Robotics and Automation, Atlanta, GA, USA, 2–6 May 1993; Volume 1, pp. 766–771.

32. Gross, R.; Dorigo, M. Evolution of solitary and group transport behaviors for autonomous robots capable of self-assembling. *Adapt. Behav.* **2008**, *16*, 285–305.

33. Chen, J.; Gauci, M.; Gross, R. A strategy for transporting tall objects with a swarm of miniature mobile robots. In Proceedings of the IEEE International Conference on Robotics and Automation, Karlsruhe, Germany, 6–10 May 2013; pp. 863–869.

34. Chen, J.; Gauci, M.; Li, W.; Kolling, A.; Gross, R. Occlusion-Based Cooperative Transport with a Swarm of Miniature Mobile Robots. *IEEE Trans. Robot.* **2015**, *31*, 307–321. [CrossRef]

35. Wang, Z.; Schwager, M. Force-Amplifying N-robot Transport System (Force-ANTS) for cooperative planar manipulation without communication. *Int. J. Robot. Res.* **2016**, *35*, 1564–1586. [CrossRef]

36. Bechlioulis, C.P.; Kyriakopoulos, K.J. Collaborative multi-robot transportation in obstacle-cluttered environments via implicit communication. *Front. Robot. AI* **2018**, *5*, 90. [CrossRef] [PubMed]

37. Vlantis, P.; Vrohidis, C.; Bechlioulis, C.P.; Kyriakopoulos, K.J. Robot Navigation in Complex Workspaces Using Harmonic Maps. In Proceedings of the IEEE International Conference on Robotics and Automation, Brisbane, QLD, Australia, 21–25 May 2018; pp. 1726–1731.

8. Vlantis, P.; Vrohidis, C.; Bechlioulis, C.P.; Kyriakopoulos, K.J. Orientation-Aware Motion Planning in Complex Workspaces using Adaptive Harmonic Potential Fields. In Proceedings of the 2019 International Conference on Robotics and Automation (ICRA), Montreal, QC, Canada, 20–24 May 2019; pp. 8592–8598. [CrossRef]
9. Loizou, S.G. The navigation transformation: Point worlds, time abstractions and towards tuning-free navigation. In Proceedings of the 2011 19th Mediterranean Conference on Control & Automation (MED), Corfu, Greece, 20–23 June 2011; pp. 303–308.
0. Loizou, S.G. The Navigation Transformation. *IEEE Trans. Robot.* **2017**, *33*, 1516–1523. [CrossRef]
1. Arslan, O.; Koditschek, D.E. Smooth extensions of feedback motion planners via reference governors. In Proceedings of the 2017 IEEE International Conference on Robotics and Automation (ICRA), Singapore, 29 May–3 June 2017; pp. 4414–4421. [CrossRef]
2. Krstic, M.; Kokotovic, P.V.; Kanellakopoulos, I. *Nonlinear and Adaptive Control Design*, 1st ed.; John Wiley & Sons, Inc.: Hoboken, NJ, USA, 1995.

Article

Globally Optimal Redundancy Resolution with Dynamic Programming for Robot Planning: A ROS Implementation

Enrico Ferrentino [1,*], Federico Salvioli [2] and Pasquale Chiacchio [1]

[1] Department of Computer Engineering, Electrical Engineering and Applied Mathematics (DIEM), University of Salerno, 84084 Fisciano, Italy; pchiacchio@unisa.it
[2] Robotic Exploration, ALTEC S.p.A., 10146 Torino, Italy; federico.salvioli@altecspace.it
* Correspondence: eferrentino@unisa.it

Abstract: Dynamic programming techniques have proven much more flexible than calculus of variations and other techniques in performing redundancy resolution through global optimization of performance indices. When the state and input spaces are discrete, and the time horizon is finite, they can easily accommodate generic constraints and objective functions and find Pareto-optimal sets. Several implementations have been proposed in previous works, but either they do not ensure the achievement of the globally optimal solution, or they have not been demonstrated on robots of practical relevance. In this communication, recent advances in dynamic programming redundancy resolution, so far only demonstrated on simple planar robots, are extended to be used with generic kinematic structures. This is done by expanding the Robot Operating System (ROS) and proposing a novel architecture meeting the requirements of maintainability, re-usability, modularity and flexibility that are usually required to robotic software libraries. The proposed ROS extension integrates seamlessly with the other software components of the ROS ecosystem, so as to encourage the reuse of the available visualization and analysis tools. The new architecture is demonstrated on a 7-DOF robot with a six-dimensional task, and topological analyses are carried out on both its state space and resulting joint-space solution.

Keywords: dynamic programming; redundancy resolution; redundant robot; inverse kinematics; ROS; industrial manipulator

Citation: Ferrentino, E.; Salvioli, F.; Chiacchio, P. Globally Optimal Redundancy Resolution with Dynamic Programming for Robot Planning: A ROS Implementation. *Robotics* **2021**, *10*, 42. https://doi.org/10.3390/robotics10010042

Received: 2 February 2021
Accepted: 1 March 2021
Published: 4 March 2021

Publisher's Note: MDPI stays neutral with regard to jurisdictional claims in published maps and institutional affiliations.

1. Introduction

When a robot is redundant with respect to its task, the inverse kinematics problem admits an infinite set of solutions almost everywhere in its workspace. It results in augmented dexterity that can be exploited to achieve several objectives, which are usually desirable for a multitude of real applications, such as obstacle avoidance and constrained motions, improvement of manipulability and local or global optimization of generic performance indices [1]. Choosing the joint-space configuration in agreement with the criteria above for each point of the given end-effector trajectory is usually referred to as *redundancy resolution*.

Redundancy resolution via global optimization of performance indices is of particular interest in real applications as it allows saving resources, such as energy and time, and consequently maximizing the use of the robotic asset. On the other hand, such a technique requires time to provide a solution, therefore it is only suited for off-line planning scenarios. In space applications of exploration and construction, the minimization of energy translates into increasing the number of operations that can be completed within the available energy budget. In manufacturing industries, redundancy can be exploited to perform tasks as fast as possible, so as to increase the plant throughput. Both applications are characterized by an off-line programming approach, where the controller references either are generated a long time before they are executed, as in space applications with windowed communications, or are computed once and executed several times, as in automated manufacturing.

If the environment in which the robotic asset operates is subject to contingencies, there still is a chance to optimize the resources by coupling the off-line planner with an on-board planner. The latter should allow for the execution of the trajectory computed off-line, so that resources are still globally optimized, until some unplanned event arises, in which case the plan has to be modified. Yet the modifications could be implemented in a neighborhood of the globally optimal solution.

In the literature, the problem of maximizing or minimizing an integral function, subject to geometric or differential constraints, is addressed by making use of calculus of variations, either through the Euler-Lagrange formulation [2] or the Pontryagin's maximum principle [3], dynamic programming (DP) [4,5], or other numerical optimization techniques as by Shen et al. [6], among the others. Calculus of variations only provides necessary conditions for optimality and is, therefore, prone to sub-optimal solutions. Such conditions are given in the form of second-order differential equations and related boundary conditions. When initial and final joint positions are not assigned, velocities are constrained at the beginning and at the end of the trajectory. This results in a *Two-Point Boundary Value Problem* (TPBVP) made of a system of non-linear differential equations. Such problems rarely have a closed-form solution and numerical methods are usually employed [7,8]. Unfortunately, on the practical side, off-the-shelf numeric solvers do not guarantee the achievement of a solution, if it exists, and the successful computation might depend on a suitable choice of an initial guess [7]. Lastly, the equations that make up the TPBVP can be derived from Euler-Lagrange or Pontryagin's Maximum Principle necessary conditions only for some specific objective functions and constraints. Real applications, foreseeing the employment of industrial robots with state and actuation limits performing complex tasks, usually require the definition of more complex objectives and constraints. Limiting our analysis to geometric and kinematic considerations, typical real-world constraints include joint mechanical limits, maximum/minimum velocities and accelerations, as well as collisions and self-collisions. There exist both unilateral and bilateral constraints, which are defined on the state variables and their derivatives (up to the second order). Fitting such constraints in a mathematical formulation based on calculus of variations is not straightforward [9]. For all these reasons, numerical approaches and, in particular, discrete dynamic programming ones, proved to be much more promising in solving redundancy in real scenarios, because of their flexibility [4,5,10–12].

Guigue et al. [4] first applied a dynamic programming algorithm to an industrial use case, where a 7-DOF manipulator is inserted in a supersonic wind tunnel and a trajectory is planned to Pareto-optimize the square norm of velocities and the aerodynamic interference. In a later work [12], by using the same use case, they confirmed the superiority of the dynamic programming method over calculus of variations in approximating the Pareto optimal set. Other examples of similar DP-inspired redundancy resolution algorithms concern applications of laser cutting and fiber placement [5,10,11]. They adopt a discrete representation, which is used to demonstrate that a formulation of the problem in terms of graph theory is possible, through which the search space can be modeled as an acyclic directed graph and the so-called dynamic programming algorithm is, in truth, an optimal path search algorithm [11]. Dynamic programming has been also successfully employed for redundancy resolution of *parallel kinematic machines* (PKM) [13].

In a previous work [14], we demonstrated that the achievement of the globally optimal solution is subject to building a complete representation of the state space and designing an algorithm capable of transiting between posture sets, crossing singularities or semi-singularities. Building upon a topological analysis of the inverse kinematics mapping, we adopted a formulation of the problem based on multiple state space grids and developed a procedure to perform a simultaneous optimal search on these grids. Although the presence of multiple grids had already been discussed in previous works, no evidence was given on the possibility of exploring them together, possibly compromising the global-optimality of the solution. On the other hand, our algorithms [9,14], which boast the capability of returning the globally optimal solution, have only been demonstrated on simple planar robots. We remark that, since they are based on a discretization of the state space, global optimality should be intended here as *resolution-optimality*, i.e., the solution is globally optimal, across all homotopy classes, within the finite set of feasible trajectories in the discretized state space. For small discretization steps, the DP redundancy resolution algorithm returns a solution that is homotopic to the real global optimum. As clarified next, because of the characteristics of the problem, a resolution-optimal solution can be found even though the local optimization problems are not convex.

The objective of this communication is to provide an implementation for complex robotic structures, effectively turning in practice the outcomes of our topological analysis [14]. At the same time, a unified architecture, built upon the *Robot Operating System* (ROS) [15], is proposed so as to respect the usual requirements of maintainability, re-usability, modularity and flexibility of a robotic software library. To the best of our knowledge, ROS is not currently equipped with a globally optimal inverse kinematics library for redundant manipulators: our aim is to fill this gap. Furthermore, since previous implementations were application-specific, thereby requiring specific coding for accommodating generic constraints and objective functions, we aim at overcoming this limit by the design of suitable interfaces allowing for an easy integration of custom requirements. Being the underlying methodology introduced in previous works [9,14], the main contribution of this paper is on the application. First, a topological analysis is carried out of a realistic 7-degrees-of-freedom redundant industrial manipulator, then, we design an efficient trajectory planner/inverse kinematics solver for spatial robots, based on dynamic programming, considering actuation constraints. We show that the algorithm is able to compute resolution-optimal solutions in a time frame that is compatible with the considered off-line applications.

In Section 2 we recall the fundamental traits of the problem formulation based on multiple state space grids and dynamic programming. Therein, we give some notions of topology that are necessary to the comprehension of the problem. In Section 3, we first describe how redundancy resolution is addressed in ROS and what the extension points are for the design of our modules. Then, we present a modular architecture of a resolution-optimal DP-inspired redundancy solver integrated in ROS, where the user can plug custom objective functions and constraints. In Section 4, we demonstrate the capabilities of the solver on a 7-DOF robot and give a topological interpretation of the results. We discuss about the advantages of the proposed architecture in Section 5 and also highlight its limitations. The conclusions of the work are drawn in Section 6, where we also propose some lines of development for the future.

2. Materials and Methods

2.1. Discrete Dynamic Programming

Although a continuous time formulation of the dynamic programming problem is possible, this communication is limited to the discrete time systems, as the objective here is to propose a solution that can be directly implemented on digital hardware.

A trajectory $x(t) \in \mathbb{R}^m$, with $t \in [0, T]$, is given in the task space. Assume to discretize $[0, T]$ such that $t = i\tau$, where τ is the sampling interval, $i = 0, 1, 2..N_i$ and $N_i = \frac{T}{\tau}$. The following discrete time system is given with its initial conditions:

$$\mathbf{q}(i+1) = \mathbf{f}(\mathbf{q}(i), \mathbf{u}(i)), \quad \mathbf{q}(0) = \mathbf{q}_0, \tag{1}$$

where $\mathbf{q} \in \mathbb{R}^n$, i.e., the joint positions, represents the state vector of the system, $\mathbf{u}$ is the input vector, and $\mathbf{f}$ is a generic discrete-time first-order inverse kinematics expression containing the constants $\mathbf{x}(i)$. The dimension of $\mathbf{u}$ depends on the particular inverse kinematics model and it is $\mathbf{u} \in \mathbb{R}^{n-m}$ for minimal representations [16]. The objective is to find the optimal sequence of inputs that minimizes or maximizes a given cost function defined, in general, on both the state and input vectors and their derivatives.

Usually, $\mathbf{u}$ is not free, but constrained to belong to a certain time-variant domain $\mathcal{A}_i$. Its time derivative $\dot{\mathbf{u}}$ may also be limited to a given domain $\mathcal{B}_i(\mathbf{u}(i))$, which is, in principle, time-variant as well as input-variant [4]:

$$\mathbf{u}(i) \in \mathcal{A}_i,$$
$$\dot{\mathbf{u}}(i) \in \mathcal{B}_i(\mathbf{u}(i)). \tag{2}$$

Thus, at each i, the set of admissible values of $\mathbf{u}(i)$, from which it is possible to reach $\mathbf{u}(i+1)$, is given by the intersection between $\mathcal{A}_i$ and the set of $\mathbf{u}$-values respecting the constraint on the derivative, i.e.,

$$\mathcal{C}_i = \mathcal{A}_i \cap \left\{ \mathbf{u}(i) : \frac{\mathbf{u}(i+1) - \mathbf{u}(i)}{\tau} \in \mathcal{B}_i(\mathbf{u}(i)), \; \text{with } \mathbf{u}(i+1) \in \mathcal{A}_{i+1} \right\}, \tag{3}$$

where the Euler approximation has been used in place of $\dot{\mathbf{u}}$. More in general, the definition above could be extended to represent the intersection of all the constraints on $\mathbf{u}$ and its higher-order derivatives.

Let the objective function to optimize be

$$I(0) = \psi(\mathbf{q}(N_i)) + \sum_{j=0}^{N_i-1} l(\mathbf{q}(j), \dot{\mathbf{q}}(j), \mathbf{u}(j), \dot{\mathbf{u}}(j)), \tag{4}$$

where ψ is the cost of the final configuration. The assumption is made that the cost function computed locally l only depends on the states, on the inputs and on their first-order derivatives, but in general, more complex functions could be defined.

Rewrite the cost function in a recursive form and use the Euler approximation for $\dot{\mathbf{q}}$ and $\dot{\mathbf{u}}$:

$$I(N_i) = \psi(\mathbf{q}(N_i))$$
$$I(i) = I(i+1) + l(\mathbf{q}(i), \mathbf{q}(i+1), \mathbf{u}(i), \mathbf{u}(i+1)). \tag{5}$$

Assume that the optimization criterion is to minimize $I(0)$. By using the *Bellman* principle, we could then write:

$$I(N_i) = \psi(\mathbf{q}(N_i))$$
$$I_{opt}(i) = \min_{\mathbf{u} \in \mathcal{C}_i} \left[l(\mathbf{q}(i), \mathbf{q}(i+1), \mathbf{u}(i), \mathbf{u}(i+1)) + I(i+1) \right], \tag{6}$$

where $I_{opt}(i)$ is the *optimal return function* and $I_{opt}(0)$ the *optimal cost*.

2.2. State Space Grids

In the majority of the works mentioned in Section 1, the input vector $\mathbf{u}$ is defined using the *joint selection* method (or *joint space decomposition*), which foresees the selection of $r = n - m$ variables from the joint position vector. Alternatively, the *joint combination* method requires the definition of r functions of the joint positions, which is a more generic setup that also includes joint selection [16]. For instance, for Guigue et al. [12], $\mathbf{u}$ is of dimension one and corresponds to the sum of two joint variables. If $\mathbf{k} : \mathbb{R}^n \rightarrow \mathbb{R}^m$ is the direct kinematics function, the inclusion of r additional functions $\mathbf{k}_u(\mathbf{q})$ yields the *augmented kinematics* $\mathbf{k}_a : \mathbb{R}^n \rightarrow \mathbb{R}^n$, i.e.,

$$\begin{bmatrix} \mathbf{x} \\ \mathbf{u} \end{bmatrix} = \mathbf{k}_a(\mathbf{q}) = \begin{bmatrix} \mathbf{k}(\mathbf{q}) \\ \mathbf{k}_u(\mathbf{q}) \end{bmatrix}. \tag{7}$$

In this case, when the redundancy parameter $\mathbf{u}$ is given, the remaining joint positions can be computed as

$$\mathbf{q}(i) = \mathbf{k}_a^{-1}\big(\mathbf{x}(i), \mathbf{u}(i)\big). \tag{8}$$

To deal with a problem that is computationally feasible, it is convenient to discretize the state space, instead of the input space, so that the admissible inputs are only those that, once plugged into the dynamic system (1), yield states in the discrete domain. However, if (8) is considered, the discretization of the state space can be obtained through the discretization of the input space. Furthermore, let us assume to adopt a joint selection parametrization, whose practical advantages will be clarified in Section 3. The selected joint domains are thus discretized according to $\mathbf{u} = \mathbf{j} \circ \Delta\mathbf{q}_u$, where '$\circ$' represents the Hadamard product and $\Delta\mathbf{q}_u = [\Delta q_{u,1}, ..., \Delta q_{u,r}]^T$ is the vector of the sampling intervals, defined on each axis of the domain and $\mathbf{j} \in \mathbb{N}^r$. Because of (8), the state space is also discretized. This allows building a grid of $r+1$ dimensions in t_i and $\mathbf{u_j}$, where i is the *stage index*, or *waypoint index*, and $\mathbf{j}$ is the vector of the *redundancy parameters indices*. Each node $(i, j_1, j_2, ..., j_r)$ in the grid contains a configuration computed as

$$\mathbf{q_j}(i) = \mathbf{k}_a^{-1}\big(\mathbf{x}(i), \mathbf{u_j}\big), \tag{9}$$

with $i \in [0, N_i] \subset \mathbb{N}$ and $\mathbf{j} \in \{[1, N_{u,1}] \times ... \times [1, N_{u,r}]\} \subset \mathbb{N}^r$. A pictorial view of this process is given in Figure 1 for the case $\mathbf{j} = j \subset \mathbb{N}$ (one degree of redundancy, i.e., $r = 1$): each column in the grid corresponds to a waypoint on the path and the nodes in a column span the self-motion manifold(s) lying in the null-space. Grids combine self-motions with path tracking and are, therefore, the pre-image of the workspace path in the configuration space.

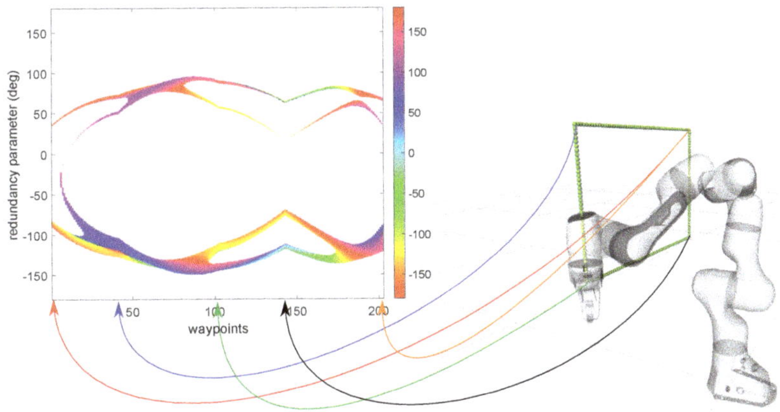

Figure 1. Mapping of the workspace path in the joint space yielding the state space grids.

The "augmented" kinematic function $\mathbf{k}_a^{-1}$ in (9), also termed *node function*, is equivalent to a standard inverse kinematic function of a non-redundant manipulator, meaning that $\mathbf{q_j}(i)$ is not unique, but a finite set of solutions exists. The number of such solutions N_g only depends on the mechanical characteristics of the manipulator and is equal to 2, 2, 4 and 16 for planar, spherical, regional and spatial manipulators respectively [17,18]. However, for some specific kinematic structures, as well as for some specific trajectories [19], the actual number of solutions could be less than the maximum theoretical value. For instance, for most six-axis industrial manipulators, the number of distinct configurations is equal to 8 [11]. Multiple inverse kinematic solutions imply the existence of multiple grids. Therefore, let us rewrite Equation (9) to make the *grid index g* explicit:

$$\mathbf{q_j}^{(g)}(i) = \mathbf{k}_a^{-1}\big(\mathbf{x}(i), \mathbf{u_j}\big), \quad \text{with } g = 1..N_g. \tag{10}$$

All the inverse kinematic solutions along the assigned path can be classified so as to ensure continuity for each single grid, determining a continuous posture set. For a redundant planar manipulator with Denavit-Hartenberg reference frames, such a classification reduces to testing the sign of the "elbow" joint, so that one grid represents the *elbow-down* configurations and the other the *elbow-up* ones. If the kinematic chain is made of many joints, identifying the elbow could be not immediate. Indeed, the elbow itself depends on the chosen redundancy parameter: it will be the joint variable nullifying the extended (or augmented) Jacobian's determinant; its sign will mark the separation between *elbow-down* and *elbow-up* inverse kinematic solutions [14]. In general, if the redundancy parameter changes, the extended Jacobian changes too, and the joint variable nullifying its determinant could be different, as well as the criterion to distinguish between continuous posture sets. For a generic redundant manipulator, the extended Jacobian determinant could be made of several factors. From the topological point of view, all the solutions determining the same signs for such factors are said to belong to the same *extended aspect* [20] and constitute a continuous posture set. If a grid contains solutions from one and only one extended aspect, it is said to be *homogeneous* [14]. This concept will be useful in the discussion that follows in Section 2.3.

As a concluding remark on state space grids, it is important to highlight that, although an extended Jacobian is virtually defined once the redundancy parameter is selected, it is never used for kinematic inversion. In agreement with (10), inverse kinematics is always positional, so that the optimization is immune to singularities and algorithmic singularities. As will be clarified next, transitions through singularities are nominal in a discrete dynamic programming approach and, indeed, this is a relevant advantage of this technique over others, including calculus of variations.

2.3. DP-Inspired Search Algorithm

If all the states by which the manipulator can transit are available in the grids, the optimization problem reduces to the selection of the nodes that provide the minimum cost. In this case, the problem is equivalent to that of finding the optimal path on an acyclic directed graph [11], which means that an optimal solution can be found regardless of the Bellman optimality principle, which would rather be necessary if the state set was continuous. Thus, we keep the problem decomposition typical of dynamic programming, where the local optimization problem is simply a node selection problem, which is typical of search algorithms on trees and graphs [21]. With this formulation, several implementations are possible. For instance, we proposed a *forward* (i.e., from the first waypoint to the last) iterative implementation [14], but alternatives exist to proceed *backward* (i.e., from the last waypoint to the first) and/or using recursive approaches. It has to be noted that if a *forward* implementation was chosen, Equation (6) should be rewritten as

$$I(0) = \psi\big(\mathbf{q}(0)\big)$$
$$I_{opt}(i) = \min_{\mathbf{u}_{i-1} \in \mathcal{C}_{i-1}} \Big[I(i-1) + l\big(\mathbf{q}(i), \mathbf{q}(i-1), \mathbf{u}(i), \mathbf{u}(i-1)\big) \Big]. \tag{11}$$

In this case it is convenient to redefine $\dot{\mathbf{u}}(i)$ as:

$$\dot{\mathbf{u}}(i) = \frac{\mathbf{u}(i) - \mathbf{u}(i-1)}{\tau} \tag{12}$$

and $\mathcal{C}_{i-1}$ as:

$$\mathcal{C}_{i-1} = \mathcal{A}_{i-1} \cap \left\{ \mathbf{u}(i-1) : \frac{\mathbf{u}(i) - \mathbf{u}(i-1)}{\tau} \in \mathcal{B}_i(\mathbf{u}(i)), \text{ with } \mathbf{u}(i) \in \mathcal{A}_i \right\}. \tag{13}$$

The choice between *forward* and *backward* implementation is not, in general, arbitrary, as it often depends on considerations about performance and on the hardware architecture used, as well as on the boundary conditions of the problem.

To better understand how boundary conditions affect the choice, we could highlight that the optimal cost function $I_{opt}(0)$ (*backward*) or $I_{opt}(N_i)$ (*forward*) are conditioned by the sequence of inputs enabled $\mathcal{A}_i$ at each i. In many practical cases, unless the environment in which the robot moves is particularly constrained, applications require that either the initial joint positions or the final ones or both are assigned or otherwise constrained (e.g., cyclic joint trajectories). The optimal solution and the value of the cost function will then vary together with the initial or final set of inputs. So we may write I_{opt} as a function of such sets [4], having $I_{opt}(0, \mathcal{A}_{N_i})$ or $I_{opt}(N_i, \mathcal{A}_0)$.

Assume to run our *forward* dynamic programming algorithm once, starting with inputs in $\mathcal{A}_0$ and ending with inputs in $\mathcal{A}_{N_i}$. One execution of the algorithm is sufficient to provide the solution together with its cost for the optimal joint-space paths ending in each single element of $\mathcal{A}_{N_i}$. From the practical standpoint, the upside is that one may decide to select a sub-optimal solution if its cost does not vary too much from the optimal cost, while the final joint positions are much more favorable for the particular task the robot has to execute.

On the other hand, if one asked for a solution starting from a specific $\mathbf{u}(0)$, this may require an additional execution of the algorithm, either proceeding *backward* or by explicitly forcing the initial condition at the moment $\mathcal{A}_0$ is defined. In other words, one execution of the *forward* algorithm with free initial conditions does not guarantee the computation of a solution for each input in $\mathcal{A}_0$, as well as one execution of the *backward* algorithm with free final conditions does not guarantee the computation of a solution for each input in $\mathcal{A}_{N_i}$.

Considering the grid representation of Section 2.2, and assuming a *forward* implementation, we may rewrite Equation (11) as

$$
\begin{aligned}
I_{\mathbf{j}}^{(g)}(0) &= \psi\big(\mathbf{q}_{\mathbf{j}}^{(g)}(0)\big) \\
I_{\mathbf{j}}^{(g)}(i) &= \min_{\mathbf{k},h} \Big[I_{\mathbf{k}}^{(h)}(i-1) + l_{\mathbf{k}\to\mathbf{j}}^{(h\to g)}\big(\mathbf{q}(i), \mathbf{q}(i-1), ..., \mathbf{q}(i-d)\big) \Big], \text{ for } i > 0 \\
I_{opt}(N_i) &= \min_{\mathbf{j},g} \Big[I_{\mathbf{j}}^{(g)}(N_i) \Big],
\end{aligned}
\tag{14}
$$

where $l_{\mathbf{k}\to\mathbf{j}}^{(h\to g)}$ is the local cost to move from node $(i-1, \mathbf{k})$ on grid h to the node $(i, \mathbf{j})$ on grid g, while $d \in \mathbb{N}$ is the maximum order derivative for which a constraint is defined; for instance, if acceleration constraints are imposed, $d = 2$. A pictorial view is given in Figure 2 for $r = 1$. For each enabled node at $i+1$ (in blue), an optimal predecessor is selected among those that belong to $\mathcal{C}_i$ (in green) by solving the local optimization problem in (14). Higher-order constraints yielding $\mathcal{C}_i$ are checked by using the chain of d predecessors represented by the red arrows, which allows computing the discrete approximations of derivatives.

In case Pareto-optimal solutions have to be found, the cost comparison cannot be performed, for each pair of nodes at subsequent stages, on the basis of a scalar cost function. The minimization in (14) is replaced by the dominance rule [4]. Said z the number of performance indices in the Pareto-optimal setup, an objective vector $\mathbf{I}^*$ is Pareto-optimal if there does not exist another objective vector $\mathbf{I}$ such that $I_i \leq I_i^* \; \forall i = 1, \ldots, z$ and $I_j < I_j^*$ for at least one index j. This definition allows defining the *dominance rule*: given two nodes $(i-1, \mathbf{k}_1)$ and $(i-1, \mathbf{k}_2)$, for which the cumulative cost is computed in transiting towards the same node $(i, \mathbf{j})$, $(i-1, \mathbf{k}_1)$ is said to *dominate* $(i-1, \mathbf{k}_2)$ if $(i-1, \mathbf{k}_1)$ improves the vectorial cost function of $(i-1, \mathbf{k}_2)$ for at least one performance index, without worsening the others. As a consequence, the optimal predecessor of a given node at i is not a single node at $i-1$ (the predecessor is not unique), as each node might improve the objective vector in a different direction. Every time a dominating node is identified, it enters the list of the optimal predecessors of $(i, \mathbf{j})$ and the dominated nodes are removed from the same list. As for the scalar objective case, the process is repeated for each i and $\mathbf{j}$.

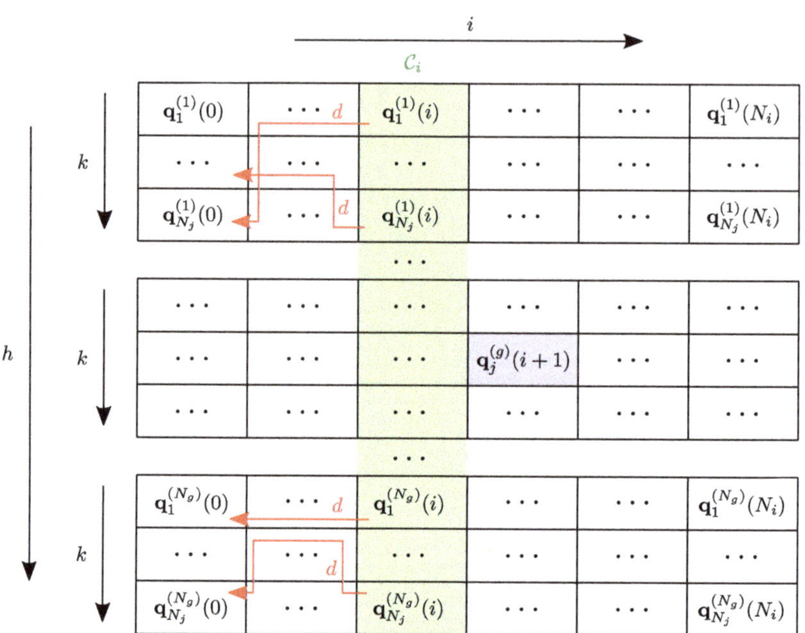

Figure 2. Pictorial representation of the local optimization problem.

With respect to previous implementations, our algorithms [9,14] provide the evidence that the resolution-optimal solution can be achieved by applying (14) onto all the N_g grids at the same time, i.e., exploring the whole configuration space, and transiting between such grids when necessary, through singularities or semi-singularities of the kinematic chain, using configurations from different extended aspects. They differ in that the former does not make any specific assumption on the homogeneity of the grids, while the latter assumes to work with homogeneous grids to reduce the computational complexity. Both algorithms adopt a *forward* iterative implementation since, in practical situations, it is more convenient to compute an optimal solution for each single final configuration, starting from one initial configuration, corresponding to the current state of the system. In this communication, we do not make any prior assumption on the grids [9]. However, if we can ensure or detect that grids are homogeneous, we can exploit this information to speed up the search [14].

The computational complexity of the algorithm [9], assuming that the same number of samples N_u is chosen for all the redundancy parameters, is $O(N_i N_u^{2r} N_g^2)$. This is due to the fact that for each waypoint (N_i waypoints overall) and for each node in the grids ($N_u^r N_g$ nodes for each waypoint), a comparison shall be made with each node at the next waypoint ($N_u^r N_g$ overall). If the grids are homogeneous, transition points (i.e., nodes corresponding to singularities or semi-singularities where the transition between grids/extended aspects is possible) are clearly identified and the comparisons with the nodes at the next waypoint are only performed within the same grid, lowering the computational complexity to $O(N_i N_u^{2r} N_g)$. For a given manipulator, N_g is constant, regardless of the trajectory and, in general, $N_g << N_u$. Thus, the computational complexity can be approximated, in most of the cases, to $O(N_i N_u^{2r})$. Also, homogeneous grids are characterized by continuity of solutions, meaning that, if a certain node does not satisfy the velocity constraints, a farther node will not satisfy them either: this is the assumption at the basis of our optimization [14]. In other words, the velocity constraints allow reducing the number of comparisons for each couple of waypoints to a limited window of nodes, whose cardinality is N_w^r, with $N_w << N_u$. Therefore, the computational complexity reduces to $O(N_i N_u^r N_w^r)$.

Since all the nodes satisfying the constraints are tested to find the optimum at each stage, resolution-optimality is guaranteed even though the problem in (14) is not convex. Indeed, our method is based on the Bellman optimality principle, ensuring that the solution of the lowest cost is returned for a given discretization of the state space. The compliance of our method to the necessary conditions of calculus of variations has been verified in [9], for a much simpler use case (a planar robot), for which a formulation based on calculus of variations can be obtained straightforwardly.

While this approach might sound demanding in terms of CPU time and memory occupancy, the results of the computational complexity analysis, as well as the specific topological features of spatial robots subject to real-world constraints suggest that solving problem (14) is doable within a time horizon that is compatible with the off-line planning applications mentioned in Section 1. This is shown in Section 4 with an example.

3. ROS Implementation

3.1. Designing an Extension for MoveIt!

MoveIt! [22] is the planning framework of ROS, including several libraries for motion planning, manipulation, 3D perception, kinematics, control and navigation. To the purpose of extending it to perform resolution-optimal inverse kinematics along a specified workspace trajectory, we focus on the analysis of three concepts of interest, which are *capabilities*, *planners* and *inverse kinematics*.

For a robotic manipulator, the MoveIt! user can plan joint-space trajectories and perform several other actions through *capabilities*, exposed by the *move_group* node. For instance, the *MoveGroupCartesianPathService* is used to plan Cartesian paths (straight lines) passing by pre-defined waypoints, the *MoveGroupPlanService* performs the point-to-point trajectory planning in the joint space, the *MoveGroupKinematicsService* computes direct and inverse kinematics, and so on.

Except for the *MoveGroupCartesianPathService*, the *move_group* node does not allow for any other form of planning in the joint space along a constrained end-effector trajectory. MoveIt! provides several *planners*, such as OMPL and STOMP [23], just to mention some, which are typically employed in a point-to-point planning scenario, i.e., move the end-effector to a new location along an arbitrary path. Additional constraints can be specified in the *Motion Plan Request* [24] for any link in the kinematic chain, including the end-effector, but planning for complex paths may not be straightforward and just a few possibilities exist to tune the process to work with generic (possibly multiple) objective functions and application-specific constraints. As far as the *MoveGroupCartesianPathService* is concerned, the user defines the workspace waypoints, then the end-effector trajectory is simply calculated by first order interpolation. The joint-space planning consists of computing the inverse kinematics for each single waypoint in the interpolated set, but, for a redundant manipulator, no mean exists to control the optimality of the solutions in the joint space along the whole trajectory.

On the other hand, since the dynamic programming algorithm presented in Section 2 is no more than a resolution-optimal inverse kinematic procedure, we may think to extend the *inverse kinematics* capabilities of MoveIt!. Nonetheless, the resolution-optimal planning is defined on a pre-defined set of waypoints, so that the objective function and constraints can regard the derivatives of the joint position variables, which will depend on the time law defined at the end-effector.

The algorithm we aim to extend MoveIt! with is a planner in that it computes a joint-space trajectory from infinite possible solutions but, at the same time, it is an IK solver, as it works with an assigned workspace trajectory. Because resolution-optimal inverse kinematics is different from other capabilities of the framework, we believe that providing the functionality as a new *move_group* capability is the most seamless solution.

3.2. Requirements

On the basis of the considerations above, let us consider the following requirements for our extensions:

Req. 1: to allow for a seamless integration with the ROS ecosystem, so as to reuse as much as possible, the available technologies (e.g., visualization and analysis tools);

Req. 2: to support the generation of multiple homogeneous grids;

Req. 3: to perform a search on such grids to find the resolution-optimal joint-space solution [9];

Req. 4: to support the optimization for homogeneous grids [14];

Req. 5: to allow for the addition of user-defined constraints and objective functions;

Req. 6: to allow for the topological analysis of the state space and the resolution-optimal trajectory.

3.3. Context

Our *moveit_dp_redundancy_resolution* package, publicly available on the Internet [25] constitutes an additional *move_group* capability, which is offered to the MoveIt! users through a *ros::ServiceServer*. Specific messages, called *GetOptimizedJointsTrajectory*, are exchanged through the service interface. The user can be any ROS node that implements a *ros::ServiceClient* interface and is able to assembly and send a *GetOptimizedJointsTrajectoryRequest*.

To exploit, as much as possible, the available visualization and analysis tools, i.e., Req. 1:, the results of the computation performed by the resolution-optimal planner are published through *ros::Publisher* objects on specific pre-existing topics, so that other nodes from the ROS ecosystem, such as *RViz* and *rqt_multiplot*, can be used to animate the robot along the assigned trajectory and to plot the resulting joint position, velocity and acceleration curves respectively.

Furthermore, in order to satisfy Req. 6:, several functions are provided to export the data structures of interest to the filesystem, to be later imported by additional analysis tools, such as MATLAB functions/scripts or replayed through *ROS bagfiles*.

The *moveit_dp_redundancy_resolution* context is reported in Figure 3, where the extensions are drawn in red. Together with the main ROS developments, additional analysis functions have been developed in MATLAB to perform the off-line topological analysis of both the state space and the resulting resolution-optimal joint-space trajectory. State space grids are exported to custom binary formats, while optimal trajectories are exported to bagfiles and imported in MATLAB through the *rosbag* interface.

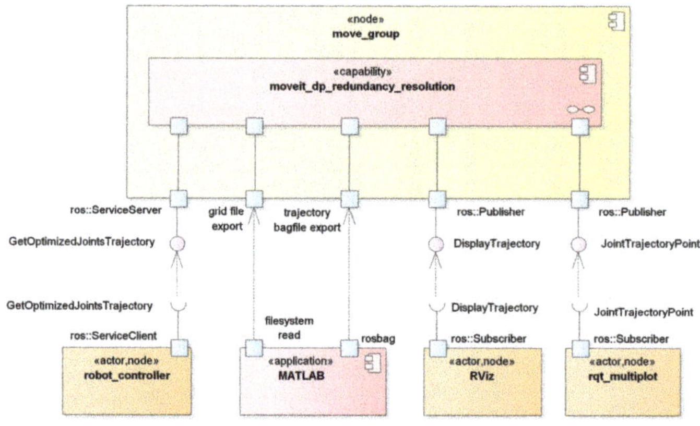

Figure 3. Context diagram.

3.4. Architectural Design

Internally, the *ros::ServiceServer* is hosted in the *MoveGroupDPRedundancyResolutionService* class, which constitutes the capability plugin. It is in charge of receiving requests, calling the lower level functions, building up responses and disseminating them through publishers, as well as generating bagfiles. It is instantiated at run-time, depending on the configured *move_group* capabilities and usually stands next to other default capabilities, as mentioned in Section 3.1. Its client counterpart, the *robot_controller* node, can call its service or other *move_group* capabilities, covering a broad range of planning scenarios. For example, one may request a point-to-point planning to OMPL and then use the generated workspace trajectory to issue a resolution-optimal redundancy resolution request. The same could be done with a workspace trajectory generated with the *MoveGroupCartesianPathService* capability, as will be shown in Section 4.

Behind the *MoveGroupDPRedundancyResolutionService*, several other objects interact to satisfy the requirements of Section 3.2. In particular, the *StateSpaceGrid* is the class implementing the data structure representing the joint space along the assigned trajectory. It provides import/export functions for the grid's custom binary file format as well as the generation of colormaps in the form of raster images that can be directly interpreted by the human user for quicklook purposes. To deal with homogeneous grids, state space grids cannot be generated independently, as it is necessary that the multiple IK solutions are classified per extended aspect, as observed in Section 2.2. For this reason, multiple grids are created by a single execution of the IK solver, supervised by a *StateSpaceMultiGrid* object. Its primary objective is to control the non-redundant IK solver implementing (10) and to classify the solutions, so as to satisfy Req. 2:.

To speed up the calculation of the state space grids, it is convenient to adopt an analytic inverse kinematic solver that is several orders of magnitude faster than numeric solvers. In the ROS framework, a possibility is given by *IKFast*, which can find all the IK solutions on the order of 6 microseconds, while most numeric solvers may require even 10 milliseconds or longer, and convergence is not certain [26]. *IKFast* performs an off-line analytic kinematic inversion and generates a C++ library containing the algebraic IK solver, able to return all the solutions for given end-effector pose. The off-line process may require several minutes, but is independent from the assigned trajectory and, thus, needs to be executed only once for a given kinematic chain. Currently, *IKFast* is able to manage open kinematic chains with one degree of redundancy. The value of the redundancy parameter has to be provided at the time the algebraic solver is called, which is the case of the DP grids considered in this communication. Nonetheless, it is worth observing that, in this context, it is not necessary that the IK solver natively supports redundant inverse kinematics, as the redundancy parameters are given for each single grid node and the inverse kinematics always involve a non-redundant kinematic chain. This means that our solution is scalable with respect to an arbitrary redundancy degree and *IKFast* can always be used, provided a suitable definition of the redundant (i.e., including the redundant joints) and non-redundant (i.e., excluding the redundant joints) planning groups.

The equations in (14) are implemented by the *DynamicProgrammingSolver* class, supporting multi-grid search, thereby satisfying Req. 3:. Since state space grids are added to the solver one by one, specific posture sets can be excluded by not passing them to the solver. To satisfy Req. 4:, the optimization can be explicitly enabled/disabled, depending on whether homogeneous grids can be obtained.

The *moveit_dp_redundancy_resolution* package provides two extension points, which are the abstract classes *PerformanceIndex* and *StateSpaceNode*. The former allows for the definition, through a specific XML-based language, of custom performance indices that can be combined together in an *ObjectiveFunction*. This is done offline, through configuration files. Optionally, performance indices can be characterized by a weight, used for weighted optimization. Otherwise, they are inserted in a vector and the solution will be Pareto-optimal (in a discrete sense, as discussed above). On the other hand, the *StateSpaceNode* class allows for the definition of application-specific semantics (even beyond resolution-optimal inverse kinematics) and related constraints. For instance, one may think of using the same classes for time-optimal planning along specified paths, provided that a specific *StateSpaceNode* implementation is given. These two classes allow satisfying Req. 5:.

The structure of the described classes is reported in the hybrid decomposition/class diagram of Figure 4, while its dynamic behavior, for the operations yielding the dynamic programming redundancy resolution, is represented in the sequence diagram of Figure 5

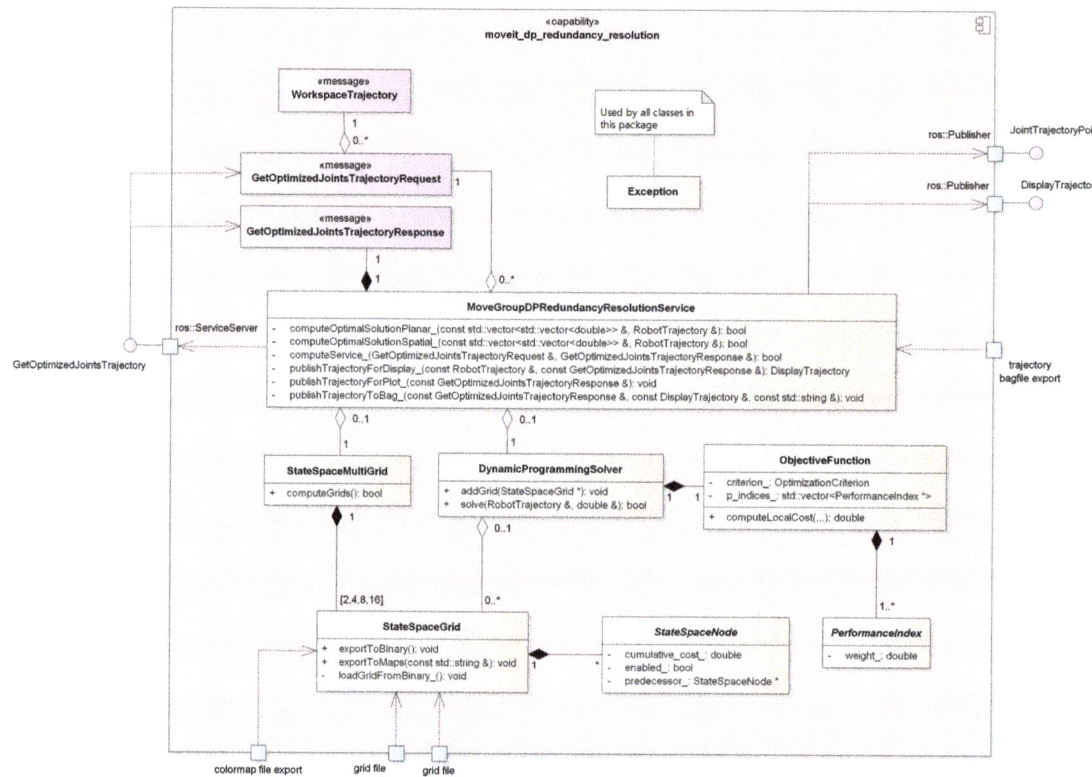

Figure 4. Hybrid decomposition/class diagram.

3.5. Use of Numeric Solvers

IKFast is characterized by the nice property of returning all the IK solutions in a single call to the solver. Furthermore, under certain circumstances, such solutions could be returned in the same order with respect to the extended aspects. When this happens, no post-processing is required on the IK solutions to generate homogeneous grids. Otherwise, either an explicit analytic factorization of the Jacobian is available, or numeric techniques are employed in order to ensure continuity in the state space.

If an analytic solution cannot be computed at all, or is hard to obtain, numeric solvers can be used instead. They are characterized by the property of returning one or no solution for each single call to the solver and their processing is notoriously time-consuming. To still have some control on the execution time, it is extremely important that the conditions characterizing the extended aspects are known beforehand, so that the search space of the solver can be drastically reduced. If this is not the case, the employment of numeric solvers practically rules out the usage of dynamic programming, as multiple IK searches do not guarantee the achievement of solutions in different extended aspects, but, worse, the same solution may be returned multiple times, making inverse kinematics a time-consuming try-and-error process.

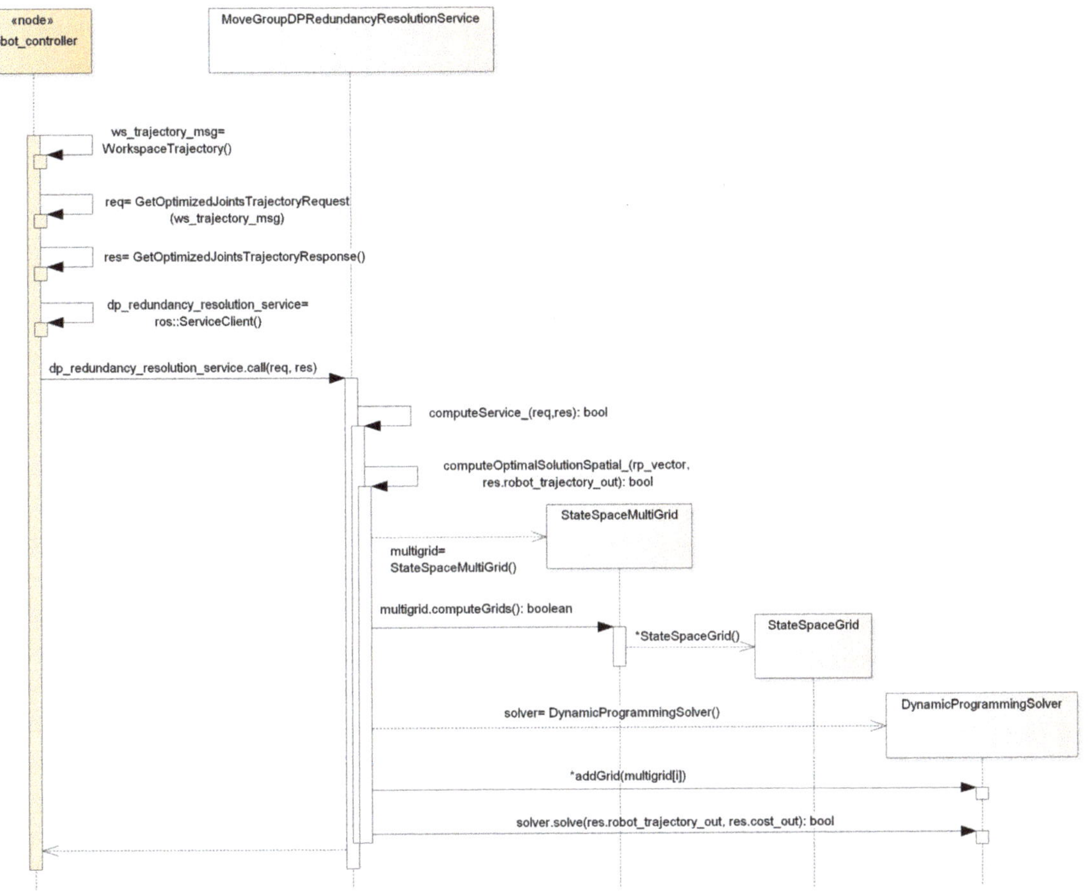

Figure 5. Sequence diagram representing dynamic programming redundancy resolution.

In our implementation, we also provide the possibility of using the numeric solver KDL [27], but with planar manipulators only, where we know the condition separating the (two) extended aspects. This capability is transparently provided by the *MoveGroupDPRedundancyResolutionService* class. In our experience, for more complex cases where the Jacobian's factorization is not available, the grids generation time becomes much greater than the time needed for the DP search itself and, consequently, the algorithm becomes unusable in any case of practical relevance.

We have been using the KDL-based implementation for planar robots to validate our implementation against a known use case employed in previous works [2,14,28] and have kept this capability for further similar cases involving planar manipulators.

4. Results

4.1. Use Case Description

To demonstrate that the methodology developed in the previous sections can be effectively applied to a real scenario, let us consider a real robotic arm with 7 degrees of freedom, to which a task constrained in position and orientation is assigned, with a time law. In a first trial, the objective is to reduce the energy consumption indirectly through the minimization of the square norm of joint velocities. Sub-optimal solutions are not of interest, hence the globally optimal one must be found. In a second trial, while minimizing the energy consumption as before, in a Pareto-optimal setup, the distance between the elbow (corresponding to the fourth joint) and an obstacle in the workspace is maximized so as to find a collision-free joint-space trajectory. Since the Panda robot by Franka Emika [29] has 7 degrees of freedom and is the flagship robot of MoveIt!, it is a convenient choice for the experiment at hand.

The modified Denavit-Hartenberg parameters [30] of the Panda robot [31] are reported in Table 1. Let us set the joint position, velocity and acceleration limits according to the datasheet. Limits on the jerk are not considered, but, as noticed by Gao et al. [5], they contribute to generating smoother solutions, which is necessary in all the cases where the resulting trajectory has to be executed on real hardware.

Table 1. Panda modified Denavit-Hartenberg parameters.

	d_i	θ_i	a_i	α_i
J1	0.333	q_1	0	0
J2	0	q_2	0	$-\pi/2$
J3	0.316	q_3	0	$\pi/2$
J4	0	q_4	0.0825	$\pi/2$
J5	0.384	q_5	-0.0825	$-\pi/2$
J6	0	q_6	0	$\pi/2$
J7	0	q_7	0.088	$\pi/2$
Flange	0.107	0	0	0

The workspace path is defined in terms of position and orientation and is depicted in Figure 6, together with the base reference frame and the obstacle. The axes x, y and z are in red, green and blue respectively. The planning is performed for the end-effector's flange that has to visit five waypoints, in the order x_A, x_B, x_C, x_D and x_E, describing the corners of a rectangle in the y-z plane, with variable orientation. Their values with respect to the base reference frame, considering a roll-pitch-yaw representation for the orientation, are

$$x_A = \begin{bmatrix} 0.3 \\ -0.3 \\ 0.8 \\ 0 \\ -\pi/2 \\ 0 \end{bmatrix} \quad x_B = \begin{bmatrix} 0.3 \\ -0.3 \\ 0.4 \\ 0 \\ -\pi/2 \\ 0 \end{bmatrix} \quad x_C = \begin{bmatrix} 0.3 \\ 0.3 \\ 0.4 \\ 0 \\ -\pi \\ 0 \end{bmatrix} \quad x_D = \begin{bmatrix} 0.3 \\ 0.3 \\ 0.8 \\ 0 \\ \pi/2 \\ 0 \end{bmatrix} \quad x_E = \begin{bmatrix} 0.3 \\ -0.3 \\ 0.8 \\ \pi/2 \\ \pi/2 \\ 0 \end{bmatrix}. \tag{15}$$

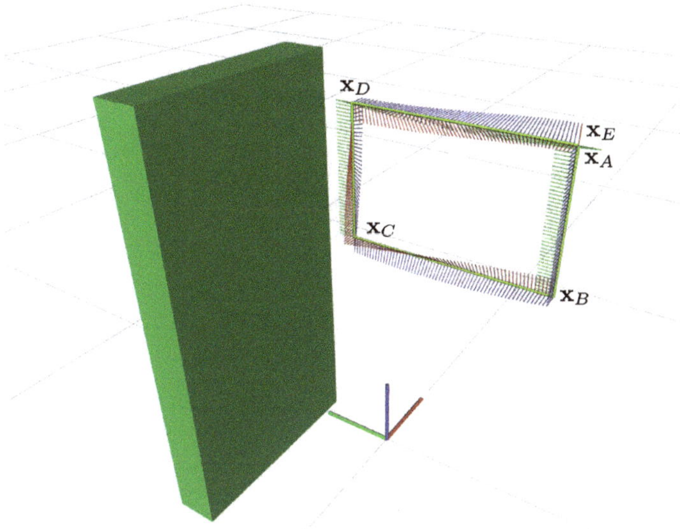

Figure 6. Workspace path assigned to the Panda arm, together with the base reference frame and obstacle.

All the points in between each pair of waypoints are obtained by linear interpolation, with a linear resolution not exceeding 0.01 m. A time law is defined so as to complete the whole trajectory in 60 s, with a constant time offset between consecutive points. The total number of points is $N_i = 203$.

4.2. Grids Computation

As mentioned above in Section 3.4, grids are computed through the *StateSpaceMulti-Grid* class, making use of the *IKFast* kinematic plugin. The plugin is based on a C++ solver generated off-line, which requires to select a redundant joint with respect to which inverse kinematics expressions are computed. In general, the choice of the redundancy parameter is not arbitrary, for two reasons:

- inverse kinematics in (10) foresees the availability of an analytical IK solver: in order for the solver to exactly implement (10), it must be parametrized with respect to the same parameter **u**;
- when fixing the redundant joints to specific values, the manipulator must be no longer redundant in order for (10) to return a finite number of solutions: at algorithmic singularities, i.e., configurations nullifying the determinant of the extended Jacobian, this is not the case.

Both issues are beyond the scope of this paper, but we note that solutions exist to select redundant joints in view of performing inverse kinematics, as well as correctly representing internal motions [32]. In our case, we can obtain an *IKFast* solver by selecting the redundancy parameter $u = q_4$. In fact, since joint 4 is in the middle of the kinematic chain and its axis does not intersect any other joint axis, we minimize the chances of encountering degenerate cases and of handling more complicated expressions [26]. A posteriori, we verify that the selected parameter is representative of the internal motion for the assigned path, i.e., state space grids do not degenerate to lines.

The redundancy parameter can be discretized so that $N_u = 2880$, either between -180 deg and 180 deg, which yields a resolution of 0.125 deg, or between its physical limits, i.e., -176 deg and -4 deg, which yields a resolution of about 0.06 deg. The Panda manipulator has 8 IK solutions, i.e., $N_g = 8$, for all the points on the trajectory, but in practice, because of joint limits, some points have less. The "slices" corresponding to q_1 of the grids computed with *IKFast* are reported in Figure 7, while those computed neglecting joint limits, for comparison purposes, are reported in Figure 8.

The first interesting thing to notice about these grids is that they are homogeneous as evident from those of Figure 8. The extended Jacobian $\mathbf{J}_a$, obtained from the 6×7 rectangular Jacobian by adding the row $[0\,0\,0\,1\,0\,0\,0]$, cannot be easily factorized, implying that we are not provided with analytic conditions to classify the solutions of *IKFast*. For this reason, the following three conditions are used, obtained from an a-posteriori numerical analysis of the solution sets:

- $|\mathbf{J}_r^{(4)}| > 0$;
- $q_2 > 0$;
- $q_5 > 0$.

Each of the grids in Figures 7 and 8 corresponds to a different combination of the conditions above, providing an homogeneous classification of the solutions. It is possible to demonstrate that both q_2 and q_5 are factors of $|\mathbf{J}_a|$ and, being the "augmented" Panda manipulator of type 1, according to Wenger [19], they are sufficient conditions for classifying the solutions.

The second trait of interest is that there might exist redundancy parameters other than q_4 that are more representative of the internal motion, for the trajectory assigned. In fact, by looking at the grids of Figure 8 (without joint limits), a large portion of the joint domain does not contain any solution. This means that large variations of the other joints shall be expected for little variations of the redundancy parameter: a fine discretization of the redundancy parameter is needed for the dynamic programming algorithm to provide a smooth solution. We recall that, in our case, the selection of q_4 is unavoidable for *IKFast* to produce an anlytical IK solver.

Lastly, it is worth noting that joint limits, in real scenarios, notably reduce the search space, giving a chance to the dynamic programming algorithm to find the resolution optimal solution in a short time. Also, because of joint limits, the Panda is not able to track the assigned trajectory remaining in the same extended aspect, as none of the grids admits a feasible joint-space path from $\mathbf{x}_A$ (corresponding to $i = 0$) to $\mathbf{x}_E$ (corresponding to $i = 203$). Hence, the robot will need to reconfigure its posture on the way by passing through singularities of its kinematic subchains.

4.3. Globally Optimal and Pareto-Optimal Solution

Since grids are homogeneous, the search can be optimized. Thus, both our algorithms [9,14] can be executed to find the resolution-optimal solution on the grids of Figure 7. Table 2 reports the execution time of both algorithms and different discretization steps of the redundancy parameter, together with the associated cost function, for the case where only energy minimization is considered. Tests have been executed on a 64-bit Ubuntu 16.04 LTS OS running on an Intel® Core™ i7-2600K CPU @ 3.40GHz × 8. No multi-core execution model has been used in the tests.

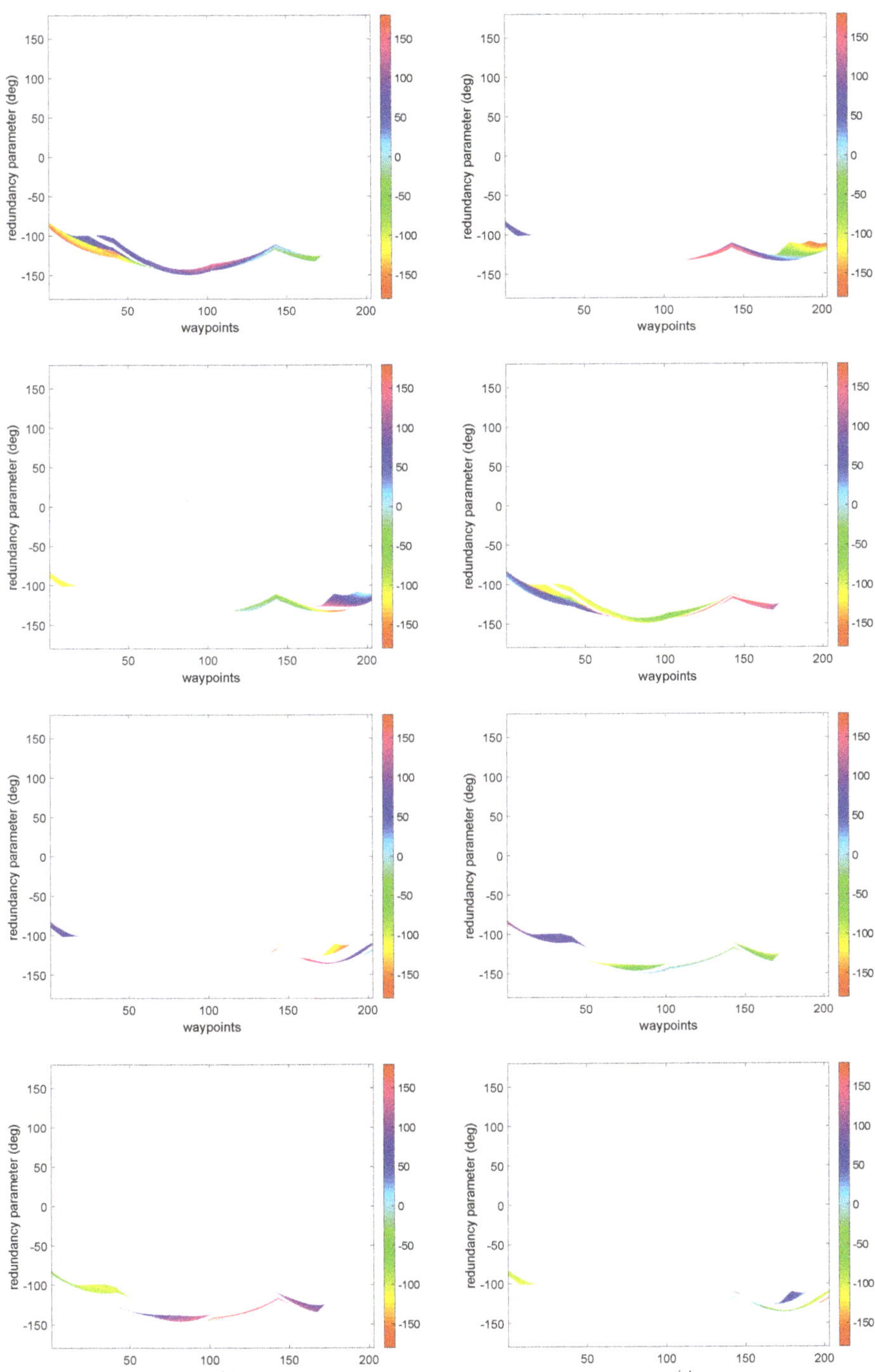

Figure 7. Panda grids (each corresponding to a different extended aspect) representing q_1 for the trajectory described in Section 4.1 considering joint limits.

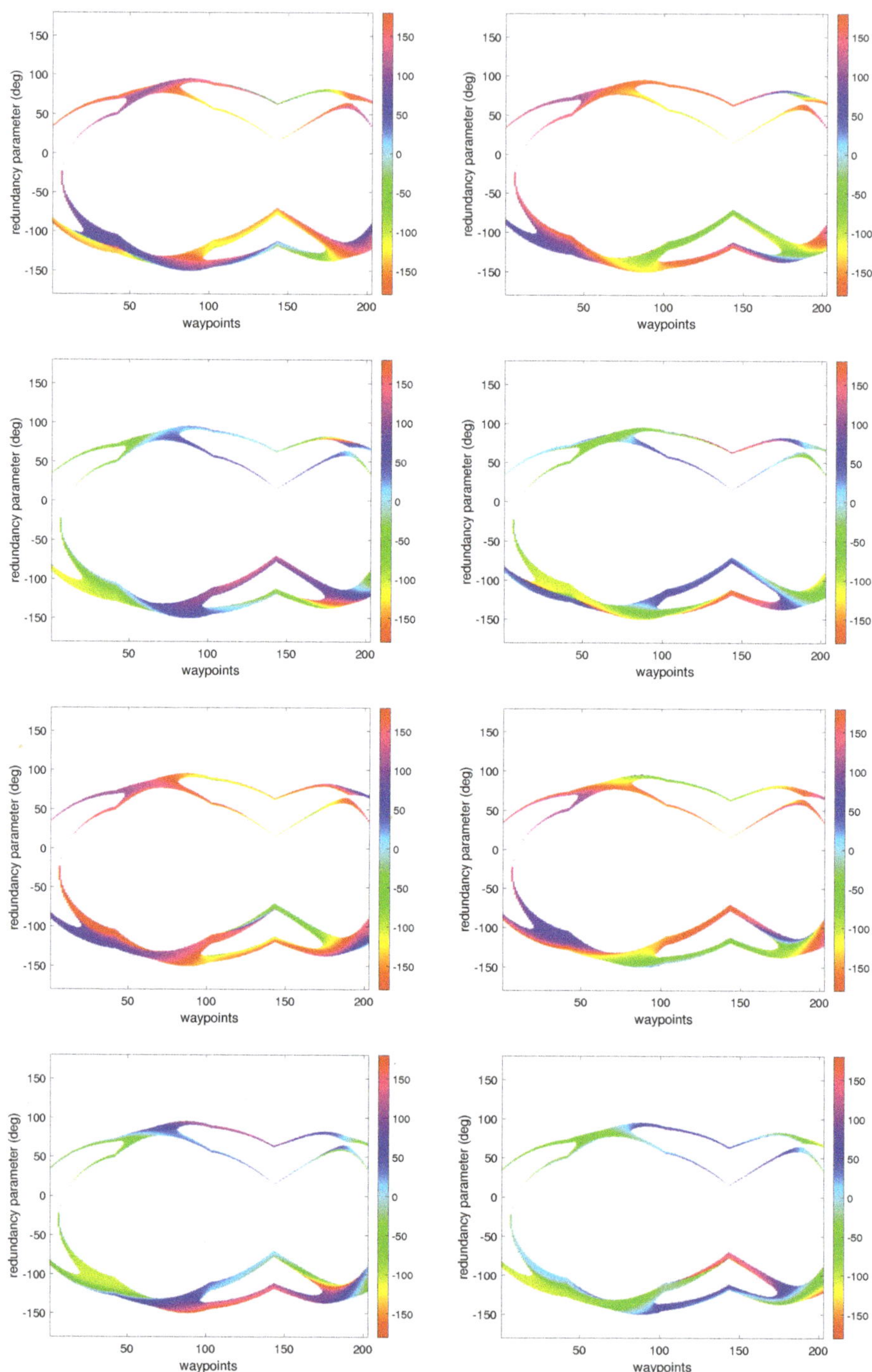

Figure 8. Panda grids (each corresponding to a different extended aspect) representing q_1 for the trajectory described in Section 4.1 neglecting joint limits.

Table 2. Cost function and performance of DP redundancy resolution algorithm for the Panda example, minimizing the square norm of velocities.

N_u	Redundancy Parameter Resolution	Non-Optimized Algorithm [9]	Optimized Algorithm [14]	Cost
360	0.48 deg	11 s	11 s	4.27
720	0.24 deg	54 s	54 s	2.76
1440	0.12 deg	4 min	4 min	2.44
2880	0.06 deg	14 min	13 min	2.16
4000	0.04 deg	27 min	26 min	2.04

It is interesting to notice that there is not any considerable improvement in the performance by using the optimized algorithm in place of the unoptimized algorithm. This means that either position or acceleration limits are almost everywhere stricter than velocity limits for the assigned trajectory. This is in contrast to the use cases where the existence of much less unfeasible cells (i.e., white regions) allows velocity constraints activate first [14].

The convergence rate that we may estimate from the values of the cost function, compared to our previous use case [14], is a confirmation that q_4 is very sensitive for the considered trajectory, meaning that small variations of q_4 yield considerable changes in the solution for the other joints and, as consequence, in the final value of the cost function.

The solution obtained for $N_u = 4000$ is reported in Figure 9 (left). It starts from grid 5 (i.e., $|\mathbf{J}_r^{(4)}| < 0, q_2 < 0, q_5 > 0$), then, at $t = 3.3$ s ($i = 12$), it jumps to grid 6 (i.e., $|\mathbf{J}_r^{(4)}| > 0$, $q_2 < 0, q_5 > 0$) and, at $t = 14.6$ s ($i = 50$), to grid 1 (i.e., $|\mathbf{J}_r^{(4)}| < 0, q_2 < 0, q_5 < 0$). For the majority of the trajectory, up to $t = 48.7$ s ($i = 165$), the solution lies on grid 1. Afterwards, it transits to grid 2 (i.e., $|\mathbf{J}_r^{(4)}| > 0, q_2 < 0, q_5 < 0$) and terminates, achieving 3 posture reconfigurations in total and visiting 4 different extended aspects. As commented in Section 2.3, posture reconfigurations always happen on the boundaries of the feasible (non-white) regions, where two or more of the maps have the same color for all the joints (only q_1 is shown in Figure 7). It is easy to verify that this is the case for the sequence of grids visited by the algorithm and transitions at the stages mentioned above.

In the Pareto-optimal setup, in order to provide the solver with a unique optimization criterion (minimization or maximization), the distance from the obstacle is maximized by minimizing the distance between joint 4 and the point $\mathbf{p} = [0, -2, 0.5]$, lying on the opposite side with respect to the robot, with the effect of "pulling" it away from the obstacle. Among the solutions in the Pareto set, we select the one that minimizes the norm of the objective vector. For $N_u = 4000$, the cost is $[3.20, 191.33]$, corresponding to square norm of velocities and distance from $\mathbf{p}$ respectively, while the joint-space solution is shown in Figure 9 (right). As shown in the video [33], the dynamic programming algorithm achieves the computation of an obstacle-free trajectory, while the robot collides if only the square norm of velocities is minimized.

As far as the solutions of Figure 9 are concerned, the reader may clearly notice the discontinuities in the derivative of the joint positions at each of the three intermediate corners of the trajectory. In between these points the curves are not everywhere smooth. As commented by Gao et al. [5], either a post-processing step or the introduction of jerk constraints would be desirable to allow for the execution on real hardware.

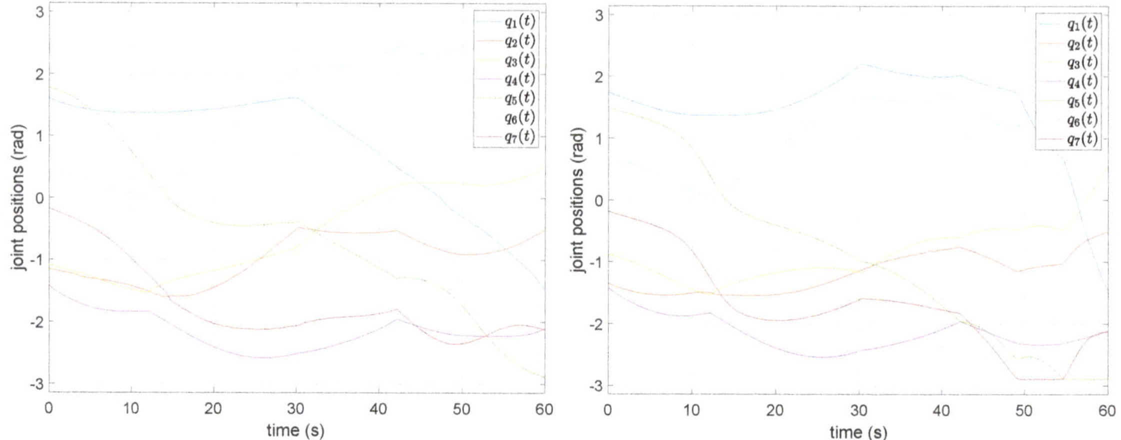

Figure 9. Discrete globally optimal (**left**) and Pareto-optimal (**right**) solution for the Panda example.

5. Discussion

With respect to previous works using a similar DP-based problem formulation, we focused on the design of a maintainable, re-usable, modular and flexible ROS extension that can be generic enough to be employed in a broad range of applications. We provided clear extension points to adapt the software to specific scenarios and introduce custom requirements, such as specific constraints and objective functions. Our solution foresees the development of a minimal amount of code to introduce such modifications, because when possible, they are enforced through configuration files.

The whole extension is provided in the form of a new *move_group* capability, meaning that the already existing interfaces are reused as much as possible, so that the user can benefit from already available analysis and visualization tools. Also, we implemented several file export functions, relying as much as possible on existing formats, e.g., bagfiles, so that further tools can be developed even in environments outside of ROS, such as MATLAB.

We emphasized the capability of our algorithm of exploring multiple grids at the same time, which was not evident in previous works, thus ensuring the achievement of the resolution-optimal solution. In our treatment, we did not renounce to the topological analysis of both the manipulator's null space along the trajectory and the resulting reslution-optimal joint-space solution. Although this is more complicated for spatial manipulators than planar ones, some key features of the problem can be highlighted, as the feasibility of the task and the necessity of traversing singularities or semi-singularities to complete it. Attention was paid to the computational complexity and, in fact, we extended previous algorithms [9,14] to be applied to real robots and demonstrated that, depending on the chosen redundancy parameter and joint limits, not necessarily homogeneous grids yield lower computation times. However, on the other hand, we showed that for real manipulators, dynamic programming is perfectly suited for redundancy resolution as the constraints characterizing real applications drastically reduce the search space and yield a fast convergence.

In our implementation, inverse kinematics plays an important role. First of all, it only concerns position kinematics, thus the robot is free to pass through its singularities as no Jacobian inversion is performed. Second, if an analytic IK solver is available, state space grids can be computed without any knowledge of the extended aspects, and their homogeneity can be imposed numerically. On the contrary, if an analytic solver is not available, and the extended Jacobian cannot be factorized, finding all the possible solutions could be cumbersome, especially in the presence of joint limits. If there is not any certainty that the state space is completely represented, the global optimality of the solution could be affected as well.

Beside possible application-specific extensions, several improvements are possible for our ROS-based implementation. For instance, because the state space is discretized, it might be possible that the solution resulting from the application of the algorithms described in Section 2.3 is not feasible on real hardware. Rather, in other circumstances, it might happen that the trajectory is feasible, but it is not smooth enough to be repeated over and over again without damaging the mechanical parts on the long run. The proposed formulation is straightforward, but, in practice, is not enough to ensure that the motion is always feasible and smooth. In fact, on one hand, the output joint trajectory could exceed joint torque capacities and, on the other, could result in oscillations of the joints because of its non-smoothness. Indeed, while the usage of acceleration constraints allows for smooth joint position functions, it might not be enough to guarantee smoothness at velocity level. In such cases, it might be suitable considering additional constraints on the derivative of the acceleration, which could also be provided in the robots datasheets [5].

Together with the imposed constraints, the discretization step of the redundancy parameter also plays an important role in the generation of smooth joint-space trajectories. It is clear that the finer the discretization is, the smoother the trajectory can be, but this comes to the detriment of time. Indeed, some redundancy parameters have a lower sensitivity with respect to the motion to be performed, meaning that for large changes of their value, all the other variables in play, such as the joint position variables, change less. If this is the case, a coarser discretization can be used for the redundancy parameter, as it is very representative of the motion, resulting in a smooth trajectory, still at a reasonable computation time. Alternatively, an iterative approach can be used, where a finer discretization is performed in the neighborhood of a solution obtained with a coarser discretization at the previous iteration [5]. This technique yields satisfactory results, but may compromise the optimality of the solution if the first discretization is too coarse.

In some other works [10,11], the trajectory smoothness has also been explicitly included in the performance index to optimize. Specific smoothness measures can be suitably combined with other performance indices of interest for the specific application, but the result will always be a sub-optimal solution with respect to each of the indices. Gao et al. [5] consider a different approach, which is based on the post-processing of the solution. In particular, the redundancy parameter curve is smoothened by applying a fifth-order polynomial approximation. Then, in order to guarantee that the trajectory is exactly tracked, inverse kinematics is solved again with the new values of the redundancy parameter.

6. Conclusions

In this paper we proposed a novel architecture to perform redundancy resolution through the global optimization of performance indices, employing a dynamic programming formalism. In particular, the problem formulation foresees the discretization of the state space and its representation in the form of multiple grids. Then, a DP-inspired graph search algorithm is used to ensure the achievement of the resolution-optimal solution.

The developed software components extend the open-source framework ROS, and integrate seamlessly with the existing packages so as to promote the reuse of the available visualization and analysis tools. On the other hand, they provide clear extension points that can be used to introduce user-specific requirements, so that the new capability can be easily adopted in a broad range of applications, with a minimum development effort, even beyond redundancy resolution.

If the underlying state space grids are characterized by continuity, i.e., they are homogeneous, the developed algorithm can exploit this feature to optimize the multi-grid search. This is particularly advantageous when the velocity limits are stricter than other constraints. Moreover, the proposed architecture provides the means to analyze the intermediate and final products of the computation from the topological point of view, and further analysis tools can be developed in MATLAB or other languages.

Nonetheless, our architecture does not guarantee that the resolution-optimal joint-space solution can be directly sent to a real robot controller, although additional constraints and/or processing steps can be defined either through our extension points or the ROS ecosystem.

Future work may concern the introduction of parallel computational models, which would further improve the performance of the algorithm, as well as the extension toward other problems/semantics, such as time-optimal planning along specified paths.

Author Contributions: Conceptualization, E.F. and P.C.; methodology, E.F.; software, E.F. and F.S.; validation, E.F.; investigation, E.F.; resources, F.S. and P.C.; data curation, E.F. and F.S.; writing—original draft preparation, E.F.; writing—review and editing, E.F., F.S. and P.C.; visualization, E.F.; supervision, F.S. and P.C.; project administration, P.C.; funding acquisition, P.C. All authors have read and agreed to the published version of the manuscript.

Funding: This research was partly funded by Italian Ministry of Education, University and Research (MIUR) grant number CUP D49D17000250006. The APC was funded by the University of Salerno.

Institutional Review Board Statement: Not applicable.

Informed Consent Statement: Not applicable.

Data Availability Statement: The source code presented in this study is openly available in GitHub/Zenodo [25], version v0.1.0. More data are available on request from the corresponding author.

Conflicts of Interest: The authors declare no conflict of interest.

References

1. Siciliano, B.; Khatib, O. *Springer Handbook of Robotics*; Springer International Publishing: Berlin/Heidelberg, Germany, 2016. [CrossRef]
2. Kazerounian, K.; Wang, Z. Global versus local optimization in redundancy resolution of robotic manipulators. *Int. J. Robot. Res.* **1988**, *7*, 3–12. [CrossRef]
3. Nakamura, Y.; Hanafusa, H. Optimal redundancy control of robot manipulators. *Int. J. Robot. Res.* **1987**, *6*, 32–42. [CrossRef]
4. Guigue, A.; Ahmadi, M.; Hayes, M.J.D.; Langlois, R.; Tang, F.C. A dynamic programming approach to redundancy resolution with multiple criteria. In Proceedings of the IEEE International Conference on Robotics and Automation, Rome, Italy, 10–14 April 2007; pp. 1375–1380. [CrossRef]
5. Gao, J.; Pashkevich, A.; Caro, S. Optimization of the robot and positioner motion in a redundant fiber placement workcell. *Mech. Mach. Theory* **2017**, *114*, 170–189. [CrossRef]
6. Shen, Y.; Huper, K. Optimal trajectory planning of manipulators subject to motion constraints. In Proceedings of the 12th International Conference on Advanced Robotics, Seattle, WA, USA, 17–20 July 2005. [CrossRef]
7. Ascher, U.M.; Mattheij, R.M.M.; Russell, R.D. *Numerical Solution of Boundary Value Problems for Ordinary Differential Equations*; Society for Industrial and Applied Mathematics: Philadelphia, PA, USA, 1995. [CrossRef]
8. Keller, H.B. *Numerical Methods for Two-Point Boundary-Value Problems*; Courier Dover Publications: Garden City, NY, USA, 2018.
9. Ferrentino, E.; Chiacchio, P. A topological approach to globally-optimal redundancy resolution with dynamic programming. In *ROMANSY 22—Robot Design, Dynamics and Control*; Arakelian, V., Wenger, P., Eds.; Springer: Cham, Switzerland, 2018; Volume 584, pp. 77–85. [CrossRef]
10. Pashkevich, A.P.; Dolgui, A.B.; Chumakov, O.A. Multiobjective optimization of robot motion for laser cutting applications. *Int. J Comput. Integr. Manuf.* **2004**, *17*, 171–183. [CrossRef]
11. Dolgui, A.; Pashkevich, A. Manipulator motion planning for high-speed robotic laser cutting. *Int. J. Prod. Res.* **2009**, *47*, 5691–5715. [CrossRef]
12. Guigue, A.; Ahmadi, M.; Langlois, R.; Hayes, M.J.D. Pareto optimality and multiobjective trajectory planning for a 7-DOF redundant manipulator. *IEEE Trans. Robot.* **2010**, *26*, 1094–1099. [CrossRef]
13. Cavalcanti Santos, J.; Martins da Silva, M. Redundancy Resolution of Kinematically Redundant Parallel Manipulators Via Differential Dynamic Programing. *J. Mech. Robot.* **2017**, *9*. [CrossRef]
14. Ferrentino, E.; Chiacchio, P. On the optimal resolution of inverse kinematics for redundant manipulators using a topological analysis. *J. Mech. Robot.* **2020**, *12*. [CrossRef]
15. What is ROS? Available online: http://www.ros.org/ (accessed on 3 March 2021).
16. Ferrentino, E.; Chiacchio, P. Redundancy parametrization in globally-optimal inverse kinematics. In *Advances in Robot Kinematics 2018*; Lenarcic, J., Parenti-Castelli, V., Eds.; Springer: Cham, Switzerland, 2018; Volume 8, pp. 47–55. [CrossRef]
17. Burdick, J.W. On the inverse kinematics of redundant manipulators: Characterization of the self-motion manifolds. In *Proceedings of the 4th International Conference on Advanced Robotics, Scottsdale, AZ, USA, 14–19 May 1989*; Waldron, K.J., Ed.; Springer: Berlin/Heidelberg, Germany, 1989; Volume 4, pp. 25–34. [CrossRef]

8. Burdick, J.W. On the Inverse Kinematics of Redundant Manipulators: Characterization of the Self-Motion Manifolds. In Proceedings of the IEEE International Conference on Robotics and Automation, Scottsdale, AZ, USA, 14–19 May 1989; pp. 264–270. [CrossRef]

9. Wenger, P. A New General Formalism for the Kinematic Analysis of All Nonredundant Manipulators. In Proceedings of the IEEE International Conference on Robotics and Automation, Nice, France, 12–14 May 1992; pp. 442–447. [CrossRef]

10. Pámanes, J.A.; Wenger, P.; Zapata, J.L. *Motion Planning of Redundant Manipulators for Specified Trajectory Tasks*; Advances in Robot Kinematics; Springer: Dordrecht, The Netherlands, 2002; pp. 203–212. [CrossRef]

11. Sniedovich, M. Dijkstra's algorithm revisited: The dynamic programming connexion. *Control. Cybern.* **2006**, *35*, 599–620.

12. MoveIt. Available online: https://moveit.ros.org/ (accessed on 3 March 2021).

13. MoveIt Planners. Available online: https://moveit.ros.org/documentation/planners/ (accessed on 3 March 2021).

14. MoveIt Concepts. Available online: https://moveit.ros.org/documentation/concepts/ (accessed on 3 March 2021).

15. Ferrentino, E.; Salvioli, F. ROS/MoveIt! Extension for Redundancy Resolution with Dynamic Programming. GitHub/Zenodo. Available online: https://zenodo.org/record/3236880#.YAsFiehKhPY (accessed on 3 March 2021).

16. Diankov, R. Automated Construction of Robotic Manipulation Programs. Ph.D. Thesis, Carnegie Mellon University, Pittsburgh, PA, USA, 26 August 2010. Available online: http://www.programmingvision.com/rosen_diankov_thesis.pdf (accessed on 3 March 2021).

17. KDL Wiki. Available online: http://www.orocos.org/kdl (accessed on 3 March 2021)

18. Ferrentino, E.; Chiacchio, P. Topological analysis of global inverse kinematic solutions for redundant manipulators. In *ROMANSY 22—Robot Design, Dynamics and Control*; Arakelian, V., Wenger, P., Eds.; Springer: Cham, Switzerland, 2018; Volume 584, pp. 69–76. [CrossRef]

19. Panda Powertool. Available online: https://www.franka.de/panda (accessed on 3 March 2021).

20. Khalil, W.; Kleinfinger, J. A new geometric notation for open and closed-loop robots. In Proceedings of the IEEE International Conference on Robotics and Automation, San Francisco, CA, USA, 7–10 April 1986; pp. 1174–1179. [CrossRef]

21. Franka Control Interface Documentation. Available online: https://frankaemika.github.io/docs/ (accessed on 3 March 2021).

22. Zaplana, I.; Basanez, L. A novel closed-form solution for the inverse kinematics of redundant manipulators through workspace analysis. *Mech. Mach. Theory* **2018**, *121*, 829–843. [CrossRef]

23. Ferrentino, E.; Salvioli, F. Redundancy Resolution for Energy Minimization and Obstacle Avoidance with Franka Emika's Panda Robot. Available online: https://youtu.be/AxL755_t3_o (accessed on 3 March 2021).

Article

A Recursive Algorithm for the Forward Kinematic Analysis of Robotic Systems Using Euler Angles

Fernando Gonçalves [1,2], Tiago Ribeiro [3,4], António Fernando Ribeiro [3,4,*], Gil Lopes [5] and Paulo Flores [1,2]

1 Department of Mechanical Engineering, University of Minho, 4800-058 Guimarães, Portugal; id8699@alunos.uminho.pt (F.G.); pflores@dem.uminho.pt (P.F.)
2 Center for Microelectromechanical Systems (CMEMS), University of Minho, 4800-058 Guimarães, Portugal
3 Department of Industrial Electronics, University of Minho, 4800-058 Guimarães, Portugal; id9402@alunos.uminho.pt
4 Centro ALGORITMI, University of Minho, 4800-058 Guimarães, Portugal
5 Department of Communication Sciences and Information Technologies, University Institute of Maia—ISMAI, 4475-690 Maia, Portugal; alopes@ismai.pt
* Correspondence: fernando@dei.uminho.pt

Abstract: Forward kinematics is one of the main research fields in robotics, where the goal is to obtain the position of a robot's end-effector from its joint parameters. This work presents a method for achieving this using a recursive algorithm that builds a 3D computational model from the configuration of a robotic system. The orientation of the robot's links is determined from the joint angles using Euler Angles and rotation matrices. Kinematic links are modeled sequentially, the properties of each link are defined by its geometry, the geometry of its predecessor in the kinematic chain, and the configuration of the joint between them. This makes this method ideal for tackling serial kinematic chains. The proposed method is advantageous due to its theoretical increase in computational efficiency, ease of implementation, and simple interpretation of the geometric operations. This method is tested and validated by modeling a human-inspired robotic mobile manipulator (CHARMIE) in Python.

Keywords: forward kinematics; computational mechanics; robot manipulator kinematics; 3D robot modeling

Citation: Gonçalves, F.; Ribeiro, T.; Ribeiro, A.F.; Lopes, G.; Flores, P. A Recursive Algorithm for the Forward Kinematic Analysis of Robotic Systems Using Euler Angles. *Robotics* **2022**, *11*, 15. https://doi.org/10.3390/robotics11010015

Academic Editor: António Paulo Moreira, Félix Vilariño and Pedro Neto

Received: 26 November 2021
Accepted: 6 January 2022
Published: 14 January 2022

Publisher's Note: MDPI stays neutral with regard to jurisdictional claims in published maps and institutional affiliations.

1. Introduction

The control of robotic manipulators is strongly linked to the study of their motion. Forward kinematics refers to the process of determining the position and orientation of a robotic end effector with known joint parameters [1]. Although by definition, the position and orientation of all links are not required to solve a forward kinematics problem, in this paper, the goal is to obtain the complete definition of all the link's orientations and positions to fully describe the robot's 3D configuration.

The most used method for the kinematic analysis of robotic manipulators is the Denavit–Hartenberg parameters [1]. This approach concisely allows the characterization of each link using four parameters, providing a compact definition of a robot's kinematic structure. However, this methodology has two drawbacks. The first is fixing the choice of axes, which is defined by the orientation of the joints. This prevents researchers from picking a more natural axes orientation based on the configuration of the kinematic links, where each axis could be associated with a specific physical meaning (for example, using the z-axis for heights, or the length of parts). The second is that calculations are made based on homogeneous transformations. These $[4 \times 4]$ matrices define rotations and translations in a single operation, however, the last line of the matrix does not contain any relevant information, being constituted by 0 s and 1 s to allow algebraic operations. These additional multiplications reduce the computational efficiency of the forward kinematics analysis.

A known solution to this problem is to divide translations and rotations into different operations [2].

This paper presents an alternative generalizable methodology that intends to deal with both of these limitations. This method is based on a recursive algorithm that builds a 3D model of the robot from its base, to its end-effector. The algorithm progresses along the kinematic chain, determining the rotation matrix R_i^0 that defines the orientation of each link i. This matrix is obtained from the orientation of the preceding link R_{i-1}^0, and the relative orientation between the current link and its predecessor R_i^{i-1}. The rotation between consecutive links is defined using Euler Angles. After the orientation of a link is determined, its position is obtained from the joint coordinates resulting from the 3D modeling of its predecessor $i-1$. This process provides the necessary information for the definition of the geometry of each of the robot's links, the determination of the position and orientation of these links, and their three-dimensional representation.

This method was implemented in Python using the numpy library for matrix and trigonometry operations, and Matplotlib for the 3D plotting of the robot's points to allow the observation of its behavior in a 3D graphical interface. For validation, the algorithm was used to analyze CHARMIE [3], a human-inspired mobile manipulator robot (Figure 1). This robot also serves as an example throughout the paper to better explain the developed algorithm.

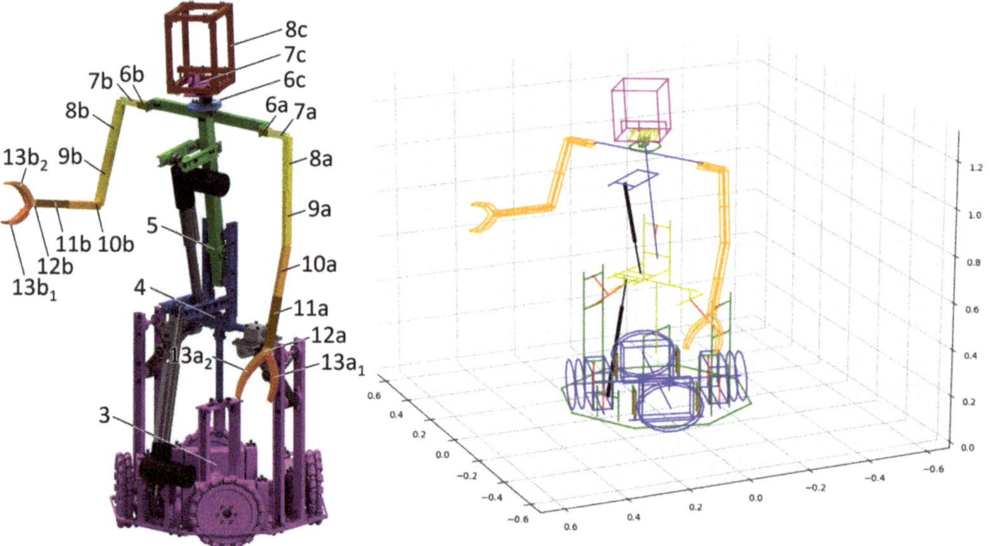

Figure 1. 3D model of the CHARMIE mobile manipulator in a CAD software (**left**) and in the developed kinematics simulation environment (**right**). On the left, the robot's kinematic links are color coded and named.

In Figure 1 the robot's base is shown as a single kinematic link. At this stage, the behaviour of the omnidirectional locomotive system was not considered (the wheels and suspension system are represented, but they do not move in relation to the robot). Instead, as a simplification, the robot was modeled as having its base sliding (in the x and y axes) about the floor using two linear actuators, and rotating (around the z-axis) using a revolute actuator. The arms and end-effector shown in the CAD model are mere placeholdes, since one of the results from this kinematic analysis is the study responsible for dimensioning them.

The main contribution of this paper is the presentation and explanation of a methodology for the study of forward kinematics. This methodology produces a computationally defined 3D model of a robot, which can be used for a set of relevant analyses in the field of robotics. Two examples of these applications are the studies being conducted in CHARMIE, where this kinematic model is being used not only as the starting point to build a simulation environment for the training of a neural network to control the robot's motion and trajectories, but also for multibody dynamics analysis, where the recursive Newton Euler algorithm described in [4] is used to compute the robot's inverse dynamics. The cited Newton Euler algorithm is fully compatible with the methodology presented in this paper, directly using the information obtained from it (positions, configuration of each link, and orientations between consecutive bodies in the form of rotation matrices).

To fully describe and validate this methodology, the paper is structured as follows: in Section 2 a Literature Review is provided, where several methods for the 3D representation of rotations are listed and described, followed by a justification of the choice of method for this paper; Section 3 presents, formulates, explains and describes the recursive algorithm developed in this paper, dividing it into three simple steps; Section 4 provides an example of application of this methodology, using it to build a 3D model of the CHARMIE mobile manipulator, defining the robot and explaining how the modeling of some of its particularities was dealt with; Section 5 finishes the paper discussing results and commenting on possible future works.

2. Literature Review

The forward kinematic analysis can be tackled as a matter of obtaining the 3D configuration of a group of bodies (links) from a set of known conditions (joint parameters). It is possible to fully define a rigid body in 3D space with its orientation and the coordinates of one of its points. The position of a point, in Euclidian space, is easily described using three coordinates: x, y, and z, however, defining three-dimensional rotations, often denoted SO(3), is a more complex topic. Several formalisms have been developed for this purpose, they can be used together, and the conversion between them is a well-studied process. In this Literature Review, some of the main notations used in the field of robotics for the description of 3D rotations are listed and described, followed by a few examples of works that use them. This section finishes by presenting and justifying the method chosen for this paper.

Rotation Matrices are [3×3] matrices commonly used to define the rotation between two coordinate frames i and j. A rotation matrix R_j^i describes the rotation from frame i to frame j. These matrices represent the dot product between the basis vectors $[\hat{x}_i, \hat{y}_i, \hat{z}_i]$ and $[\hat{x}_j, \hat{y}_j, \hat{z}_j]$ of the two considered frames [2]. When multiplying the coordinates of a point P with the rotation matrix R_j^i, the result is a transformation which follows the rotation defined between frames i and j. This can be used to convert the coordinates of points between references with different orientations. Rotation matrices can be combined by simple matrix multiplication ($R_k^i = R_j^i\, R_k^j$), allowing the representation of a limitless sequence of rotations. Due to the simplicity of their manipulation, they are often used to describe rotations obtained from the application of different formalisms, such as Euler Angles [5].

The Denavit—Hartenberg convention is one of the most used notations for the kinematic analysis of serial manipulators. Four parameters are used to describe the transformations between each consecutive element of the kinematic chain: a_i and α_i describe the link's length and twist; d_i and θ_i describe the joint's offset and angle [6]. To apply this method, a set of reference axes are attached to the links of the kinematic chain. The definition of these references is based on the orientation and position of the joints and follows a set of rules and conventions usually described by a set of steps (such as presented in [1]), to guarantee the cohesion of the resulting Denavit–Hartenberg parameters. From these four parameters, a [4×4] homogeneous transformation matrix is constructed containing information regarding the rotation and translation between each consecutive pair of links. These matrices can be combined, multiplied, and easily manipulated like the aforementioned rotation matrices. Mostly used for serial manipulators, this method is highly advantageous due to: representing robot kinematics in a compact form; producing consistent results thanks to the rigid and detailed methodology; the vast amount of algorithms and works already developed for it. Examples of papers using this notation are available in [7–9].

Euler angles represent any rotation of a three-dimensional object as a sequence of three consecutive rotations. These rotations can be extrinsic (around the fixed motionless original *xyz* axes) or intrinsic (around the rotating coordinate axes of the considered body). The sequence of rotations uses proper Euler angles if the first and third rotations are around the same axis, or Tait–Bryan angles if all three rotations are around different axes. Depending on the research field, different authors use different names, and axes, to define Euler angles so it is important to verify the nomenclature used in each work. There are two well-known limitations related to the use of Euler angles. The first is the Euler Angle singularity, which occurs when the angle of the second rotation is $\pi/2$ or $-\pi/2$. In these cases, the first and third Euler Angles can vary independently, both controlling the same degree of freedom resulting in an infinite number of possible combinations for defining a single orientation. The second problem, gimbal lock, also occurs for the same values of the second rotation. Due to two rotation axes being aligned, a degree of freedom is lost, which prevents the system from immediately doing determined motions. These limitations can be corrected or become severe problems, depending on the intended applications. An explanation of these limitations, and ways to address them, is available in [10]. Some examples of works using the Euler-Angles notation are [11–13].

A quarternion is a four-dimensional vector, represented by 4 scalar entities, which can be harnessed to compute rotations on points and vectors in three dimensions. They are one of the major alternatives to rotation matrices, and are commonly used due to their high efficiency in computer calculations and their ease of interpolation. They also avoid both previously described problems related to Euler Angles. It should be noted that quarternions have their limitations, such as a reduce in efficiency when calculating the rotation of a vector [14]. The formulas required for the use of quaternions are well-known, but the understanding of these formulas, and underlying principles, is complex [14]. Some works allow a deeper understanding of quaternions, such as the paper [14], and the books [15,16]. A survey is presented in [17] which reviews and compares methods for the computation of quaternions from rotation matrices. Quaternions are used in the following works: [18,19].

Another possible method for the computational analysis of multi-body kinematics is screw theory (usually alongside Lie groups). In screw theory, two three-dimensional vectors are used to represent: the position and orientation of a rigid body; the linear and angular velocity of a rigid body; a force and a couple [20]. The two vectors define the Plücker coordinates of a line in space (the position and direction of the screw axis), the magnitude of the screw, and its pitch. These four factors completely define a screw [20]. Using screw theory in conjunction with Lie algebra se(3), associated with Lie group SE(3) [21], it is possible to develop recursive algorithms that solve the kinematics of multi-body problems with high computational efficiency [22]. Some examples of works that use Screw Theory are [23–25].

Regarding the methodology used in this paper, the rotation between two consecutive bodies is defined using ZXZ intrinsic Euler angle rotations. These rotations can be easily converted into other formalism [5] (such as rotation matrices that can be more conveniently manipulated), the comprehension of their geometry is straightforward, and they allow a free choice of local axes for each link (it may become beneficial to program a constant rotation to guarantee convenient axes orientation). Since most used joints in robotics rotate around a single axis (revolute joints), problems regarding singularities and the gimbal lock are easily avoided with the choice of local orientation. More complex joints (such as spherical joints) can be modeled using a single complex Euler angle rotation, or as a sequence of rotations with one degree of freedom around the same point. The rotations obtained from the Euler angles are converted into rotation matrices, and all following mathematical operations are made using said matrices.

3. Recursive Algorithm for the Computation of Forward Kinematics

The structure of the developed recursive forward kinematics algorithm is illustrated in Figure 2. This algorithm computes the positions of the robot's links from known joint configurations. After running this algorithm from link 0 (the global reference) to the robot's end-effector, the coordinates and orientations of all bodies in the kinematic chain are fully defined. The calculations are made sequentially, using information regarding both the current link i and the previous link $i-1$.

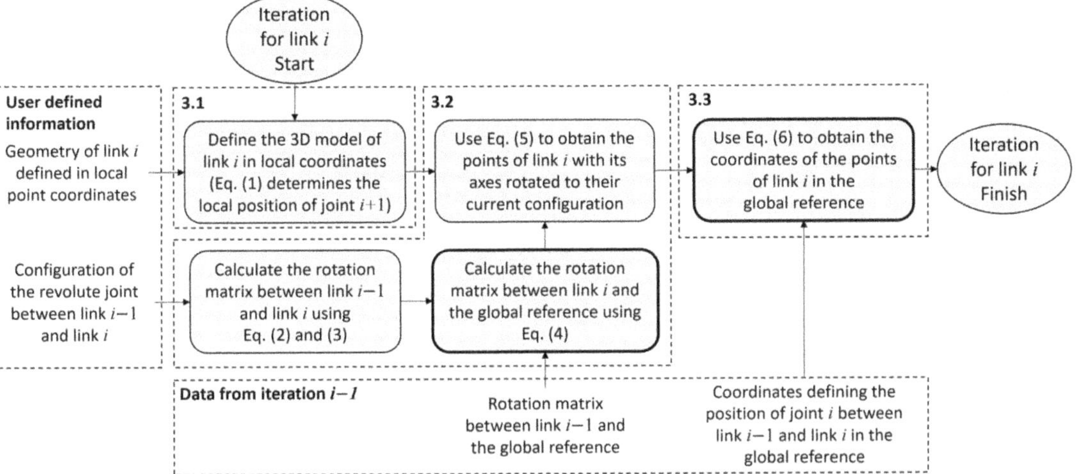

Figure 2. Flowchart of the developed recursive algorithm for the computation of forward kinematics.

In Figure 2, the boxes with stronger outlines represent the steps where the orientation (in the form of a rotation matrix) and position (in the form of coordinates) of the link are defined. With this information, additional data (such as linear and angular velocities and accelerations) can be calculated.

The algorithm is divided into three main steps, which are described in greater detail—accompanied by an example—in the following subsections:

1. Definition of the geometry of link i;
2. Rotation of link i into its current orientation;
3. Translation of link i into its current position.

The following terms are used to define and name points and axes in this paper:

- $x_i' y_i' z_i'$—The local coordinate axis of link i;
- $x_i'' y_i'' z_i''$—The coordinate axis of link i after considering the link's rotation to its actual orientation;
- xyz—The global reference axis;
- A_i—The origin of link i, placed on the connection point between link i and link $i-1$;
- B_i—The connection point between link i and link $i+1$;
- C_i—The link's center of mass (important for posterior dynamics calculations);
- D_i—The position of the joint between link i and link $i+1$ considering a linear joint displacement d_{i+1} of 0.
- P_i—Refers to all points of link i.

If the joint after a link is revolute, point B_i' is fixed and equal to D_i'. However, if this joint is prismatic, this point will move based on the position of the linear actuator. These calculations must be made in local coordinates when analyzing link i, so that when iteration $i+1$ begins, the position of the origin of link $i+1$ in global coordinates is already known. This is done using the equation:

$$B'_i = D'_i + d_{i+1} v'_{d_{i+1}}, \tag{1}$$

where d_{i+1} is the linear displacement of the prismatic joint between link i and link $i+1$ and $v'_{d_{i+1}}$ the unit vector defining the orientation of this prismatic joint in relation to the $x'_i y'_i z'_i$ local axes.

The rotation matrix associated with an intrinsic ZXZ Euler rotation, defined by the angles (Z_1, X_2, and Z_3), is obtained from [5]:

$$ZXZEuler(Z_1, X_2, Z_3) = \begin{bmatrix} c_1 c_3 - s_1 c_2 s_3 & -c_1 s_3 - s_1 c_2 c_3 & s_1 s_3 \\ s_1 c_3 + c_1 c_2 s_3 & c_1 c_2 c_3 - s_1 s_3 & -c_1 s_2 \\ s_2 s_3 & s_2 c_3 & c_2 \end{bmatrix}, \tag{2}$$

where c represents a cosine function, s a sine function, and the indexes 1, 2, and 3 the corresponding angles (angle of the 1st rotation around the z-axis, angle of the 2nd rotation around the new x-axis, angle of the 3rd rotation around the z-axis after the first two rotations are applied).

If the local coordinate axes for two consecutive links are aligned when the associated joint rotation angle θ_i is 0, the rotation matrix R_i^{i-1} between them can be directly determined using:

$$R_i^{i-1} = \begin{cases} ZXZEuler(0, \theta_i, 0) & \text{if rotation axis is } x \\ ZXZEuler(\pi/2, \theta_i, 0) & \text{if rotation axis is } y \, , \\ ZXZEuler(0, 0, \theta_i) & \text{if rotation axis is } z \end{cases} \tag{3}$$

in this equation, the orientation of the joint axis and the joint rotation angle θ_i are used to define the inputs for Equation (2).

The orientation of link i, defined by R_i^0, is then determined using:

$$R_i^0 = R_{i-1}^0 R_i^{i-1}, \tag{4}$$

where the rotation matrix R_{i-1}^0, which defines the orientation of link $i-1$ in relation to reference 0 (already determined in iteration $i-1$ of the algorithm), is multiplied by the rotation matrix R_i^{i-1} determined in the previous Equation (3).

The P''_i points of link i rotated into its current orientation (expressed in the $x''_i y''_i z''_i$ axes) are obtained using:

$$\begin{bmatrix} P''_i x \\ P''_i y \\ P''_i z \end{bmatrix} = R_i^0 \begin{bmatrix} P'_i x \\ P'_i y \\ P'_i z \end{bmatrix}, \tag{5}$$

where the rotation matrix R_i^0 is used to rotate the P'_i points in the local axes of link i around point A'_i.

A last equation then determines the coordinates of the points P_i of link i expressed in the xyz global axes:

$$\begin{bmatrix} P_i x \\ P_i y \\ P_i z \end{bmatrix} = \begin{bmatrix} P''_i x \\ P''_i y \\ P''_i z \end{bmatrix} + \begin{bmatrix} B_{i-1} x \\ B_{i-1} y \\ B_{i-1} z \end{bmatrix}, \tag{6}$$

where the coordinates of point B_{i-1} (determined in the previous iteration of the algorithm) are used to translate the P''_i points of the rotated axes of link i into their correct position and orientation.

In particular cases, a joint may not be directly actuated, and additional calculations are required based on the geometry of the robot. Examples of this are given in Section 4.2.

This simple algorithm can model robots with different configurations and purposes. Besides CHARMIE, studied in the following sections, Figure 3 shows three examples of kinematic models obtained using this method: (a) a mobile quadruped robot similar to

SPOT from Boston Dynamics [26]; (b) a fixed serial manipulator similar to KUKA KR 500-3 [27]; and (c) a mobile hexapod similar to the one presented in [28].

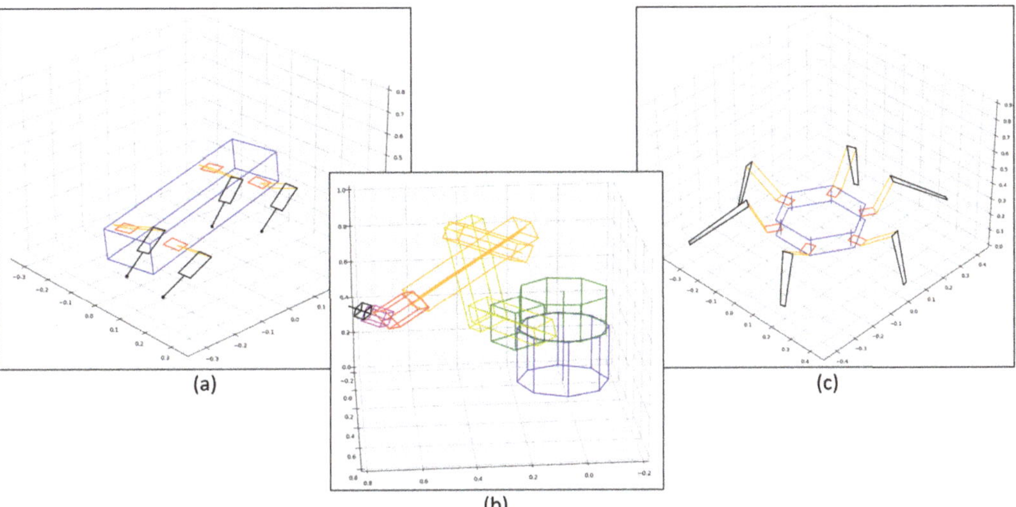

Figure 3. Examples of robots modeled using the proposed recursive algorithm: (**a**) a quadruped robot; (**b**) a fixed serial manipulator; (**c**) a hexapod robot.

These three models were made as follows:

- Quadruped robot—First, the main body of the robot was created. Then, a single leg was modeled. The leg model was replicated four times, and each of them was placed in its respective connection point to the body (a constant rotation altered the origin orientation between legs on the left and the right side of the robot). This resulted in a fully defined kinematic model controlled by 12 joint angles (3 for each leg). To study the robot's locomotion, the model can be placed in a simulation environment that considers the dynamic interactions between the feet and the floor.
- Fixed serial manipulator—To build this model, each body was created in local coordinates, and then the robot was assembled with the proposed algorithm. This model is controlled by the angles of the 6 revolute joint. This analysis is similar to problems commonly tackled using Denavit-Hartenberg parameters.
- Hexapod robot—Modeling the hexapod was similar to the quadruped robot, with the main difference being the use of a $\pi/3$ rotation between each leg in relation to the body. The resulting kinematic model is controlled by 18 joint angles (3 for each leg). Similar to the quadruped robot, after interaction with the floor is defined, this model can be used to study locomotion.

The use of this algorithm provides the same advantages for the study of these three robots as for CHARMIE. Besides the resulting models being compatible with other methods, they are inherently parametric and modular. As an example, this allows doing a parametric study of the limb length for the mobile robots to minimize actuator torque or maximize locomotion velocity. All shown models can also be used for machine learning applications.

The algorithm's behaviour will now be explained and illustrated by using it to model the head (link 8c) of the CHARMIE robot. At this iteration, the algorithm has already finish running for all links leading to link 8c, which produces the model shown in Figure 4.

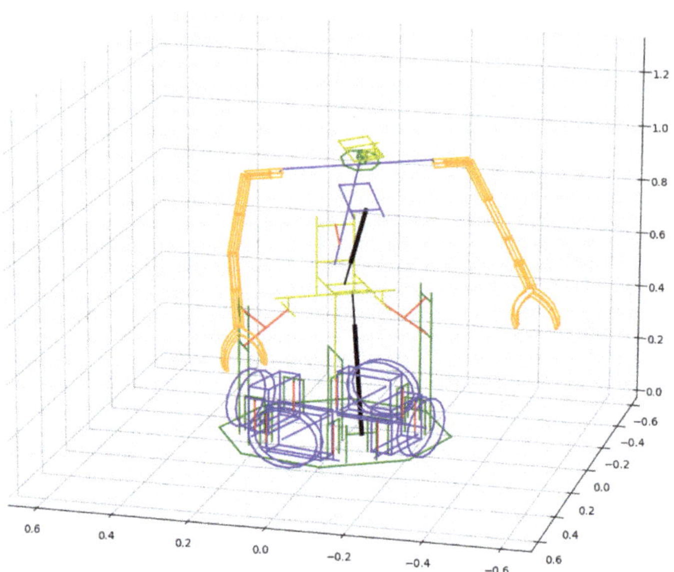

Figure 4. Starting configuration for the application example of the recursive algorithm. The robot has been completely modeled with the exception of its head (link 8*c*).

3.1. Modeling Link i in Its Local Coordinates Axis

The first step is the construction of a 3D model of the link being analyzed in its local coordinates. The origin of each link, identified as point A'_i, is placed on the point where link i is connected to link $i-1$. The geometry of each link can be defined by any number of points (or other methods, such as surfaces or equations).

A coherent choice of local axes orientation for the 3D models facilitates the implementation of the algorithm. In this example, z'_i represents the positive height of complex parts or the length of tubes, the positive y'_i points to the front of the part, and the x'_i axis points from left to right.

To demonstrate this step, Table 1 shows the local coordinates of the CHARMIE robot's head (link 8*c*). Since this is the end of a kinematic chain, point B'_i corresponds to the end-effector (in this example, it is a point in the top of the robot's head, but the position of a camera could also be considered). Equation (1) defines the position of B'_{8c} being the same as D'_{8c} because the joint after link 8*c* has no motion (in this case, it is a fixed point).

Table 1. Coordinates (in millimeters) of the relevant points of the robot's head (link 8*c*) in the $x'_{8c}y'_{8c}z'_{8c}$ local axes.

Point	x	y	z	Point	x	y	z	Point	x	y	z
A'_{8c}	0	0	0	G'_{8c}	−84.75	0	41	M'_{8c}	−84.75	110	−9
B'_{8c}	−14.75	20	201	H'_{8c}	55.25	0	−9	N'_{8c}	55.25	110	201
C'_{8c}	−11.83	15.23	75.68	I'_{8c}	−84.75	0	−9	O'_{8c}	55.25	−70	201
D'_{8c}	−14.75	20	201	J'_{8c}	55.25	110	−9	Q'_{8c}	−84.75	−70	201
E'_{8c}	0	0	41	K'_{8c}	55.25	−70	−9	R'_{8c}	−84.75	110	201
F'_{8c}	55.25	0	41	L'_{8c}	−84.75	−70	−9				

Figure 5 shows the 3D model obtained from the points of Table 1. The head is represented in a 3D plot by drawing lines between the relevant points. Points B'_{8c} to D'_{8c} were not illustrated since they are only relevant for internal calculations, not for the graphical representation of the head.

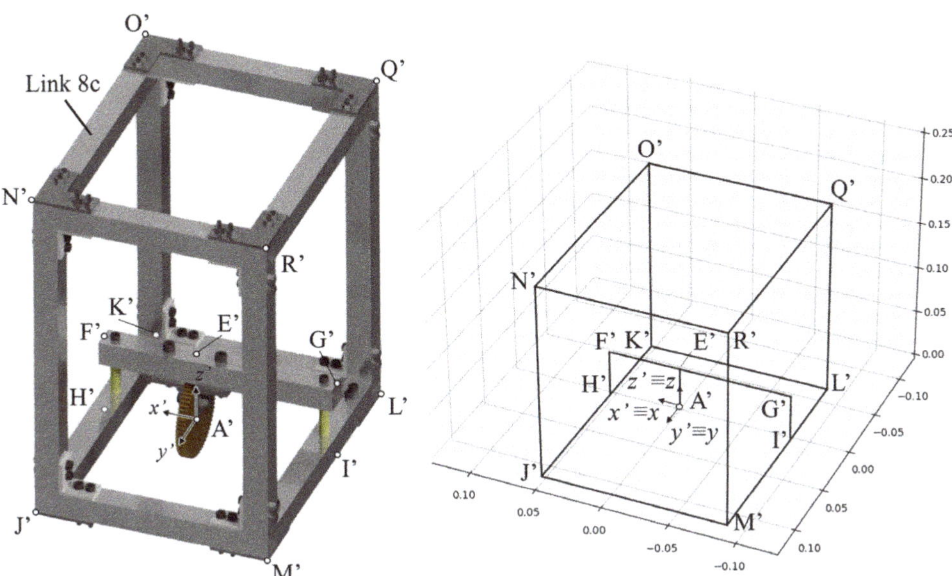

Figure 5. 3D Model of the robot's head (link $8c$) in CAD (**left**), and in the computational simulation in the $x'_{8c}y'_{8c}z'_{8c}$ local axes (**right**). With the exception of the axes of the global reference (xyz), represented points and axes have sub-index ($8c$) but this notation was omitted for simplification.

It should be noted that if the goal is to obtain the solution of the forward kinematics with maximum computational efficiency, each link can be simplified to only contain points B'_i and D'_i (when the following joint is revolute, only B'_i is needed).

3.2. Rotating Link i to Its Current Orientation

In the second step of the algorithm, the orientation of link i is determined. In this work, intrinsic proper Euler rotations along the ZXZ axes are used to describe the rotation between each pair of consecutive links.

Since the local axes were chosen fulfilling the conditions for Equation (3), it can be used together with Equation (2) to obtain the rotation matrix R^{7c}_{8c}. If the axes were not aligned, constant angles could be added to Equation (3) to include the change of orientation.

Knowing the orientation of link $7c$ about the global reference R^0_{7c}, which was determined in the previous iteration of the algorithm, and the rotation between link $7c$ and link $8c$, the orientation R^0_{8c} of link $8c$ is obtained using Equation (4).

With the orientation of link $8c$ determined, it is now possible to rotate it to its current configuration. A new auxiliary reference, $x''_{8c}y''_{8c}z''_{8c}$, is created to represent link $8c$ after its rotation. This rotation is applied to all P'_{8c} points of the link using Equation (5).

Figure 6 shows the robot's head, and its corresponding points, after being rotated to a specific configuration. This orientation is calculated using data resulting from all iterations until the current one is reached, indirectly utilizing information from the rotation of all revolute joints along the kinematic chain.

3.3. Moving Link i to Its Current Position

With link $8c$ in its correct orientation, the only step left is the translation to its current position. The coordinates of the connecting point B_{7c} between link $7c$ and link $8c$ in the xyz global axes were already determined in iteration $7c$ of the algorithm. Since the model of link $8c$ was built around this same connection point in its local reference (A'_{8c}), the global coordinates are obtained by adding the coordinates of point B_{7c} to all previously rotated points P''_{8c} using Equation (6).

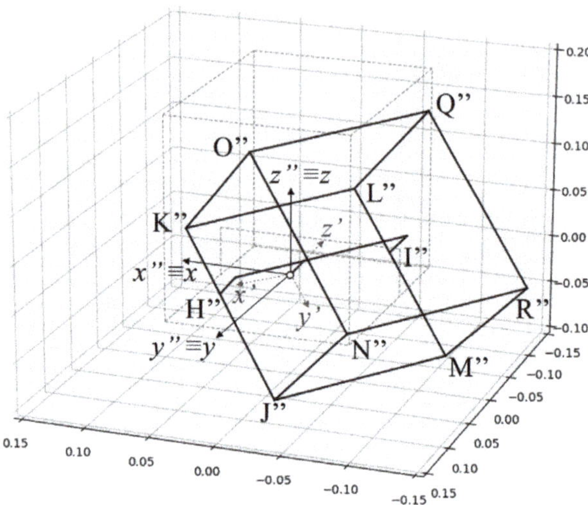

Figure 6. 3D Model of the robot's head in the computational simulation in the rotated $x''_{8c}y''_{8c}z''_{8c}$ axes. The labeling of points A–G was omitted. The non-rotated position of the head is shown in dotted grey lines. With the exception of the axes of the global reference (xyz), all points and axes represented have sub-index ($8c$) but this notation was omitted for simplification.

The coordinates of B_{7c} are used for all P_{8c} points since they all follow the same translation. With this step finished, link $8c$ becomes modeled in its current position with its current orientation, as shown in Figure 7.

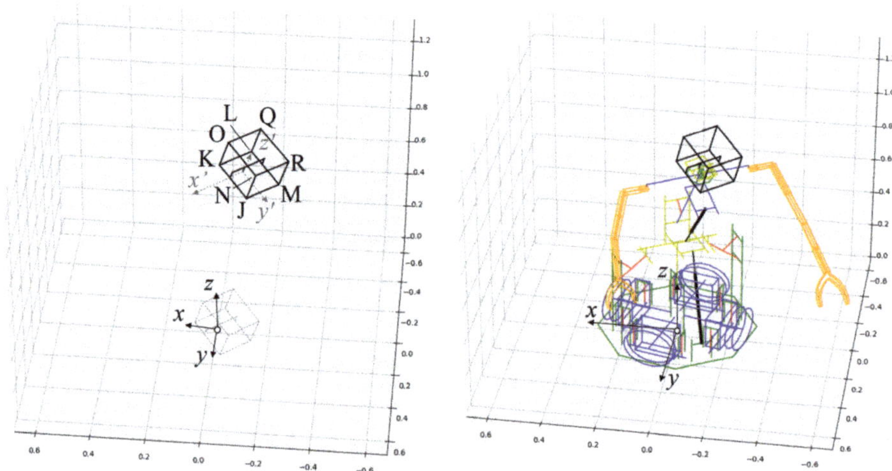

Figure 7. 3D Model of the robot's head in the computational simulation in the xyz global axes. The labeling of points A–I was omitted. The head after rotation, but before translation, is shown in dotted grey lines (**left**). On the (**right**), the head is shown with the whole robot also being visible. With the exception of the axes of the global reference (xyz), all points and axes represented have sub-index ($8c$) but this notation was omitted for simplification.

4. Application of the Algorithm for the CHARMIE Robot

To validate the proposed algorithm, and provide a further understanding of how it is applied to a practical example, this section describes the process of using it to model CHARMIE [3], a human-inspired mobile manipulator robot (Figure 1).

The application of the algorithm to an ongoing project also proved its usefulness by successfully serving as a basis for two studies:

- Develop a 3D environment for multibody dynamic analysis—Using the recursive Newton-Euler algorithm for the inverse dynamics analysis presented in [4], together with the algorithm described in this paper, a 3D multibody dynamic analysis of the CHARMIE robot has been built. This study was fundamental for the mechanical project of the robot and the choice of actuators. Results from this methodology were validated both using benchmarks from literature, and comparing results to other multibody simulation software (WorkingModel4D and CoppeliaSim).
- Analyse and train neural network solutions for trajectory control—The forward kinematic analysis defines the robot's configuration as a function of its joint positions. After defining limits for the joints, and a set of conditions that determine the success and failure of a trajectory generation (example: a collision is a failure, and getting into an intended position is a success), a neural network can be trained to control the joint actuators to find optimal trajectories. By hiding the visual interface of the developed algorithm, high computational efficiency is achieved, optimal for neural network training. By also including contact between bodies (using, for example, the mathematical models in [29]) this method will be used to train CHARMIE for manipulation tasks, such as picking and placing of objects.

In the following subsections, CHARMIE's kinematic structure is described, followed by an explanation of the auxiliary calculations required to both correctly model the robot's forward kinematics, and to extract data regarding the configuration of components that were not modeled directly. This section finishes by presenting a comparative study, made using WorkingModel4D, to validate the obtained results.

4.1. Definition of the Robot's Kinematic Chain

The first step for analyzing CHARMIE was organizing its kinematic structure by defining and naming its links and joints. The result from this process is shown in Figure 8, which schematically represents the kinematics of the robot. The manipulator was modeled as a single serial kinematic chain from the global reference to its upper body and then split into three different serial kinematic chains, one for the left arm (chain a), one for the right arm (chain b), and one for the head (chain c). The end of both arms also splits into the two halves of the end effectors claws. The motion of both halves of each claw is controlled by a single actuator (joint $13a$ and joint $13b$).

Next, all information required to run the algorithm was prepared and computed. This includes the coordinates of the points considered for each link in their local coordinates, and the R_i^{i-1} matrices obtained using Equation (3). This information is listed in Table 2. For simplicity, only the coordinates of points B_i' (enough to model the robot's kinematics) are shown. Since link 5 is connected to three different links, it has three different B_i' associated. Despite being connected to two links, links $12a$ and $12b$ have a single B_i' due to both halves of the end effector having the same origin. Links $13a_1$, $13a_2$, $13b_1$, $13b_2$, and $9c$ have no B_i' because they are end effectors, not connected to any following link.

This information (and the coordinates of the other points to draw each link) was enough to build the model of Figure 1. The only exceptions were Joints 4 and 5 which, due to not being directly actuated, required additional calculations.

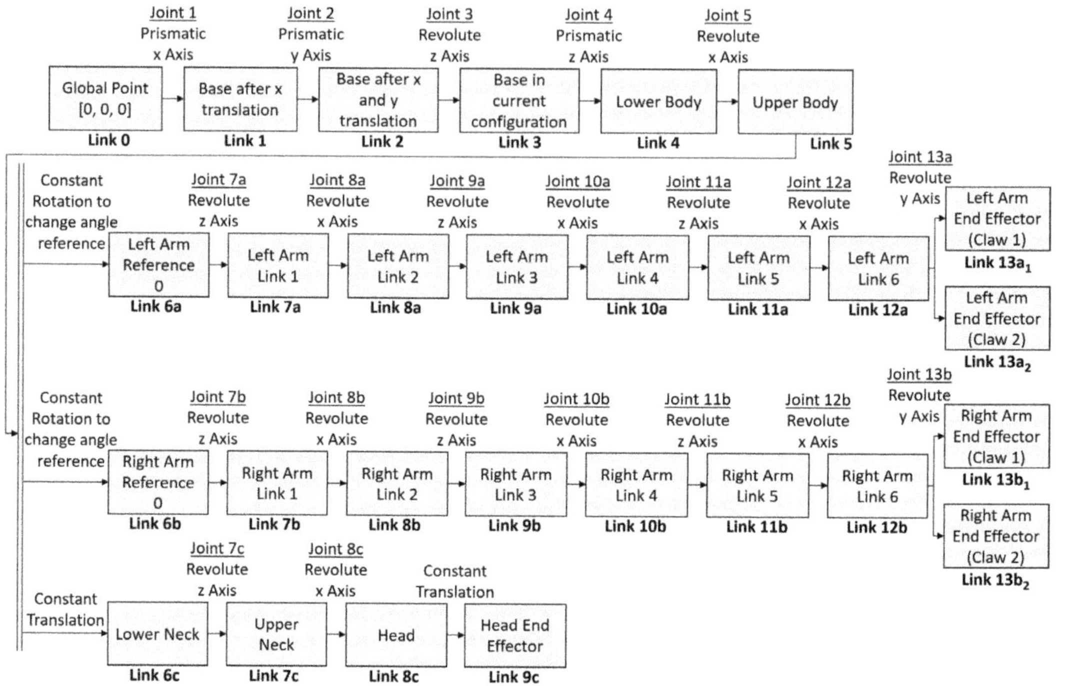

Figure 8. Schematic representation of the kinematic chain of the CHARMIE mobile manipulator.

Table 2. Information regarding the coordinates of the B_i' points (in millimeters) and the R_i^{i-1} rotation matrices for the kinematic model of the CHARMIE robot. Variables marked with * were determined indirectly using the methods described in Section 4.2.

Iteration (i)	B_i'	R_i^{i-1}
0	$[0;0;36.8] + d_1[1;0;0]$	$ZXZEuler(0,0,0)$
1	$[0;0;0] + d_2[0;1;0]$	$ZXZEuler(0,0,0)$
2	$[0;0;0]$	$ZXZEuler(0,0,0)$
3	$[0;0;290] + d_4{}^*[0;0;-1]$	$ZXZEuler(0,0,\theta_3)$
4	$[0;0;434]$	$ZXZEuler(0,0,0)$
5	$\begin{array}{l} B_{5a} = [-200;3;460]; \\ B_{5b} = [200;3;460]; \\ B_{5c} = [0;0;495] \end{array}$	$ZXZEuler(0,\theta_5{}^*,0)$
6a	$[0;0;30]$	$ZXZEuler(\pi/2,-\pi/2,0)$
7a	$[0;0;61.5]$	$ZXZEuler(0,0,\theta_{7a})$
8a	$[0;0;145]$	$ZXZEuler(0,\theta_{8a},0)$
9a	$[0;0;145]$	$ZXZEuler(0,0,\theta_{9a})$
10a	$[0;0;140]$	$ZXZEuler(0,\theta_{10a},0)$
11a	$[0;0;140]$	$ZXZEuler(0,0,\theta_{11a})$
12a	$[0;0;15]$	$ZXZEuler(0,\theta_{12a},0)$
13a_1	X	$ZXZEuler(\pi/2,\theta_{13a},0)$
13a_2	X	$ZXZEuler(\pi/2,-\theta_{13a},0)$
6b	$[0;0;30]$	$ZXZEuler(\pi/2,\pi/2,0)$
7b	$[0;0;61.5]$	$ZXZEuler(0,0,\theta_{7b})$
8b	$[0;0;145]$	$ZXZEuler(0,\theta_{8b},0)$
9b	$[0;0;145]$	$ZXZEuler(0,0,\theta_{9b})$
10b	$[0;0;140]$	$ZXZEuler(0,\theta_{10b},0)$
11b	$[0;0;140]$	$ZXZEuler(0,0,\theta_{11b})$
12b	$[0;0;15]$	$ZXZEuler(0,\theta_{12b},0)$

Table 2. *Cont.*

Iteration (*i*)	B'_i	R^{i-1}_i
$13b_1$	X	$ZXZEuler(\pi/2, -\theta_{13b}, 0)$
$13b_2$	X	$ZXZEuler(\pi/2, \theta_{13b}, 0)$
$6c$	$[0; 0; 46.3]$	$ZXZEuler(0, 0, 0)$
$7c$	$[14.8; -5.3; 18.3]$	$ZXZEuler(0, 0, \theta_{7c})$
$8c$	$[-14.8; 20; 201]$	$ZXZEuler(0, \theta_{8c}, 0)$
$9c$	X	$ZXZEuler(0, 0, 0)$

4.2. Auxiliary Calculations for Complex Joints

The recursive algorithm presented in Section 3 can model any link from known joint values. However, in some practical examples, joints are not directly actuated, and it is necessary to establish the correspondence between the actuator position and the joint value. This happens in two joints of CHARMIE: joint 4, a prismatic joint actuated indirectly by a linear actuator; and joint 5, a revolute joint controlled by a linear actuator. Since these calculations can be applied to similar situations in other robots, they are explained in detail.

4.2.1. Joint 4

Joint 4 of CHARMIE is a prismatic joint indirectly actuated by a linear actuator. The goal is to establish a relation between the length of the linear actuator c_3 and the linear displacement of joint 4 d_4. The relevant geometric features for this calculation are shown in Figure 9.

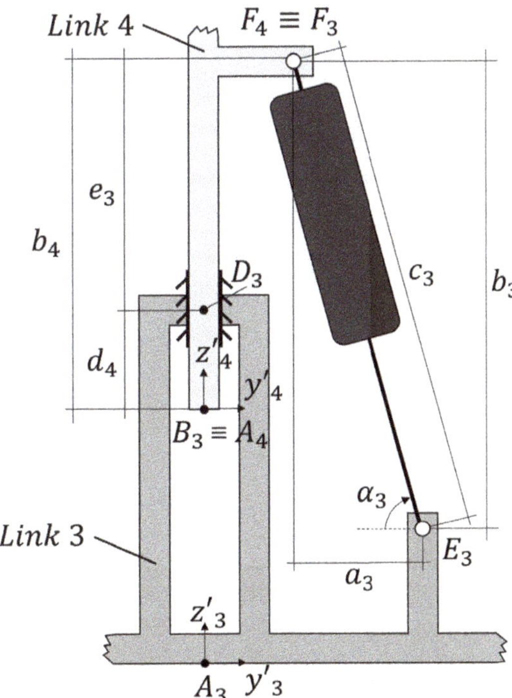

Figure 9. Schematic representation of the geometry of joint 4 between link 3 and link 4 of the CHARMIE robot.

For these calculations, global coordinates from link 4 cannot be used (because at this stage of the algorithm, this link is not modeled yet), but the local coordinates can since they only depend on the link's geometry, which is constant. First, an auxiliary point F_3' was included in link 3, which corresponds to the endpoint of the linear actuator. The y' coordinate of this point is fixed, and can be used to determine a_3 using the expression:

$$a_3 = F_3'y - E_3'y. \tag{7}$$

With this value and the current lenght of the actuator c_3, the auxiliary angle α_3 can be determined:

$$\alpha_3 = acos\left(\frac{a_3}{c_3}\right). \tag{8}$$

This angle allows the height b_3 to be obtained:

$$b_3 = c_3 sin(\alpha_3), \tag{9}$$

which then allows the height of the auxiliary point $F_3'z$ to be determined:

$$F_3'z = E_3'z + b_3. \tag{10}$$

With $F_3'z$ determined, and D_3' being a known and fixed point, e_3 is calculated using:

$$e_3 = F_3'z - D_3'z. \tag{11}$$

The height b_4 is only dependant on the geometry of link 4, and is calculated using:

$$b_4 = F_4'z - A_4'z. \tag{12}$$

It then becomes possible to calculate d_4 using:

$$d_4 = b_4 - e_3. \tag{13}$$

By combining Equations (7)–(13), the relation between d_4 and c_3 is obtained, resulting in equation:

$$d_4 = F_4'z - E_3'z - c_3 sin\left(acos\left(\frac{F_3'y - E_3'y}{c_3}\right)\right) + D_3'z. \tag{14}$$

Point $A_4'z$ was removed from this equation since its value is 0. To allow reproducibility of results, the required geometric values for this step are shown in Table 3.

Table 3. Coordinates (in millimeters) of the relevant points for the calculations of joint 4 in their respective local coordinates.

Point	x	y	z	Point	x	y	z
D_3'	0	0	290	F_3'	0	162.5	-Calculated-
E_3'	0	257.5	99.5	F_4'	0	162.5	383

4.2.2. Joint 5

Joint 5 of CHARMIE is a revolute joint actuated by a linear actuator between links 4 and 5. The goal is to establish the relation between the configuration of the linear actuator c_4 and the rotation of the joint θ_5. The geometric features used in this calculation are represented in Figure 10.

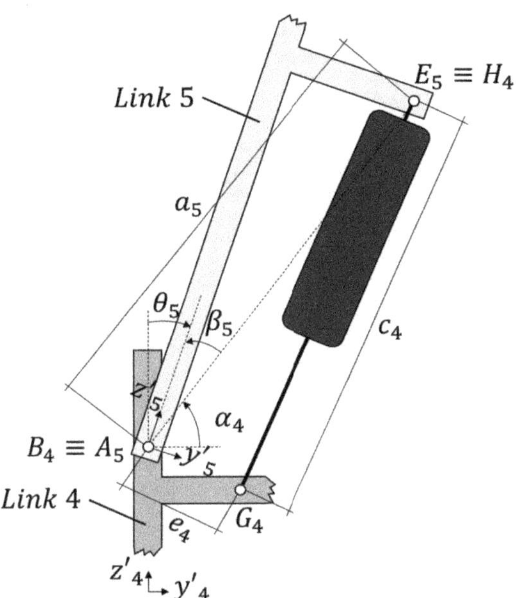

Figure 10. Schematic representation of the geometry of joint 5 between link 4 and link 5 of the CHARMIE robot.

As in the previous example, only local coordinates (related to the parts' geometry) or lengths (constant regardless of the chosen reference) are used. The angle θ_5 can be determined from two auxiliary angles, α_4 and β_5, using the expression:

$$\theta_5 = \alpha_4 + \beta_5 - \pi/2, \tag{15}$$

where α_4 can be calculated using:

$$\alpha_4 = atan\left(\frac{H_4'z - B_4'z}{H_4'y - B_4'y}\right), \tag{16}$$

and β_5 using the expression:

$$\beta_5 = atan\left(\frac{E_5'y - A_5'y}{E_5'z - A_5'z}\right). \tag{17}$$

The points required for obtaining β_5 are only dependant on the geometry of link 5, therefore, they can be determined directly. However, α_4 requires the coordinates of point H_4', which are dependant on the geometry of link 4, the geometry of link 5, and the length of the actuator c_4.

The coordinates of H_4' were calculated as the intersection between two circumferences, one with its center on point G_4' and radius c_4, and the other centered on point B_4' with radius a_5. Radius c_4 is obtained from the actuator length, and a_5 from the expression:

$$a_5 = \sqrt{(E_5'y - A_5'y)^2 + (E_5'z - A_5'z)^2}. \tag{18}$$

To calculate the intersection between two circumference, the distance between their centers e_4 is also needed, given by the expression:

$$e_4 = \sqrt{(B_4'y - G_4'y)^2 + (B_4'z - G_4'z)^2}. \tag{19}$$

To facilitate the formulation of the intersection, two auxiliary mathematical parameters were used. The first is l_4, given by the expression:

$$l_4 = \frac{c_4{}^2 - a_5{}^2 + e_4{}^2}{2e_4},$$ (20)

and the second is h_4, calculated from the expression:

$$h_4 = c_4{}^2 - l_4{}^2,$$ (21)

The coordinates of H_4' can then be determined using the pair of equations:

$$H_4'y = \frac{l_4}{e_4}(B_4'y - G_4'y) + \frac{h_4}{e_4}(B_4'z - G_4'z) + G_4'y,$$ (22)

$$H_4'z = \frac{l_4}{e_4}(B_4'z - G_4'z) + \frac{h_4}{e_4}(B_4'y - G_4'y) + G_4'z.$$ (23)

By combining Equations (15)–(23), the expression establishing the relation between θ_5 and e_4 is obtained. This expression is not be presented in its extended form due to its size, which would hinder its comprehension.

To allow the results to be reproducible, Table 4 shows the coordinates of all geometric features used for this calculation.

Table 4. Coordinates (in millimeters) of the relevant points for the calculations of joint 5 in their respective local coordinates.

Point	x	y	z	Point	x	y	z	Point	x	y	z
B_4'	0	0	434	G_4'	0	87.5	383	E_5'	0	122.5	380

4.3. Auxiliary Calculations to Extract Relevant Data from the Model

After the model is obtained, a set of data not modeled directly can be obtained from it.

One example for this, in the CHARMIE robot, is the configuration of the springs included between links 4 and 5. The connection points between these springs, S_4' and S_5', are placed on the local coordinates, but the global configuration of the spring is not explicitly defined. After the recursive algorithm runs, the spring length l_s can be easily obtained by calculating the distance between the two connection points S_4 and S_5 in the global coordinate reference, as shown in the following equation:

$$l_s = \sqrt{(S_5x - S_4x)^2 + (S_5y - S_4y)^2 + (S_5z - S_4z)^2}.$$ (24)

Any angle α_s can also be calculated by defining a triangle using three distances: l_s; a_s; b_s. The value of these distances can be determining using Equation (24), and the value of the angle is then calculated by applying the law of cossines in the form:

$$\alpha_s = \left(\frac{l_s^2 + b_s^2 - a_s^2}{2l_sb_s}\right),$$ (25)

where a_s is the opposite side of the triangle to the internal angle being calculated.

Using integration, it becomes possible to evaluate problems over time. A possible method for obtaining the angular velocity, angular acceleration, and linear acceleration of all links is to use the forward iterations of the recursive algorithm from chapter 7.5.2 of [4]. The inputs of this method are the information obtained in this algorithm (rotation matrices and coordinates of points A_i', B_i' and C_i'), and the known inputs from the forward kinematics analysis (joint orientation, position, velocity, and acceleration).

4.4. Validation of Results

To validate the developed algorithm, the CHARMIE robot was also modeled in the multibody simulation software WorkingModel4D. The same problem, a simple motion that requires the movement of all joints, was solved using both methods to compare and verify the results.

All actuators started with a position/angle of 0, except for d_4 and θ_5, actuated by the linear actuators c_3 and c_4, which started in the positions $c_3 = 400$ mm and $c_4 = 350$ mm. All joints were defined with a constant velocity. The angular velocity considered for the directly actuated revolute joints was $\pi/36$ rad/s, and the linear velocity for the linear actuators 5 mm/s. The three exceptions were joints θ_{7c} and θ_{8c}, both with an angular velocity of $\pi/72$ rad/s, and the linear actuator c_4, with a linear velocity of 2.5 mm/s. To analyse a single set of coordinates, the claw end effectors were replaced with a point at their center— this corresponds to changing Table 2 by: removing rotations θ_{13a} and θ_{13b}; and changing the value of B'_{12a} and B'_{12b} to $[0; 0; 140]$. The robot was simulated under these conditions for a period of 20 s. Figure 11 shows a side-by-side comparison of frames captured from both simulation environments, and Figure 12 compares the extracted coordinates of the center of the left arm end effector (link 13a).

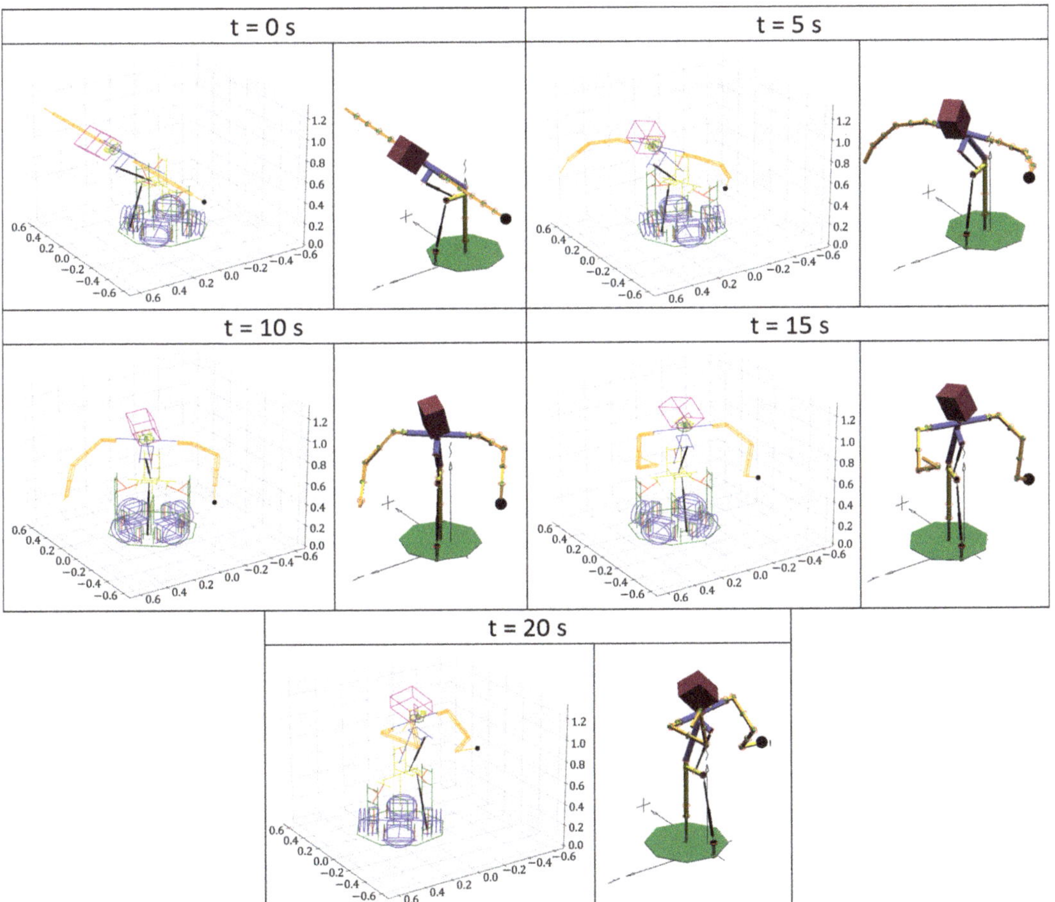

Figure 11. Comparison of the robot's configuration for the same problem analysed using the recursive algorithm presented in this paper (**left**), and WorkingModel4D (**right**).

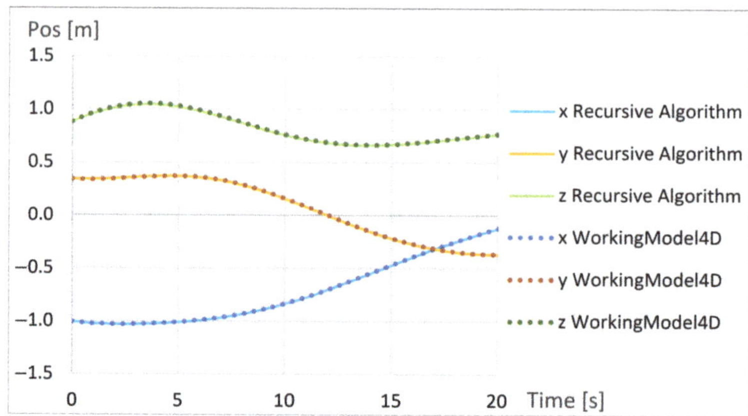

Figure 12. Comparison between the results for the end-effector position 13*a* for the same conditions analysed using both the recursive algorithm and WorkingModel4D.

The figures illustrate that the robot's behavior was identical in both environments. The results from the extracted data are indistinguishable, with a complete overlap between the graphic's lines. This data compares positions directly, not orientations (due to the many different notations available to define them), however, similar results for the positions could only be obtained with the correct definition of all the angles in the kinematic chain leading up to link 13*a*. The similar configuration of both robots, in Figure 11, further proves the cohesion between the orientations defined in the two methods.

With these two unrelated sources providing the same results for the robot's forward kinematics, the validity of the proposed algorithm is proven.

5. Conclusions

In this paper, a simple, intuitive, and theoretically computational efficient three-step recursive algorithm is presented to analyze the kinematics of robotic manipulators. From known joint values, the algorithm fully defines the 3D configuration of a kinematic chain, returning the orientation and position of each link described by rotation matrices and Cartesian coordinates. This methodology was further explained by successfully applying it to model CHARMIE, a mobile manipulator with a relatively high degree of complexity. The results were validated by comparing them with the multibody dynamics software WorkingModel4D.

All input data required to define the given example is provided with full transparency, guaranteeing its replicability. This both exposes the results to external validation and allows the presented problem to serve as a benchmark for 3D kinematic analysis.

After using this methodology, the following key advantages were identified:

- Modularity—Due to its recursive structure, models constructed with the proposed algorithm have high modularity. Entire sections in the middle of the kinematic chain can be altered with minimal effort, simplifying any changes made to both links and joints.
- Speed of implementation—After the initial equations are defined, this method allows a fast modeling of different kinematics structures. For reference, each of the models shown in Figure 3 was made in approximately four hours (including the modeling of each link in local coordinates and defining the 3D drawing of the visual interface).
- Full definition of the link's models—The algorithm contains information regarding the geometry of each of the links. This can be used for studies such as collision detection.
- Versatility—As previously stated, the implementation of this algorithm is less rigid then the Denavit–Hartenberg parameters, freeing the choice of axes orientation for each of the links. Any rotation and translation can be programmed between each of

the links, allowing the definition of any type of joint. More complex mechanisms can be analysed using a process homologous to example (Section 4.2).

- Compatibility—This algorithm outputs: the Cartesian coordinates of each point in each link in local coordinates; the Cartesian coordinates of each point in each link in global coordinates; the orientation between each pair of successive links in the form of rotation matrices; the orientation of each links in relation to the global reference in the form of rotation matrices. This information can be inputted into a set of already defined algorithms, such as the aforementioned example of inverse multibody dynamic analysis [4].

With the increasing interest in robotics, the presented method can become a powerful tool for computational analysis. It can be used to define a wide array of robots in three dimensions, and the resulting models can act as a starting point for a set of more advanced studies. In the CHARMIE project, this algorithm already functions as the basis for a multibody dynamic analysis environment and will be used for neural network training.

Theory corroborates that this algorithm increases computational efficiency in comparison to using the Denavit–Hartenberg parameters [2], however, a relevant future study would be running a set of tests to prove and quantify this improvement.

Author Contributions: Conceptualization, F.G. and G.L.; methodology, F.G.; software, F.G. and T.R.; validation, A.F.R., G.L. and P.F.; formal analysis, F.G. and T.R.; investigation, F.G. and T.R.; resources, A.F.R., G.L. and P.F.; data curation, F.G. and T.R.; writing—original draft preparation, F.G.; writing—review and editing, T.R., A.F.R., G.L. and P.F.; visualization, F.G.; supervision, A.F.R., G.L. and P.F.; project administration, G.L. and P.F.; funding acquisition, F.G., T.R., G.L. and P.F. All authors have read and agreed to the published version of the manuscript.

Funding: This work has been supported by FCT—Fundação para a Ciência e Tecnologia within the R&D Units Project Scope: UIDB/00319/2020. The author F.G. received funding through a doctoral scholarship from the Portuguese Foundation for Science and Technology (Fundação para a Ciência e a Tecnologia) [grant number SFRH/BD/145993/2019], with funds from the Portuguese Ministry of Science, Technology and Higher Education and the European Social Fund through the Programa Operacional do Capital Humano (POCH). The author T.R. received funding through a doctoral scholarship from the Portuguese Foundation for Science and Technology (Fundação para a Ciência e a Tecnologia) [grant number SFRH/BD/06944/2020], with funds from the Portuguese Ministry of Science, Technology and Higher Education and the European Social Fund through the Programa Operacional do Capital Humano (POCH).

Institutional Review Board Statement: Not applicable.

Informed Consent Statement: Not applicable.

Data Availability Statement: This study presents the data required to replicate its results.

Acknowledgments: This work has been supported by the CMEMS Research Centre and the Laboratory of Automation and Robotics (LAR) of University of Minho.

Conflicts of Interest: The authors declare no conflict of interest. The funders had no role in the design of the study; in the collection, analyses, or interpretation of data; in the writing of the manuscript, or in the decision to publish the results.

References

1. Rocha, C.R.Ã.; Tonetto, C.P.; Dias, A. Robotics and Computer-Integrated Manufacturing A comparison between the Denavit-Hartenberg and the screw-based methods used in kinematic modeling of robot manipulators. *Robot. Comput. Integr. Manuf.* **2011**, *27*, 723–728. [CrossRef]
2. Waldron, K.; Schmiedeler, J. Kinematics. In *Springer Handbook of Robotics*; Springer: Berlin/Heidelberg, Germany, 2008; pp. 9–33. [CrossRef]
3. Ribeiro, T.; Gonçalves, F.; Garcia, I.S.; Lopes, G.; Ribeiro, A.F. CHARMIE: A Collaborative Healthcare and Home Service and Assistant Robot for Elderly Care. *Appl. Sci.* **2021**, *11*, 7248. [CrossRef]
4. Siciliano, B.; Sciavicco, L.; Villani, L.; Oriolo, G. *Robotics Modelling, Planning and Control*, 1st ed.; Springer: London, UK, 2009; p. 632. [CrossRef]

5. Aye, M.M.M. Analysis of Euler angles in a simple two-axis gimbals set. *World Acad. Sci. Eng. Technol.* **2011**, *81*, 389–394. [CrossRef]
6. Corke, P. A Simple and Systematic Approach to Assigning Denavit-Hartenberg Parameters. *IEEE Trans. Robot.* **2007**, *23*, 590–594. [CrossRef]
7. Ding, F.; Liu, C. Applying coordinate fixed Denavit-Hartenberg method to solve the workspace of drilling robot arm. *Int. J. Adv. Robot. Syst.* **2018**, *15* , 1729881418793283. [CrossRef]
8. Thomas, M.J.; Sanjeev, M.M.; Sudheer, A.P.; Joy, M.L. Comparative study of various machine learning algorithms and Denavit-Hartenberg approach for the inverse kinematic solutions in a 3- PP SS parallel manipulator. *Ind. Robot. Int. J. Robot. Res. Appl.* **2020**, *47*, 683–695. [CrossRef]
9. Klug, C.; Schmalstieg, D.; Gloor, T.; Arth, C. A Complete Workflow for Automatic Forward Kinematics Model Extraction of Robotic Total Stations Using the Denavit-Hartenberg Convention. *J. Intell. Robot. Syst.* **2019**, *95*, 311–329. [CrossRef]
10. Hemingway, E.G.; Reilly, O.M.O. Perspectives on Euler angle singularities, gimbal lock, and the orthogonality of applied forces and applied moments. *Multibody Syst. Dyn.* **2018**, *44*, 31–56. [CrossRef]
11. Perrusquía, A.; Yu, W. Human-in-the-Loop Control Using Euler Angles. *J. Intell. Robot. Syst.* **2020**, *97*, 271–285. [CrossRef]
12. Ran, C.; Cheng, X. A Direct and Non-Singular UKF Approach Using Euler Angle Kinematics for Integrated Navigation Systems. *Sensors* **2016**, *16*, 1415. [CrossRef]
13. Schappler, M.; Tappe, S.; Ortmaier, T. Resolution of Functional Redundancy for 3T2R Robot Tasks using Two Sets of Reciprocal Euler Angles. In *Advances in Mechanism and Machine Science*; Uhl, T., Ed.; Springer International Publishing: Cham, Switzerland, 2019; pp. 1701–1710.
14. Goldman, R. Understanding quaternions. *Graph. Model.* **2011**, *73*, 21–49. [CrossRef]
15. Hamilton, W.R. *Elements of Quaternions*; Cambridge Library Collection—Mathematics, Cambridge University Press: Cambridge, UK, 2010; [CrossRef]
16. Kuipers, J.B. *Quaternions and Rotation Sequences: A Primer with Applications to Orbits, Aerospace and Virtual Reality*; Princeton University Press: Princeton, NJ, USA, 1999.
17. Sarabandi, S.; Thomas, F. A Survey on the Computation of Quaternions From Rotation Matrices. *J. Mech. Robot.* **2019**, *11*, 021006. [CrossRef]
18. Valverde, A.; Tsiotras, P. Spacecraft Robot Kinematics Using Dual Quaternions. *Robotics* **2018**, *7*, 64. [CrossRef]
19. Zupan, E.; Saje, M.; Zupan, D. Dynamics of spatial beams in quaternion description based on the Newmark integration scheme. *Comput. Mech.* **2013**, *51*, 47–64. [CrossRef]
20. Huang, Z.; Li, Q.; Ding, H. Basics of Screw Theory. In *Theory of Parallel Mechanisms*; Springer: Dordrecht, The Netherlands, 2013; pp. 1–16. [CrossRef]
21. Kecskeméthy, A. Lie-Group First-Order Operations in Rigid-Body Kinematics. In *On Advances in Robot Kinematics*; Springer: Dordrecht, The Netherlands, 2004; pp. 57–66. [CrossRef]
22. Müller, A. Screw and Lie group theory in multibody kinematics. *Multibody Syst. Dyn.* **2018**, *43*, 37–70. [CrossRef]
23. Wu, A.; Shi, Z.; Li, Y.; Wu, M.; Guan, Y.; Zhang, J.; Wei, H. Formal Kinematic Analysis of a General 6R Manipulator Using the Screw Theory. *Math. Probl. Eng.* **2015**, *2015*, 1–7. [CrossRef]
24. Chen, Q.; Zhu, S.; Zhang, X. Improved Inverse Kinematics Algorithm Using Screw Theory for a Six-DOF Robot Manipulator. *Int. J. Adv. Robot. Syst.* **2015**, *12*, 140. [CrossRef]
25. Su, H.J.; Dorozhkin, D.V.; Vance, J.M. A Screw Theory Approach for the Conceptual Design of Flexible Joints for Compliant Mechanisms. *J. Mech. Robot.* **2009**, *1*, 41001–41009. [CrossRef]
26. Biswal, P.; Mohanty, P.K. Development of quadruped walking robots: A review. *Ain Shams Eng. J.* **2021**, *12*, 2017–2031. [CrossRef]
27. Li, B.; Tian, W.; Zhang, C.; Hua, F.; Cui, G.; Li, Y. Positioning error compensation of an industrial robot using neural networks and experimental study. *Chin. J. Aeronaut.* **2021**, *35*, 346–360. [CrossRef]
28. Deepa, T.; Angalaeswari, S.; Subbulekshmi, D.; Krithiga, S.; Sujeeth, S.; Kathiravan, R. Design and implementation of bio inspired hexapod for exploration applications. *Mater. Today Proc.* **2021**, *37*, 1603–1607. [CrossRef]
29. Flores, P.; Lankarani, H.M. Contact Force Models for Multibody Dynamics. In *Solid Mechanics and Its Applications*; Springer International Publishing: Cham, Switzerland, 2016; Volume 226. [CrossRef]

Article

Evaluation Criteria for Trajectories of Robotic Arms

Michal Dobiš [1,*], Martin Dekan [1], Peter Beňo [2], František Duchoň [1] and Andrej Babinec [1]

1   Institute of Robotics and Cybernetics, Slovak University of Technology in Bratislava, 812 19 Bratislava, Slovakia; martin.dekan@stuba.sk (M.D.); frantisek.duchon@stuba.sk (F.D.); andrej.babinec@stuba.sk (A.B.)

2   Department of Robot Applications, Photoneo s.r.o. Company, 821 09 Bratislava, Slovakia; beno@photoneo.com

*   Correspondence: michal.dobis@stuba.sk

Abstract: This paper presents a complex trajectory evaluation framework with a high potential for use in many industrial applications. The framework focuses on the evaluation of robotic arm trajectories containing only robot states defined in joint space without any time parametrization (velocities or accelerations). The solution presented in this article consists of multiple criteria, mainly based on well-known trajectory metrics. These were slightly modified to allow their application to this type of trajectory. Our framework provides the methodology on how to accurately compare paths generated by randomized-based path planners, with respect to the numerous industrial optimization criteria. Therefore, the selection of the optimal path planner or its configuration for specific applications is much easier. The designed criteria were thoroughly experimentally evaluated using a real industrial robot. The results of these experiments confirmed the correlation between the predicted robot behavior and the behavior of the robot during the trajectory execution.

Keywords: trajectory optimization criteria; robotic arms; trajectory evaluation; path planning

Citation: Dobiš, M.; Dekan, M.; Beňo, P.; Duchoň, F.; Babinec, A. Evaluation Criteria for Trajectories of Robotic Arms. *Robotics* **2022**, *11*, 29. https://doi.org/10.3390/robotics11010029

Academic Editors: António Paulo Moreira, Félix Vilariño and Pedro Neto

Received: 17 January 2022
Accepted: 11 February 2022
Published: 15 February 2022

Publisher's Note: MDPI stays neutral with regard to jurisdictional claims in published maps and institutional affiliations.

1. Introduction

In many modern applications of robotic arms, there is a need to compute collision-free trajectories with variable start and goal positions. An example of such application is the problem of bin-picking, where the main goal is to pick randomly placed objects from within the bin. This problem is described in the literature as a problem of dynamic path planning. Several solutions to this problem have been proposed and applied over the past decade.

Many of these popular methods are based on randomized sampling, such as probabilistic roadmap methods (PRM) [1] or rapidly random trees (RRT) [2]. Another approach to path planning is using optimization-based methods. Algorithms such as covariant Hamiltonian optimization for motion planning (CHOMP) [3] or stochastic trajectory optimization for path planning (STOMP) [4] belong to this group. Most of the methods utilize random exploration of configuration space to speed up the computation. The disadvantage of such algorithms is their non-deterministic behavior. The repeated path planning computation from the same start state to the same goal state could yield different results. Furthermore, attributes of the calculated paths depend dramatically on the chosen path planning algorithm and its configuration. The trajectories can differ in the distance traveled by the endpoint of the robotic arm, energy consumption, or maximal joint accelerations achieved during the motion execution. Several metrics were proposed for the measurement of difference between two robot states.

Many path planner comparisons and path planner articles for specific applications evaluate mainly planning time and success rate. In their paper [5], Rodriguez and Suarez (2015) provide a comparison of sampling-based algorithms, which is based on planning time, success rate, and number of samples. Likewise, Paulin et al. (2015) [6] compare path planners for grape vine pruning application, but the trajectory quality is not considered. Magyar et al. (2019) proposed a modified version of the STOMP algorithm called

GSTOMP [7] and use the term "smoothness", which represents a cumulative function of the end effector's linear and angular accelerations. Similarly, in their study [8] De Maeyer and Demeester (2021) are focused on benchmarking path planning for robotic arc welding where path length is computed as the cumulative sum of joint differences. On the other hand, Larsen et al. (2017) [9] used the Euclidean distance of a tool center point for the path length calculation. In all of these algorithms factors/characteristics such as tool orientation, acceleration or jerks are neglected. In our previous research [10], we attempted to compare the quality of paths computed by various path planners and their configurations. However, we were not able to find any common framework for evaluation of the quality of the path.

It is hard to define what is the best universal trajectory, but it is possible to determine which trajectory is better with respect to a selected criterion or their combinations for specific setup and production type. It makes little sense to talk of the best universal trajectory, however it is possible to determine which trajectory is best suited depending on a selected criterion (or a combination of several criteria) for a specific setup and production type. Optimization typically aims to minimize the production cycle time, based on the robot's velocity and the total trajectory length. Other optimization criteria could be maximizing the durability of the robot gearing or decreasing energy consumption. [11]

In this article we formulate a complex trajectory evaluation framework consisting of multiple criteria. The objective of this article is not to provide a comparison of path planning algorithms, but to provide a methodology to compare the resulting paths. For the purpose of this article, the result of a path planner is a path defined as a set of robot states in the robot joint space.

In order to ensure a smooth motion execution, the robot control system needs to post-process the given path and execute its approximation. Therefore, it will not traverse all the positions precisely. Some of our criteria try to predict how the robot will behave during the execution. Therefore, in experimental evaluation of our framework, we tried to find the correlation between the planned and performed trajectory for each criterion.

The evaluation criteria are fully described in Section 2 "Criteria" and they are validated in Section 3 "Validation of criteria" with data measured using a real robot. These measurements are also used for evaluation of the aforementioned hypotheses about prediction of path execution, where the measured and predicted data are compared. Section 4 "Usage of criteria" provides criteria examples and an introduction on how to use them. The impact of environment complexity and the position of the bin were investigated and analyzed as well. Finally, Section 5, Results, summarizes and recapitulates the findings reported in previous sections.

2. Criteria

The first trivial criterion often used by manufacturers is the Cartesian distance of tool center point (TCP) between each waypoint of the path. This means that the sum of the Euclidean distance between each consecutive TCP position in R^3 should be minimal. In many cases programmers try to optimize the robotic program in order to reduce the production cycle time [12] in specific applications like spot welding [13], where the TCP path must be minimal. In the joint space, we could define the joint distance criterion, which is very closely related to the Cartesian distance criterion. The joint distance uses an accumulated summary of differences for each joint value. An interesting application is collaborative robotics, because robots must indicate appropriate proxemic behavior, and unnecessarily long and nonsensical trajectories could evoke fear [14].

$$x = \sum_{i=2}^{n} \sum_{j=1}^{m} ||q_{i,j} - q_{i-1,j}|| \tag{1}$$

$$x = \sum_{i=2}^{n} ||p_i - p_{i-1}|| \tag{2}$$

The score of the joint distance criterion is calculated using (1), where $q_{i,j}$ is joint value, i is waypoint ID of n waypoint, and j is the joint ID of m robot joints. Equation (2) is used for the calculation of the Cartesian distance score, where instead of the accumulated sum of joint values, the tool point position p_i is used.

The Cartesian distance does not include orientation, although in specific types of production the tool point orientation is fundamental e.g., in grinding, cutting, milling, or painting [15,16]. In the cases of object handling with pneumatic or magnetic grippers, picked objects could be lost during a move. Another interesting and practical example is object handling in the human environment [17,18] or liquid handling [19]. It follows then that the orientation change is the next crucial criterion. It is defined by (3), where q_{xi}, q_{yi}, q_{zi}, q_{wi} are values of tool point quaternion on i waypoints.

$$x = \sum_{i=2}^{n} \cos^{-1}\left(q_{xi} * q_{xi-1} + q_{yi} * q_{yi-1} + q_{zi} * q_{zi-1} + q_{wi} * q_{wi-1}\right) \tag{3}$$

The joint distance criterion could be modified by weights, which multiply each joint value, thus achieving optimization of selected joints [20]. In this article, we call this criterion control pseudo-cost.

$$x = \sum_{i=2}^{n} \sum_{j=1}^{m} w_j \left(||q_{i,j} - q_{i-1,j}||\right) \tag{4}$$

The definition (4) of this feature is similar to joint distance, with the added multiplication by w_j which represents the weight in the range between 0 and 1. This could simplify energy consumption in the application, where the actual energy consumption often remains unknown before the trajectory is executed [21,22]. The computation of the energy for each robot link is dependent on weight and inertia. A great example is an arm on a mobile platform where the power is limited by a battery [23]. There are many approaches to trajectory planning for energy efficiency [24,25], where this criterion can prove useful for comparison and evaluation of planned paths.

In addition to control pseudo-cost, in [20] the robot displacement metric is mentioned that could be used as a relevant criterion. It could also be applicable in collaborative robotics, similarly as the joint distance and the Cartesian distance criteria. Definition of criterion by [20]: for any two configurations q_1 and q_2, a robot displacement metric is defined as (5),

$$x = \sum_{i=2}^{n} \max_{a \in A}\{||a(q_i) - a(q_{i-1})||\} \tag{5}$$

In which $a(q_i)$ is the position of the point a in the world when the robot A is at configuration q_i.

The aforementioned criteria are solely focused on robot states and positions, but do not include any robot dynamics. Industrial programmers often teach robots smooth trajectories without any high jerks, which have the potential to damage mechanical units or stop movement. Less jerk reduces the steadfastness and increases the robot's durability, which can economize production costs. Jerk is defined as the third time-derivation of position, which is applied in research for optimal time-jerk trajectory planning, as shown in Huang et al. (2017) [26]. However our current research does not include any time-parametrization of trajectories, therefore we propose four criteria hypotheses that focus on jerk and are calculated solely using robot positions. One of these hypotheses is the joint jerk, defined as a summary of jerks on each joint (6):

$$x = \sum_{i=2}^{n} \sum_{j=1}^{m} ||q_{i,j} - q_{i-1,j}|| di^3 \tag{6}$$

Similar to the above, the next criterion: the Cartesian jerk; is also defined as the third derivation of the tool point position, and its score equals the sum of all jerks between each waypoint (7):

$$x = \sum_{i=2}^{n} ||p_i - p_{i-1}|| di^3 \tag{7}$$

The criterion is important in applications requiring gripper load, where higher jerk on the load is undesirable. Consider situations such as using soft robotic grippers [27] or a robot handling fragile materials or food [28].

Alternatively, another approach could be employed by finding the maximal jerk between waypoints, which could be applicable as a safety threshold. Therefore, this is defined as joint max jerk in the joint space, as opposed to Cartesian max jerk in the Cartesian space of the tool [29,30]. Joint max jerk is defined by (8) and Cartesian max jerk by (9) below:

$$x = \max_{q_i \epsilon Q} \left\{ \sum_{j=1}^{m} ||q_{i,j} - q_{i-1,j}|| di^3 \right\} \tag{8}$$

In which q_i is a joint state on i waypoint, and Q represents all joint states of a path.

$$x = \max_{p_i \epsilon P} \left\{ \sum_{j=1}^{m} ||p_i - p_{i-1}|| di^3 \right\} \tag{9}$$

In which p_i is tool position on i waypoint, and P represents all tool positions of a path.

3. Validation of Criteria

It is essential to verify the validity of the stated criteria as well as that of the hypothesis that criteria could predicate trajectory execution. The verification was carried out by planning and executing the trajectories using a real robot. Firstly, the paths are planned by the STOMP algorithm using FCL (Flexible Collision Library) [31]. The selection of the path planner is not relevant in this section and, therefore, the specific parameters of the path planner are not presented. Subsequently, these planned paths are executed on a real KUKA KR120 robot with gripper (see in Figure 1), controlled by KRC4. As the robot performs the motion measurements on positions and energy consumption are sampled in time and collected.

Figure 1. Kuka KR120 with gripper, used to executed planned paths in a series of criteria-validating experiments.

For the purpose of criteria validation an experiment was designed, where the robot moves from the start state [0.455, −1.506, 1.885, 0.0, 1.17, 0] to the goal state [−0.498, −1.543, 1.924, 0.0, 1.17, −1.061] with an obstacle added to its path. The path is illustrated in Figure 2. The dataset consists of 30 different paths between the start and the goal states, generated by the STOMP algorithm using FCL collision checking [31].

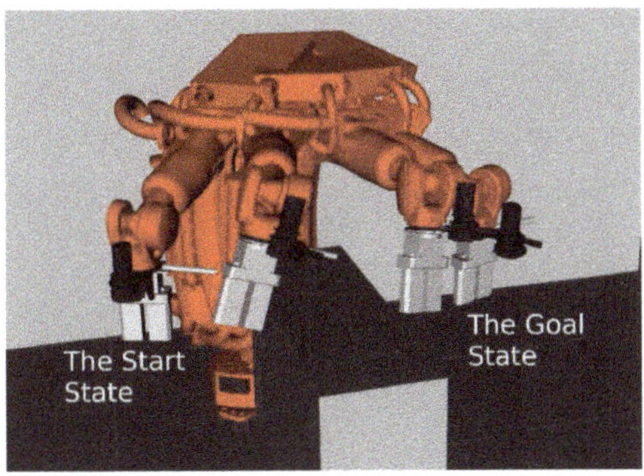

Figure 2. The path between the start state and the goal state.

The first four criteria joint distance, Cartesian distance, orientation change and robot displacement are directly related to the position of the robot in the joint space and of the Cartesian space of its tool point. As for the next set of criteria—joint jerk, joint max jerk, Cartesian jerk, Cartesian max jerk, and control pseudo-cost—these are evaluated separately, as they require the use of specific measurement procedures. The equations for criteria were applied to the entire dataset of planned paths. To provide a clear comparison, they were applied to the joint and Cartesian positions measured during planned path executions as well. Figure 3 shows the differences between planned and executed positions for all paths in the form of a histogram.

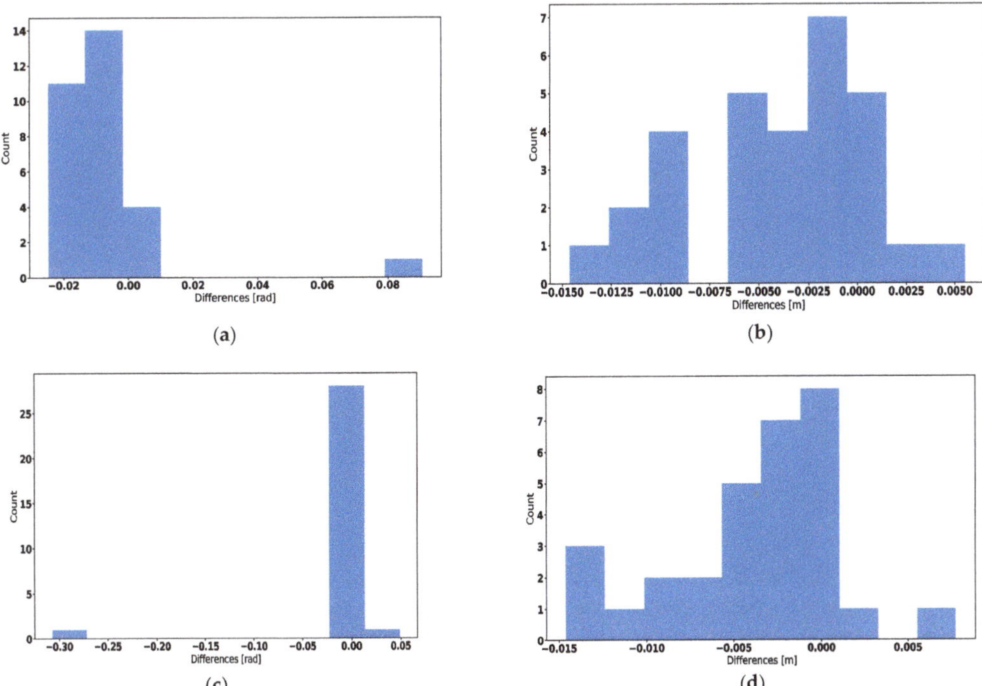

Figure 3. Histograms of differences between planned paths and executed trajectories, compared for the following criteria: (**a**) joint distance; (**b**) Cartesian distance; (**c**) orientation change; (**d**) robot displacement.

For further illustration, the averages and variances in Table 1 show that discrepancies between planned and executed trajectories are correlated.

Table 1. The average and the variance differences between planned path and executed trajectory.

Criterions	Average	Variance
Joint Distance [rad]	-5.75×10^{-3}	3.76×10^{-4}
Cartesian Distance [m]	-4.09×10^{-3}	2.22×10^{-5}
Orientation Change [rad]	-1.24×10^{-2}	3.09×10^{-3}
Robot Displacement [m]	-3.75×10^{-3}	2.52×10^{-5}

3.1. Control Pseudo-Cost

Energy consumption is an excellent example of where the control pseudo-cost criterion can play a part. It could simplify dynamic robot parameters for each arm joint, or these parameters could be replaced by criterion weights as seen in (4). However, the problem is that the weights are unknown, as they need to be calculated from the dynamic robot model. Unfortunately, not all robot vendors provide parameters such as weights, center of gravity, or an inertial matrix for each robot link. A further difficulty arises from the fact that the parameters in the whole robot workspace are not constant. They need be variable, depending on the type of movement, load, etc.

It is essential to ensure that the measuring process is not compromised by systematic error. Therefore, the robot is allowed to "warm up" as its energy consumption tends to be higher immediately after start-up. (The decreased energy consumption is illustrated in Figure 4). Variable velocity and acceleration could also affect the measurement adversely, therefore these and other robot parameters are kept constant.

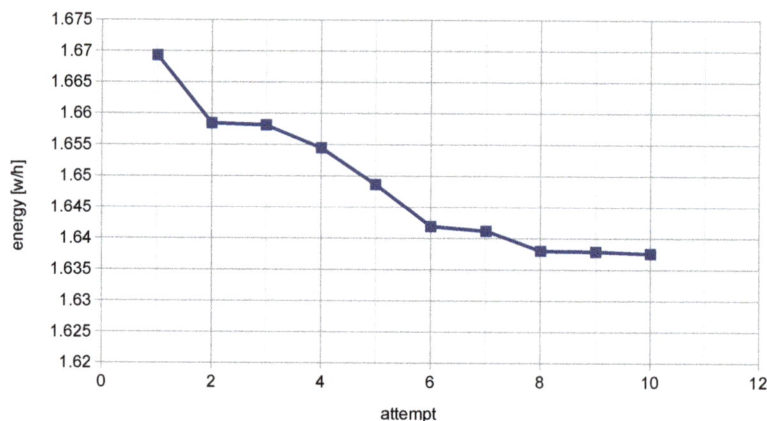

Figure 4. Energy consumption after a cold start of the robot.

The KUKA KRC4 controller allows for the measurement of consumed energy during a single trajectory execution. These measured values are used as a target for our optimization. As we do not possess all the robot parameters for the calculation of the criterion constants, we instead implement some simplifications. We employ the brute force method to determine the constants calculated on a specific set of paths, meaning that the paths will be planned between the same start and goal state. In our case, they are the positions defined in the introduction of the section "Validation of Criteria" (Figure 2). The robot will perform a "horizontal" movement (movement of the TCP of the robot will approximately correspond to the y-axis), and in this section, the experimental setup is denoted as no. 1.

Our working hypothesis is that the calculated constants will be suitable for use in similar robot motion. To prove this hypothesis, four experiments were prepared, where the first three are identical trajectories no.2 (Figure 5a "Longer" trajectories), no.3 (Figure 5b "Shorter" trajectories), and no.4 (Figure 5c TCP is further on the *x*-axis). The last trajectory, no.5. (Figure 5d "vertical" trajectory), is of a different type. A deviation between the actual energy consumption and criterion validity is increased, because the robot executes a different set of trajectories ("vertical movement" the TCP of the robot is moving along the *z*-axis).

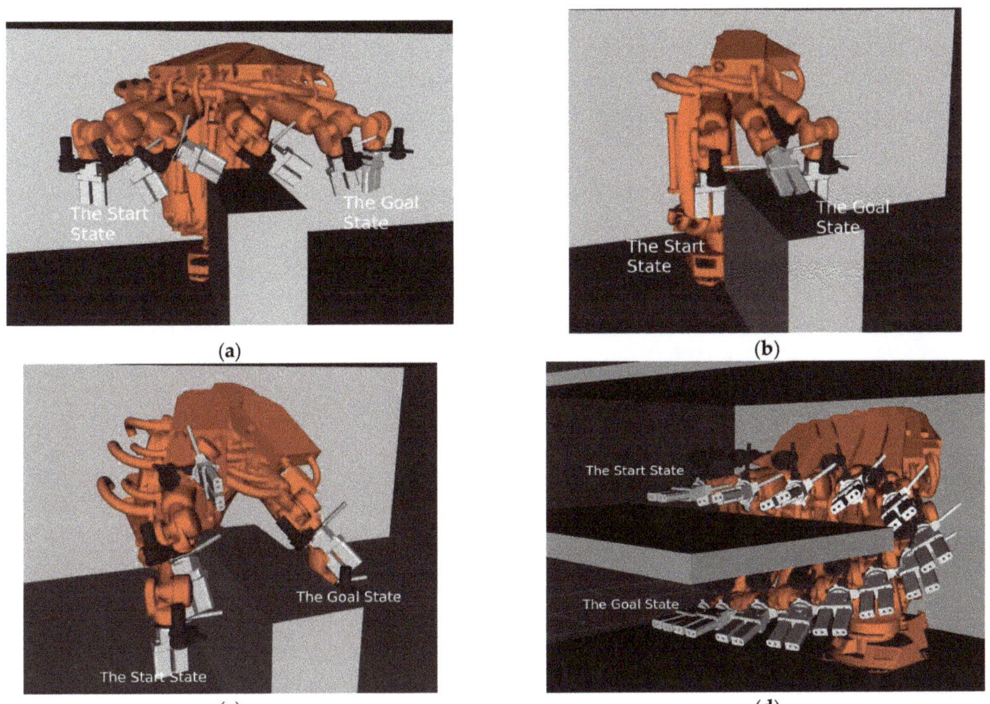

Figure 5. Illustration of one trajectory, which is executed in experiments 2–5: (**a**) experiment no. 2 the "longer" trajectory (**b**) experiment no. 3. the "shorter" trajectory; (**c**) experiment no. 4 the TCP is further on the *x*-axis trajectory; (**d**) experiment no 5. the "vertical" trajectory.

The brute force method for finding of constants requires the following steps:

1. Warm-up of the robot;
2. Generate paths between the same start and the same goal using a path planner;
3. Execute the generated trajectories on a real robot and measure energy consumption;
4. Make cycles through the whole constant space, in which constants are gradually changed:

    a. Compute the criterion score using current constants for each path;
    b. Normalize the measured energy with the criterion score;
    c. If the summary of differences between measured energy and criterion score on each trajectory is lower than the minimum (from previous cycles), store this value as a new minimum.

The output from the brute force method are the constants (0.94736842, 0.21052632, 0.42105263, 0.15789474, 0.05263158, 0.21052632) and Figure 6 shows the histogram of differences between the measured energy consumption and the criterion validity.

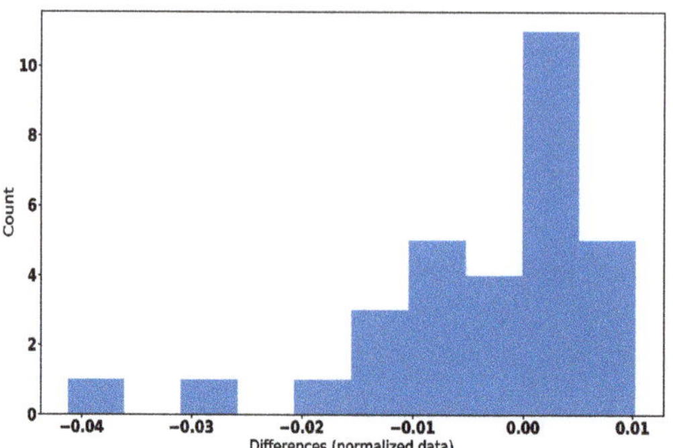

Figure 6. The histogram of differences between predicted and measured energy consumption in experiment 1, which has been used to find control cost constants.

In the next step these constants are validated in experiments 2–5. The results from experiments 2, 3, and 4 are very similar. However, the output from experiment 5 shows an increased deviation, as we anticipated. The results are reported in Figure 7. Figure 8 shows a comparison of results on the boxplot, and finally, the variances and averages for each experiment are compared in Table 2.

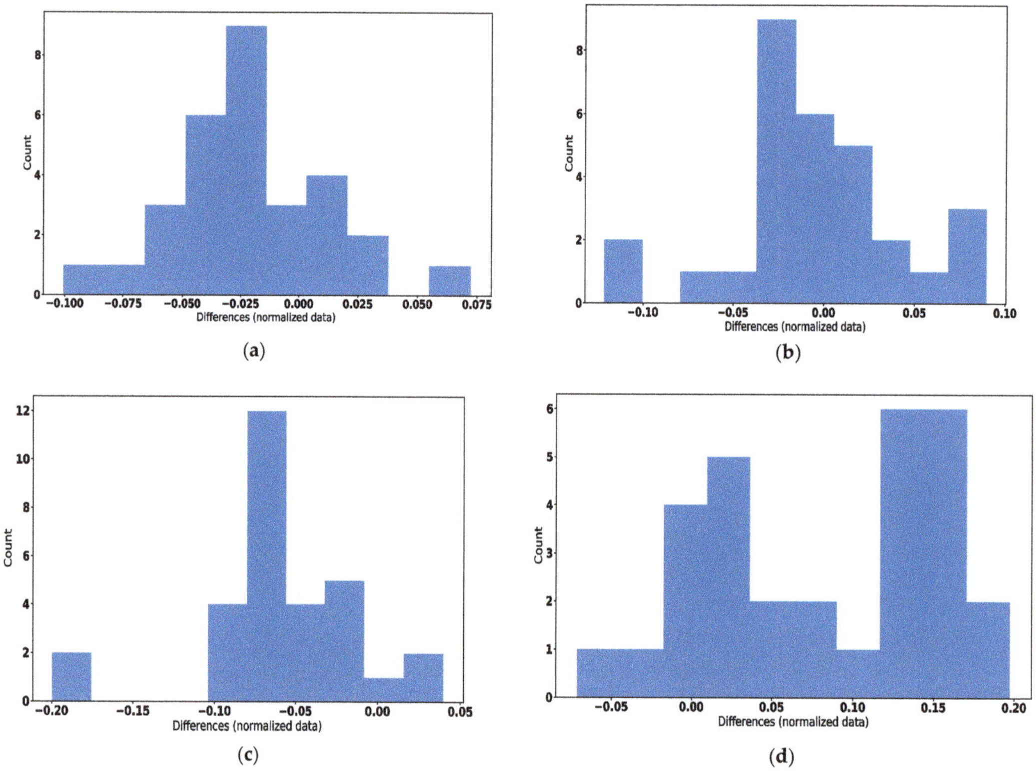

Figure 7. Histograms of the differences between predicted and measured energy consumption. (**a**) experiment no. 2; (**b**) experiment no. 3; (**c**) experiment no. 4; (**d**) experiment no. 5.

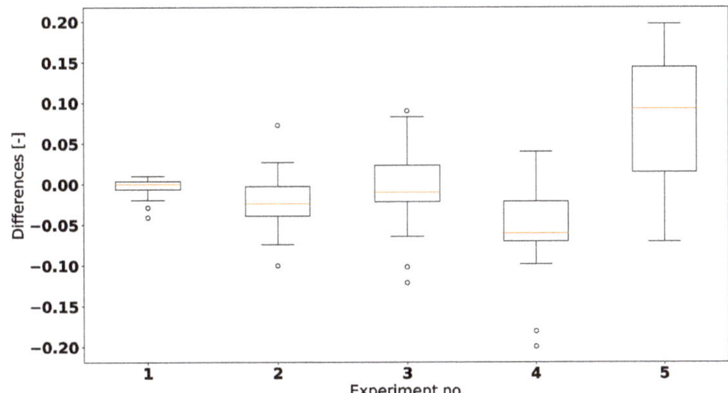

Figure 8. Boxplot comparing the results from Experiments 1–5.

Table 2. Averages and variances for each experiment.

Experiment No.	Average	Variance
1	-3.38×10^{-3}	1.25×10^{-4}
2	-2.20×10^{-2}	1.14×10^{-3}
3	-3.58×10^{-3}	2.28×10^{-3}
4	-5.67×10^{-2}	2.42×10^{-3}
5	8.08×10^{-2}	5.25×10^{-3}

3.2. Jerk-Based Criteria

The path is planned in the joint space of the robot where the path planner generates a set of waypoints. The motion between these waypoints is dependent on the robot controller. We expect that the waypoints are correctly distributed and that the path is always collision-free. Since the path planner only generates positions the jerk is unknown. It can be calculated as the third derivation of the position. This calculation returns an estimated jerk between two waypoints (Figure 9). However, it is not precisely a jerk because the planned path is not sampled in time, and therefore the unit cannot be rad/s^3 but just rad. Let us denote this jerk as a pseudo jerk. For the purpose of criterion validation, let us use the same calculation on the trajectories performed by the real robot. However, there is a drawback with this approach, namely that the measured positions are sampled in time (Figure 10). Due to the difference in sampling method and the density of waypoints the pseudo jerk calculated from the planned path and the jerk measured from the robot trajectory differ.

The real jerk is firmly dependent on the robot controller and the control parameters. In this experimental setup, we are using the KUKA KR120 robot and the KRC4 controllers. The motion command PTP_SPLINE is used; its explanation is available in the KUKA reference manual [32]. The velocity was set to 100% of the maximum speed for each joint, while acceleration was limited to 40% of maximum acceleration. The aim is to try to find a correlation between the completed movement and the expected jerk derived from the planned path. Our dataset consists of 30 different paths with different pseudo jerks.

Figure 11 shows pseudo jerk between each waypoint, calculated as joint jerk criterion by Formula (6). Between each waypoint the jerk is the Euclidean distance of the third derivation of joint positions. Therefore, the jerks are non-negative.

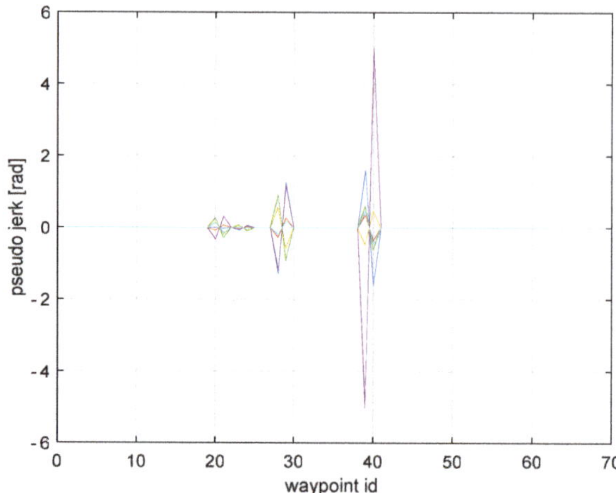

Figure 9. Pseudo jerk calculated from the planned path.

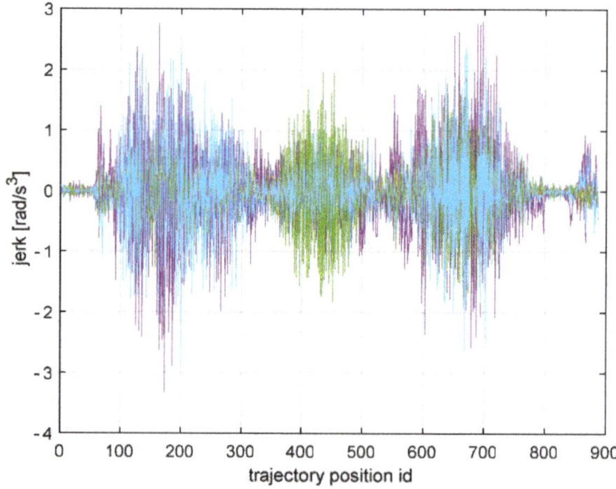

Figure 10. Jerk calculated from positions on the executed trajectory.

Interestingly, the jerk computed from measured robot positions does not correlate to the pseudo-jerk defined on the planned path. However, this highlights another aspect of the robot's motion. In each peak of a pseudo jerk (Figure 11) there is a notable change in the density of the measured robot positions around the planned waypoint. This is indicative of the evolution of the robot's velocity. Where the density of positions is higher, the robot's movement slows down and conversely, sparser density of positions signifies that the robot was moving faster.

Let us denote $X(i)$ as a count of measured robot positions around the waypoint in the planned path. When the $X(i)$ is calculated for each waypoint, then the minimum is denoted as MIN_DENSITY, and we expect that the velocity is the highest in this segment. Let us define the function:

$$f(i) = \frac{X(i)}{MIN\_DENSITY} \tag{10}$$

Function (10) represents a relation between $X(i)$ and $MIN\_DENSITY$, as shown in Figure 12. This function defines the notion of robot "slowdown" and the higher the value of the function, the higher the rate of robot "slowdown".

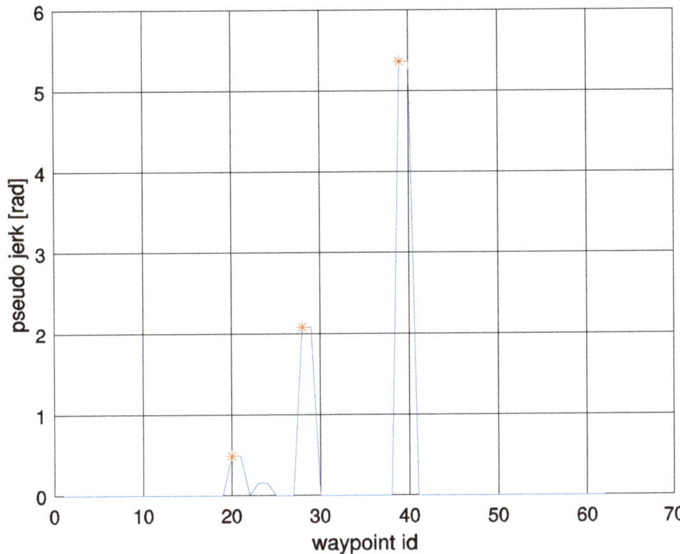

Figure 11. The calculated pseudo-jerk between each waypoint with marked local peaks.

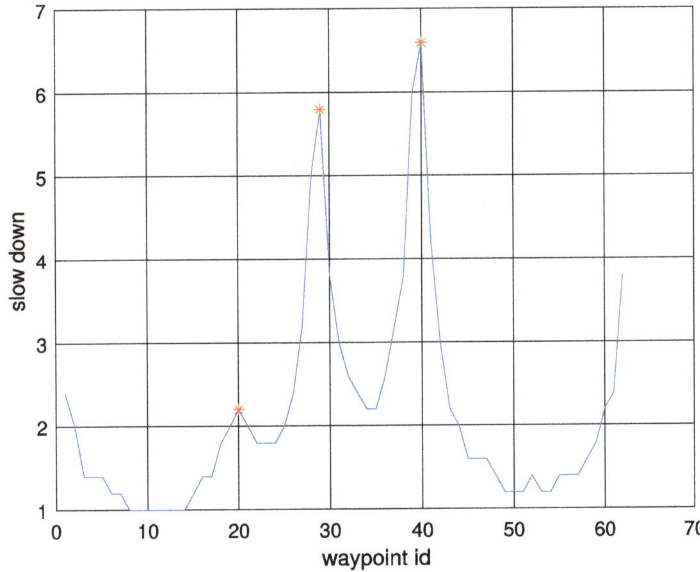

Figure 12. The "slowdown" of the robot with marked local peaks.

Our next goal is to find a relation between the calculated jerk and the robot "slowdown". The first step is to select peaks (local maximal jerks) on each path and assign a waypoint to each of these peaks. For the selection of peaks, a threshold of 0.4 rad is used, and the peaks are located where the derivation of the jerk is zero. The threshold has been found to be empirical on a series of datasets, but it can be modified depending on the indented application. When using a lower threshold the criterion is more sensitive to finding peaks. On the other hand, if the threshold is set to higher value the criterion picks out only relevant higher jerks. These peaks with their appointed waypoint id can then be assigned a value derived from Function (10) from Figure 11. Figure 13 below shows the selected peaks (blue dots) and the correlated robot "slowdown". This set of peaks is approximated by the function:

$$y = 3\,log_{10}\,x + 4 \tag{11}$$

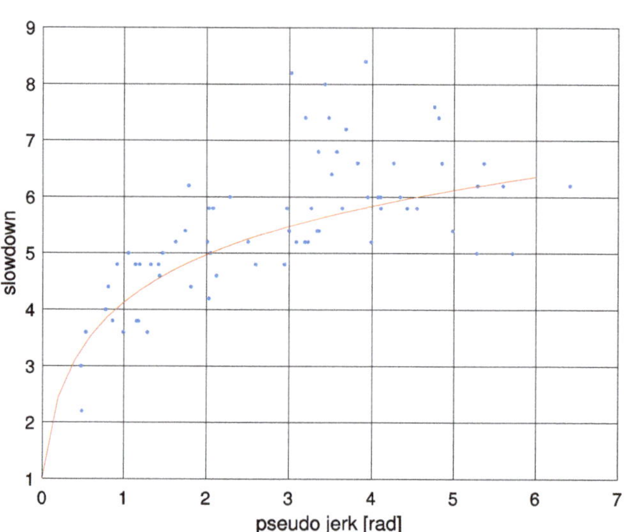

Figure 13. The dependency of the pseudo-jerk and real robot "slowdown".

The x stands for the pseudo-jerk peak on the planned path and y is the robot "slow down" in the joint space. The function and the constants have been found by regression analysis. The function describes the correlation between pseudo-jerk and the robot "slow down," which is rendered as a curved red line in Figure 13. Figure 14 shows the differences between the calculated and actual robot "slowdown."

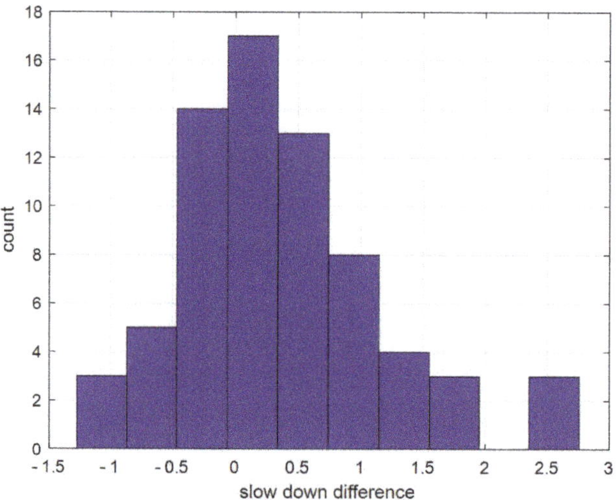

Figure 14. The histogram of differences between the predicated (function (11) and the curved red line in Figure 13) and the real measured (blue dots on Figure 13) robot "slowdown".

A similar approach is utilized for the Cartesian jerk criterion, where the Cartesian pseudo-jerk peaks were found using a threshold 0.002 m.

$$y = 1000\frac{\sqrt{2}}{2}\sqrt{x} + \sqrt[3]{4} \tag{12}$$

In Function (12) above the x stands for a Cartesian pseudo-jerk on the planned path and y is the "slowdown" of the tool point in Cartesian space. The correlation between the pseudo-jerk and the robot "slowdown" in Cartesian space rendered as a curved red line in Figure 15. Figure 16 shows differences between the predicated and real measured values.

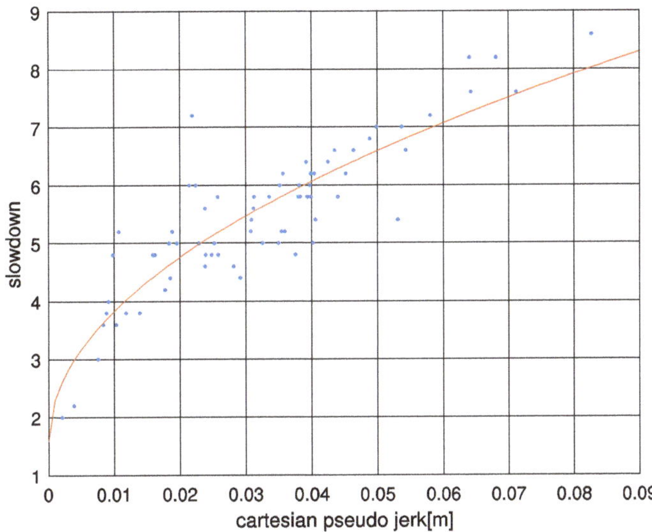

Figure 15. The dependency of the expected Cartesian pseudo-jerk and real robot "slowdown" in Cartesian space of the tool point.

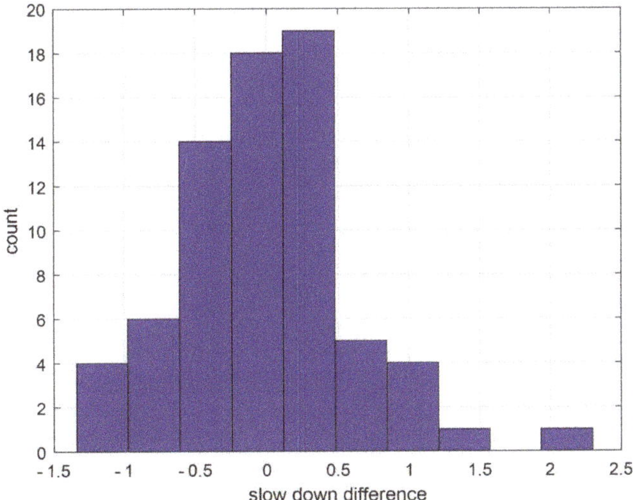

Figure 16. The histogram of differences between the predicated (function (12) and the curved red line in Figure 15) and the real measured (blue dots on Figure 15) robot "slowdown".

Our results show that the joint jerk, joint max jerk, Cartesian jerk, and Cartesian max jerk criteria cannot be compared with real jerk. For this reason, the above criteria were modified and will from now on be considered under the umbrella term joint jerk peaks. This new criterion still computes pseudo jerks, however it looks only for local pseudo jerk peaks. The score of this criterion equals the sum of all peaks, once Equations (11) and (12) have been run. The joint jerk peaks is defined as:

$$x = \sum_{i=1}^{n} 3 \, \log_{10} P_i + 4 \tag{13}$$

and the Cartesian jerk peaks is defined as:

$$x = \sum_{i=1}^{n} 1000 \frac{\sqrt{2}}{2} \sqrt{P_i} + \sqrt[3]{4} + 4 \tag{14}$$

Both in (13) and (14) i represents the peak id of n local pseudo jerk peaks and P_i is the peak value. Additionally, these modified criteria allowed us to compare whether the duration of the executed trajectory correlates with the jerk predicated in the planned path.

4. Criteria Usage

Now that the criteria are validated, our next step is a demonstration of the criteria being utilized in evaluating paths in different environmental setups. This step will be tested on a real robot. We will assess the impact an environment change will have on the path planner. Paths are generated by the STOMP path planner with the same configuration as in the section "Validation of criteria" [4]. In this section, however, we use an additional criterion "Duration", which represents measurement of computation time. The criterion "Duration" does not evaluate the quality of the path, nevertheless an assessment of the path planner's computation time is key when evaluating the impact caused by a change in the environment. Our hypothesis is that a complicated environment requires longer computation time.

4.1. The Impact of the Environment's Complexity on the Path Planner

The first setup simulates a common process in the industry—a pick and place process. The path will be computed between two robot states $(-0.527, 0.498, 0.244, -0.103, 0.983, -0.462)$ (Figure 17a) and $(0.465, 0.451, 0.312, 0.097, 0.965, 0.405)$ (Figure 17b), in both of which the robot tool is located inside a bin and the path planner must return a collision-free path.

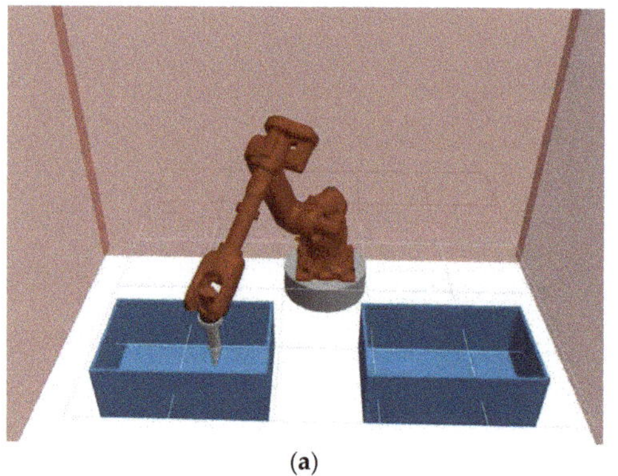
(a)

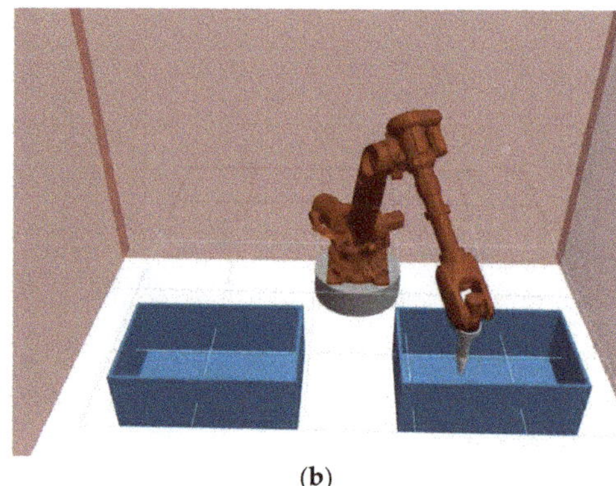
(b)

Figure 17. Illustration of the first experimental setup and robot states: (**a**) the start state; (**b**) the goal state.

In the second setup, the environment is more complicated. The robot must avoid an added simple wall embedded between the bins from the previous setup (Figure 18a). This situation is a non-standard situation in industry. However the goal of these experimental setups is to monitor the impact on path quality when the planner must avoid more obstacles on the scene. The third setup exhibits the worst traversability, as there is an extra wall inserted (Figure 18b). Start and goal states are the same as in the first setup.

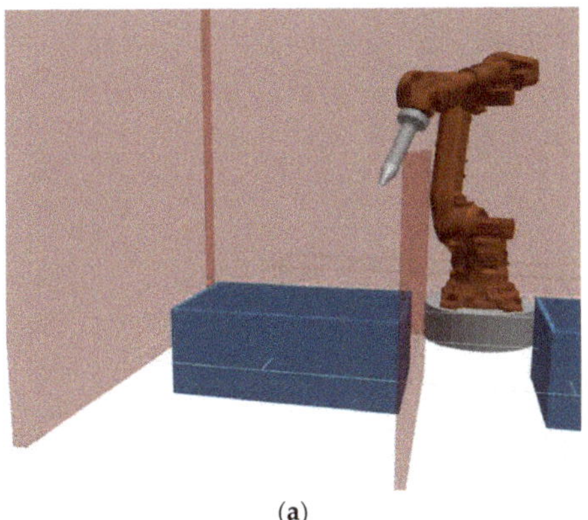

(a)

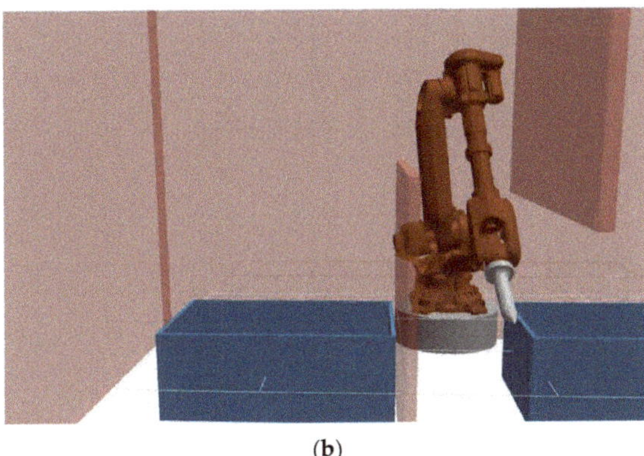

(b)

Figure 18. Experimental setups in which extra obstacles are added: (**a**) the second experiment; (**b**) the third experiment.

Our proposed criteria evaluate the environment complexity. Control pseudo-cost criteria use the same constants, calculated in Section 3.1 "Control Pseudo-Cost". In Section 3.2. "Jerk-based Criteria," the threshold for the criterion Joint Jerk Peaks 0.4 rad could not determine any peaks in experiments. Therefore, the threshold was decreased to 0.08 rad in these experimental setups, which means that the criterion is more sensitive to finding pseudo-jerk peaks. In the case of Cartesian jerk peaks, the threshold of 0.002 m is preserved. The histograms in Figure 19a–h show the score of each criterion. The results are described in the Section 5 "Results".

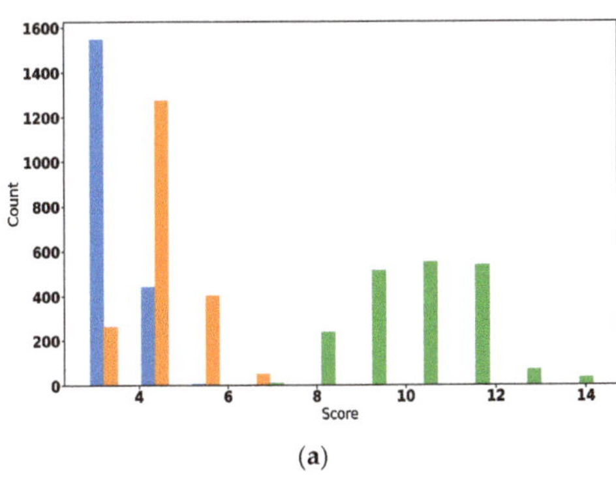

(a)

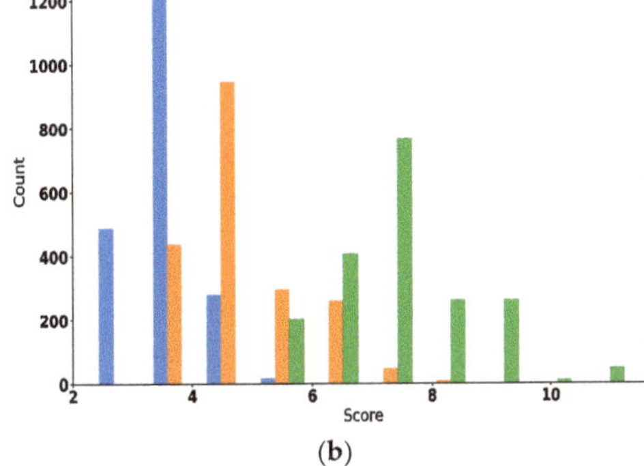

(b)

Figure 19. *Cont.*

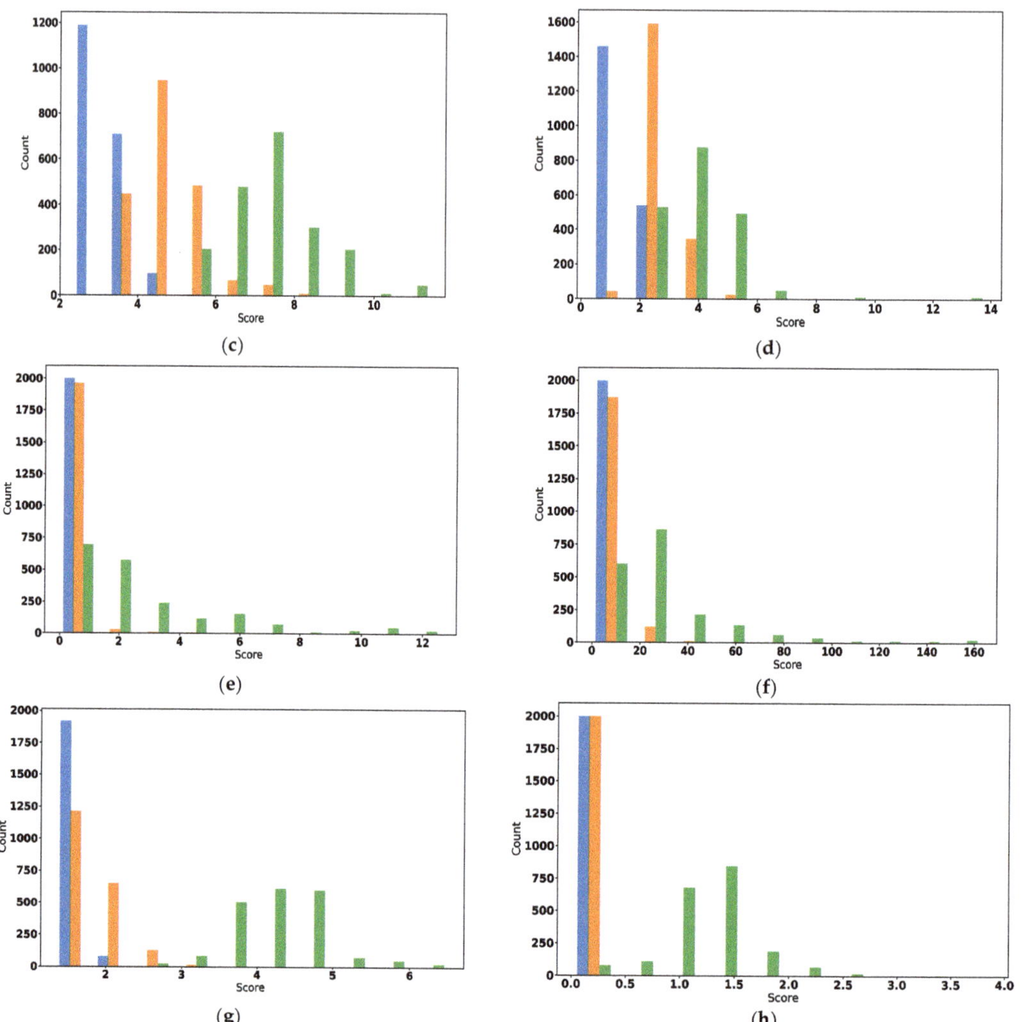

Figure 19. Histograms depicting the score for each criterion (blue—no.1, orange—no.2, green—no.3) (**a**) joint distance; (**b**) Cartesian distance; (**c**) robot displacement; (**d**) orientation change; (**e**) joint jerk peaks; (**f**) Cartesian jerk peaks; (**g**) control pseudo-cost; (**h**) duration.

4.2. The Impact on the Path Planner of the Bin's Position

This experimental setup is focused on the bin position and the behavior of the path planner as well as how the quality of the generated paths is evaluated in different workspaces. The movement is similar to the first experimental setup and the robot moves from bin A to bin B. The experiment consists of three scenarios, and in each scenario, the bins are reallocated on the x-axis of the robot's coordinate system. The layout of each scenario is illustrated in Figure 20a–c.

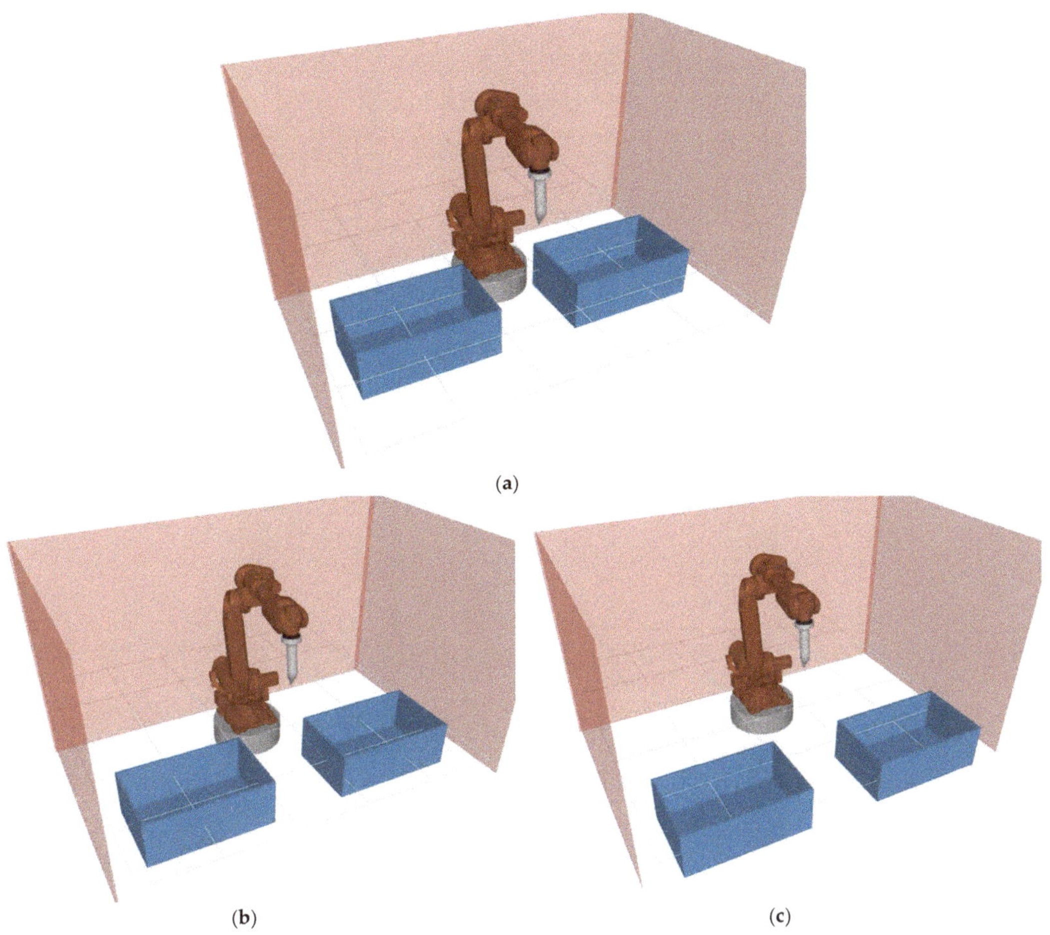

(a)

(b)

(c)

Figure 20. Experimental setups: (**a**) the first setup; (**b**) the second setup; (**c**) the third setup.

Our criteria evaluate the impact of the change in the layout and the impact of the bins' position. Control pseudo-cost, joint jerk peaks, and Cartesian jerk peaks use the same constants and parameters as were used in the previous section of this research paper. The histograms in Figure 21a–h show the score of each of these criteria. The results are described in Section 5 "Results".

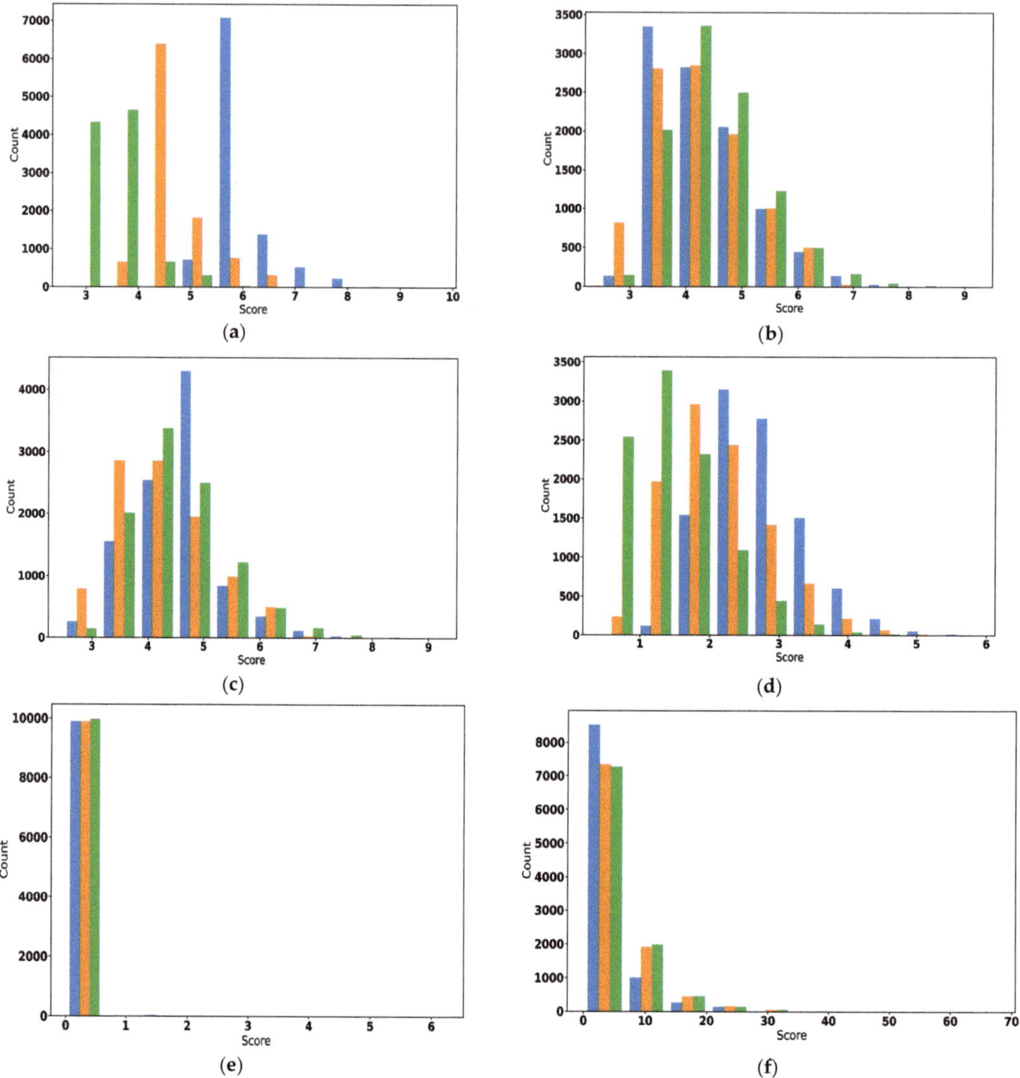

Figure 21. *Cont.*

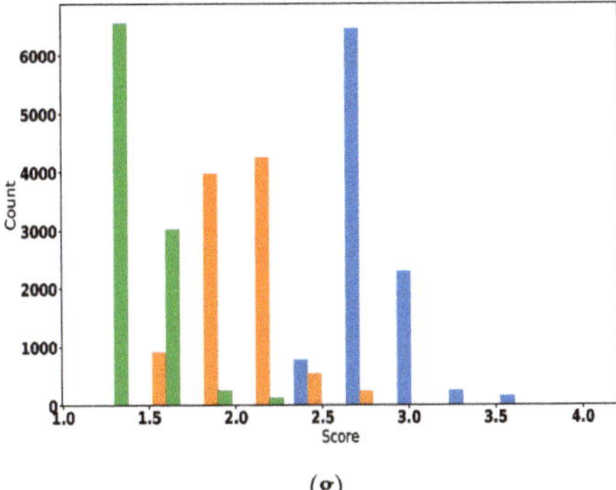
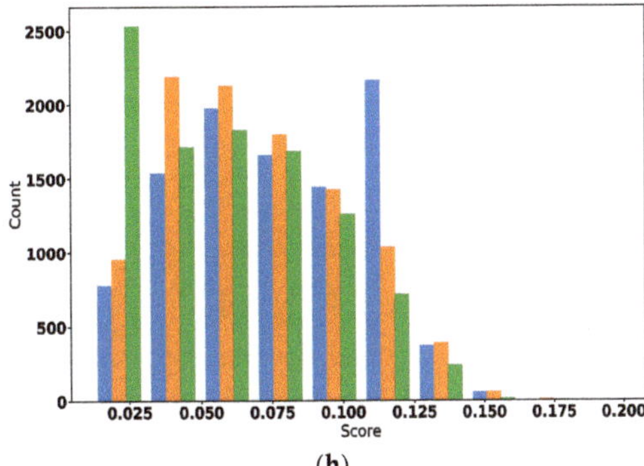

(g)   (h)

Figure 21. Histograms depicting the score for each criterion (blue—no.1, orange—no.2, green—no.3): (**a**) joint distance; (**b**) Cartesian distance; (**c**) robot displacement; (**d**) orientation change; (**e**) joint jerk peaks; (**f**) Cartesian jerk peaks; (**g**) control pseudo-cost; (**h**) duration.

5. Results

The criteria joint distance, Cartesian distance, orientation change, and robot displacement are based on the robot positions within the planned path. To validate these, they were compared with the positions on trajectory executed by a real robot. The differences between the planned and executed trajectories are recapitulated in Table 3, lines 1–4, where the averages and variances for each criterion respectively are shown. These four criteria correlate and we can, therefore, accept them as valid.

Table 3. Recapitulation of criteria validation.

Criterions	Average	Variance
Joint Distance [rad]	-5.75×10^{-3}	3.76×10^{-4}
Cartesian Distance [m]	-4.09×10^{-3}	2.22×10^{-5}
Orientation Change [rad]	-1.24×10^{-2}	3.09×10^{-3}
Robot Displacement [m]	-3.75×10^{-3}	2.52×10^{-5}
Control Pseudo Cost [1]	-2.20×10^{-2}	1.14×10^{-3}
Control Pseudo Cost [2]	8.08×10^{-2}	5.6×10^{-3}
Joint Jerk Peaks [rad]	0.3344	0.8290
Cartesian Jerk Peaks [m]	5.1×10^{-3}	0.6036

[1] The results of experiment no. 2 in Section 3.1. [2] The results of experiment no. 5 in Section 3.1.

The criterion control pseudo-cost is not directly compared with the measured robot positions, however the KUKA robot program provides the measurement of energy consumption as each trajectory is being executed. Firstly, the control pseudo-cost constants of each robot link have been calculated using the brute force method on one dataset consisting of 30 different trajectories carried out between the same start and goal states. As demonstrated on other datasets, the constants and the criteria are applicable to similar movements. The results of the second dataset are recapitulated in Table 3 below. In the table, line 5 depicts the average and the variance of differences between the consumed energy and the criterion score. Notably, the average and the variance increased significantly in line 6 when the robot was performing different movements. For more detailed results from further datasets, consult the section "Control Pseudo-Cost" above.

It was found that the jerk-based criteria—joint jerk, joint max jerk, Cartesian jerk, and Cartesian max jerk do not correlate with the executed trajectory. However we observed a different correlation, namely that the pseudo-jerk peaks calculated on a planned path corresponded to a higher density of measured robot positions in time. This correlation is discussed in more depth in the section "Jerk-Based Criteria". While the joint jerk calculates the sum of jerks along the whole path and the joint max jerk uses just one maximum jerk for the entire path, the experiments proved it would be more advantageous as combined criteria, termed joint jerk peaks and Cartesian jerk peaks respectively. These two criteria can be defined by (13) in joint space and by (14) in Cartesian space and their goal is to find all local pseudo-jerk peaks. The statistical indicators of the differences between the function's output and the jerks calculated from the measured robot positions are shown in Table 3, in lines 7 and 8.

These criteria are applied, and their usage is demonstrated in Section "Criteria Usage" in two experimental setups. In the first experiment three different environments with increasing levels of complexity are compared, and the presented criteria are used to evaluate the planned paths generated to navigate these environments. As expected, the simplest environment achieves the best scores while with increasing complexity, i.e., more obstacles we see the scores dropping. Naturally, the third and the most challenging environment has the worst scores. In several attempts, the path planner fails to find a collision-free path. The scores are compared and recapitulated in Table 4 below.

Table 4. Recapitulation of the impact of environment complexity per each criterion.

	Criterion	Experiment No.1		Experiment No.2		Experiment No.3	
		Average	Variance	Average	Variance	Average	Variance
1	Joint Distance [rad]	3.699	0.1511	4.9824	0.51754	10.456	1.544
2	Cartesian Distance [m]	3.6277	0.2882	5.107	0.7206	7.6689	1.3505
3	Orientation Change [rad]	1.6659	0.199	2.8288	0.3863	4.6727	1.64228
4	Robot Displacement [m]	3.6315	0.2800	5.1134	0.7182	7.753	1.307
5	Control Pseudo Cost	1.6349	0.0208	2.0629	0.0870	4.3652	0.3606
6	Joint Jerk Peaks [rad]	0.0022	0.0103	0.0575	0.1551	2.824	9.454
7	Cartesian Jerk Peaks [m]	2.42	10.5699	6.7953	60.34	31.2399	943.78
8	Duration [s]	8.02×10^{-2}	4.5×10^{-4}	8.6×10^{-2}	9.2×10^{-4}	1.44	0.2

In the second experimental setup, the behavior of the path planner and the quality of the generated paths are evaluated in different workspaces. On the surface, the robot performs a movement similar to that in the first experimental setup. The difference in this case comes in the position of the bins which are moved further along the X-axis of the robot's coordinate system in each scenario. The goal is to evaluate the impact of this change on the resulting paths. Each experiment has a 100% success rate of path computation. The Cartesian distance and robot displacement criteria have the same results in each scenario, and the position of bins does not change the length of the Cartesian path. The same result can be seen for the duration criterion, which means that the computation time does not dependent on the change in the reachable position of the bins. However, the criterion joint distance shows that the trajectories are longer when the bin is placed closer to the robot base. This behavior is echoed by the outcomes observed for the orientation change criterion. They display a frequent change in the orientation of the tool point in the scenario where the bins are placed closer to the robot base. The scores for the criteria joint jerk peaks and Cartesian jerk peaks prove that the change of the bin position has no impact on path jerks. Our results show that most of the paths have no significant jerk, and are therefore mostly smooth. These results are recapitulated in Table 5.

Table 5. Recapitulation of the impact of bin position per each criterion.

	Criterion	Experiment No.1		Experiment No.2		Experiment No.3	
		Average	Variance	Average	Variance	Average	Variance
1	Joint Distance [rad]	6.2858	0.3452	4.9181	0.4432	3.729	0.3086
2	Cartesian Distance [m]	4.797	0.730	4.328	0.7535	4.737	0.897
3	Orientation Change [rad]	2.974	0.4137	2.2203	0.518	1.804	0.368
4	Robot Displacement [m]	4.810	0.707	4.337	0.7435	4.7405	0.894
5	Control Pseudo-Cost	2.9122	0.0518	2.0887	0.0584	1.4711	0.0431
6	Joint Jerk Peaks [rad]	0.049	0.1156	0.0651	0.1909	0.0148	0.0347
7	Cartesian Jerk Peaks [m]	5.2656	30.308	5.43026	47.276	4.8859	38.443
8	Duration [s]	0.082	0.0011	0.078	0.0011	0.0719	0.0010

6. Discussion

In this paper, we presented a complex trajectory-evaluation framework based on seven criteria. Firstly, we introduced nine hypotheses, mostly based on well-known trajectory metrics. However, following the outcome of our experiments and subsequent modifications, we formulated a total of seven criteria to be used in comparison of trajectories. Each hypothesis was meticulously tested on a real robot, and thus proved to be a valid criterion. There is abundant research into the planning time and success rate of path planners which are viewed as important benchmarks for many industrial applications. Our research, however, focused on the quality of the path planners' output, namely a path defined as a set of robot states. When looking at the minimization of trajectory length, there are several criteria that need to be considered. A well-known approach is to assess the length of a trajectory by finding the sum of joint differences, as demonstrated in [8]. Similarly, [9] employs the tactic to compute the cumulative sum of the tool point's Euclidean distances between each waypoint. Therefore, these criteria (joint distance, Cartesian distance, orientation change, and robot displacement) were included in our trajectory-evaluation framework and were assessed experimentally using a real robot. We confirmed our expectations that these criteria are valid, given that the planned trajectories correspond with the robot-executed trajectories (Table 1). The aim of the research was to consider advanced metrics, energy consumption being one of those. The hypothesis we formulated posits that it is possible to predict which trajectory consumes more energy. Our findings show that by selecting a combination of weights for the control pseudo-cost criterion (4) we can approximate the energy consumption. We investigated the correlation between this criterion's score and the measured consumed energy during the trajectory execution on the real robot. The results confirmed this hypothesis: the higher the criterion's score, the more energy a given trajectory will consume. However, it is important to note that the set of weights is only suitable for sets of similar movements (Experiments no. 1–4 in Section 3.1). The difference between real consumed energy and the criterion's score becomes a less reliable prediction tool in case of significantly different movements (Experiment no. 5 in Section 3.1). This difference is clearly illustrated in a box plot in Figure 8. Another key metric to consider is the smoothness of a trajectory, as smooth motion puts less stress on mechanical units. Smoothness can be computed as the accumulated sum of linear and angular accelerations of the end effector, as presented in [7]. An optimal time-jerk trajectory planning, as suggested in [26], minimizes jerk which is defined as the cumulative sum of time-derivation of accelerations on each joint. Unlike the aforementioned studies, we presented a different approach of trajectory comparison. Our approach considers only robot positions, neglecting properties such as time-parametrization, velocities or accelerations, which are all unknown. We propose that many path planners generate trajectories primarily as positions only, while velocities or accelerations may not necessarily be computed. With this hypothesis in place we formulated the criteria joint jerk, joint max jerk, Cartesian jerk, and Cartesian max jerk which were all based on the computation of the pseudo-jerk. However, the experiment detailed in Section "Jerk-Based Criteria" showed that these criteria are not suitable for trajectory comparison without the trajectories being time-parametrized. Figures 9 and 10 demonstrate the lack of correlation between pseudo-jerk and real jerk. However, we were able to observe a different phenomenon. The local peaks of the pseudo-jerk (Figure 11) showed correlation with observed robot "slowdowns" (Figure 12). Therefore, we designed new criteria, the joint jerk peaks and Cartesian jerk peaks which can be used as a prediction tool for which trajectory will take more time to be executed. Finally, we demonstrate the possible applications of these criteria in Section 4 "Criteria Usage". Due to the proven validity of our criteria, these can now be simply applied in trajectory comparison, without the need for executions on a real robot. This presents a major advantage, as trajectories can now be compared in simulation mode, foregoing the need to collect any measurements from real robot performances. Furthermore, we applied these criteria experimentally to investigate what impact does a change in environment have on a path planner. Our assumption was a straightforward one, namely that the generated trajectories will be of a worse quality in a more complex environment (Table 4). In line with our expectations, the most complicated environment (Experiment no. 3) rendered the highest criterion scores meaning the quality of trajectories here was the worst.

7. Conclusions

Due to the non-deterministic behavior of randomized path planning algorithms, it is difficult to compare the resulting paths between multiple implementations or changes in configuration.

Path planners usually generate paths as sets of robot states in joint space. Trajectory attributes (e.g., velocity and acceleration) could not be calculated by every path planner but they could be calculated by a robot control system during their execution. In our work we focused on paths containing only position data. It is important to note that trajectory analysis with regard to dynamics would constitute an entirely different task, as this type of path does not contain all the necessary information.

In this article, we presented a novel approach for the evaluation and comparison of paths computed by randomized path planners. Nine hypotheses, mostly based on well-known trajectory metrics, were suggested and evaluated with experiments. On the basis of the test results, we formulated seven criteria that can be used for the comparison of the computed paths with respect to multiple production demands.

The evaluation of the criteria joint distance, Cartesian distance, orientation change and robot displacement confirmed that the planned paths correspond with the executed trajectories. We can confirm that these represent a valid metric for path comparison, and they will be valid in all situations except for situations when the robot executes a completely different trajectory than requested.

The criterion control pseudo-cost is used as a prediction of which paths could consume more energy. This criterion simplifies the dynamics of each robot link by replacing it with a simple constant. For every link, these constants have been calculated using the brute force method on a reference dataset, consisting of 30 different trajectories between the same start and goal states. Our assumption was that the constants are applicable to similar motions and for other groups of motions the constants should be recalculated. Two types of experiment were conducted to verify this hypothesis (described in Section 3.1 Control Pseudo-Cost). For experimental evaluation, we measured the real energy consumption of the robot during the trajectory execution and compared it with the predicted results. Our results confirmed this hypothesis. The difference between consumed energy and criterion scores is more accurate in the group of similar motions. The error is increased in the case of significantly different motions.

Criteria based on the jerk, namely joint jerk, joint max jerk, Cartesian jerk, and Cartesian max jerk, cannot be directly compared with the executed robot motions due to the missing information in trajectories such as velocity and acceleration. However, in our experiments, we noticed that the local peaks of the predicted pseudo-jerk correlate with observed "slowdowns" of the robot. For this reason, we designed the new joint jerk peaks criterion for the joint space and its corresponding counterpart for the Cartesian space, Cartesian jerk peaks. These new criteria could be used as a prediction of which path takes more time. This hypothesis was successfully confirmed in our experiments.

The proposed criteria can serve as a tool for deciding which path planner generates better paths. Likewise, the environment complexity has an impact on path planners and the criteria could assist in the decision of which layout is better or which path planner is appropriate for the chosen layout. Our experiments confirmed that if the environment is more complex, the quality of trajectories is worse.

In our future work, we will focus on in-depth comparison of path planners in different production types with the help of the criteria-based framework presented in this article. Furthermore, it must be stated that the criteria joint jerk peaks and Cartesian jerk peaks have been validated just on the one robot model and, therefore, we would like to investigate the behavior of jerk peaks on other robots as well.

Author Contributions: Conceptualization, P.B. and M.D. (Martin Dekan); methodology, M.D. (Martin Dekan); software, M.D. (Martin Dekan) and M.D. (Michal Dobiš); validation, M.D. (Michal Dobiš); formal analysis, M.D. (Martin Dekan) and M.D. (Michal Dobiš); investigation, M.D. (Martin Dekan) and M.D. (Michal Dobiš); resources, P.B.; data curation, M.D. (Michal Dobiš); writing—original draft preparation, M.D. (Michal Dobiš); writing—review and editing, P.B. and F.D.; visualization, M.D. (Michal Dobiš); supervision, P.B.; project administration, P.B.; funding acquisition, A.B. All authors have read and agreed to the published version of the manuscript.

Funding: This research was funded by VEGA, grant number VEGA 1/0049/20 and by the European Regional Development Fund, Grant Number 313012P386 and by Horizon 2020, Grant Number 824964.

Institutional Review Board Statement: Not applicable.

Informed Consent Statement: Not applicable.

Data Availability Statement: Not applicable.

Acknowledgments: This research was partially sponsored by Photoneo company. (http://www.photoneo.com, accessed on 16 January 2022). VEGA 1/0049/20 and DIH^2 supported this work. Operational Program Integrated Infrastructure created this publication for the project: "Robotic workplace for intelligent welding of small-scale production," code ITMS2014+: 313012P386, and it is co-sponsored by the European Regional Development Fund.).

Conflicts of Interest: The authors declare no conflict of interest. The funders had no role in the design of the study; in the collection, analyses, or interpretation of data; in the writing of the manuscript, or in the decision to publish the results.

References

1. Kavraki, L.E.; Svestka, P.; Latombe, J.-C.; Overmars, M.H. Probabilistic roadmaps for path planning in high-dimensional configuration spaces. *IEEE Trans. Robot. Autom.* **1996**, *12*, 566–580. [CrossRef]
2. Lavalle, S.M. Rapidly-Exploring Random Trees: A New Tool for Path Planning. 1998. Available online: http://citeseerx.ist.psu.edu/viewdoc/summary?doi=10.1.1.35.1853 (accessed on 14 February 2022).
3. Zucker, M.; Ratliff, N.; Dragan, A.D.; Pivtoraiko, M.; Klingensmith, M.; Dellin, C.M.; Bagnell, J.A.; Srinivasa, S.S. CHOMP: Covariant Hamiltonian optimization for motion planning. *Int. J. Robot. Res.* **2013**, *32*, 1164–1193. [CrossRef]
4. Kalakrishnan, M.; Chitta, S.; Theodorou, E.; Pastor, P.; Schaal, S. STOMP: Stochastic trajectory optimization for motion planning. In Proceedings of the 2011 IEEE International Conference on Robotics and Automation, Shanghai, China, 9–13 May 2011.
5. Rodriguez, C.; Suárez, R. Comparison of motion planners in an environment with removable obstacles. In Proceedings of the 2015 IEEE 20th Conference on Emerging Technologies & Factory Automation (ETFA), Luxembourg, 08–11 September 2015; pp. 1–7. [CrossRef]
6. Paulin, S.; Botterill, T.; Lin, J.; Chen, X.; Green, R. A comparison of sampling-based path planners for a grape vine pruning robot arm. In Proceedings of the 2015 6th International Conference on Automation, Robotics and Applications (ICARA), Queenstown, New Zealand, 17–19 February 2015; pp. 98–103.
7. Magyar, B.; Tsiogkas, N.; Brito, B.; Patel, M.; Lane, D.; Wang, S. Guided Stochastic Optimization for Motion Planning. *Front. Robot. AI* **2019**, *6*, 6. [CrossRef]
8. De Maeyer, J.; Demeester, E. Benchmarking framework for robotic arc welding motion planning. *Procedia CIRP* **2021**, *97*, 247–252. [CrossRef]
9. Larsen, L.; Kim, J.; Kupke, M.; Schuster, A. Automatic Path Planning of Industrial Robots Comparing Sampling-based and Computational Intelligence Methods. *Procedia Manuf.* **2017**, *11*, 241–248. [CrossRef]
10. Dobiš, M.; Dekan, M.; Beňo, P.; František, D.; Miroslav, K. The Comparison of Motion Planners for Robotic Arms. *J. Control. Eng. Appl. Inform.* **2021**, *23*, 87–94.
11. Chettibi, T.; Lehtihet, H.E.; Haddad, M.; Hanchi, S. Minimum cost trajectory planning for industrial robots. *Eur. J. Mech. A/Solids* **2004**, *23*, 703–715. [CrossRef]
12. Gandhi, P.; Jain, A.K. Optimization Performance of a Robot to Reduce Cycle Time Estimate. *Int. J. Sci. Res. (IJSR)* **2013**, *2*, 357–363.
13. Trnka, K.; Božek, P. Optimal Motion Planning of Spot Welding Robot Applications. *Appl. Mech. Mater.* **2012**, *248*, 589–593.
14. Mumm, J.; Mutlu, B. Human-robot proxemics: Physical and psychological distancing in human-robot interaction. In Proceedings of the 6th International Conference on Human-Robot Interaction—HRI'11, Lausanne, Switzerland, 6–9 March 2011. [CrossRef]
15. Farouki, R.T.; Li, S. Optimal tool orientation control for 5-axis CNC milling with ball-end cutters. *Comput. Aided Geom. Des.* **2013**, *30*, 226–239. [CrossRef]
16. Toh, C.A. study of the effects of cutter path strategies and orientations in milling. *J. Mater. Process. Technol.* **2004**, *152*, 346–356. [CrossRef]

17. Su, H.; Yang, C.; Ferrigno, G.; De Momi, E. Improved Human–Robot Collaborative Control of Redundant Robot for Teleoperated Minimally Invasive Surgery. *IEEE Robot. Autom. Lett.* **2019**, *4*, 1447–1453. [CrossRef]

18. Tölgyessy, M.; Dekan, M.; Duchoň, F.; Rodina, J.; Hubinský, P.; Chovanec, L. Foundations of Visual Linear Human–Robot Interaction via Pointing Gesture Navigation. *Int. J. Soc. Robot.* **2017**, *9*, 509–523. [CrossRef]

19. Cohen, B.; Chitta, S.; Likhachev, M. Search-based planning for dual-arm manipulation with upright orientation constraints. In Proceedings of the 2012 IEEE International Conference on Robotics and Automation, Saint Paul, MN, USA, 24–27 May 2012; pp. 3784–3790. [CrossRef]

20. Lavalle, S.M. *Planning Algorithms*; Cambridge University Press: New York, NY, USA, 2014; pp. 153–206.

21. Gregory, J.; Olivares, A.; Staffetti, E. Energy-optimal trajectory planning for robot manipulators with holonomic constraints. *Syst. Control. Lett.* **2012**, *61*, 279–291. [CrossRef]

22. Mohammed, A.; Schmidt, B.; Wang, L.; Gao, L. Minimizing Energy Consumption for Robot Arm Movement. *Procedia CIRP* **2014**, *25*, 400–405. [CrossRef]

23. Prado, J.; Marques, L. Energy Efficient Area Coverage for an Autonomous Demining Robot. In Proceedings of the ROBOT2013 First Iberian Robotics Conference, Madrid, Spain, 28–29 November 2013; Springer Science and Business Media LLC: Berlin/Heidelberg, Germany, 2014; pp. 459–471. [CrossRef]

24. Carabin, G.; Scalera, L. On the Trajectory Planning for Energy Efficiency in Industrial Robotic Systems. *Robotics* **2020**, *9*, 89. [CrossRef]

25. Vidussi, F.; Boscariol, P.; Scalera, L.; Gasparetto, A. Local and Trajectory-Based Indexes for Task-Related Energetic Performance Optimization of Robotic Manipulators. *J. Mech. Robot.* **2021**, *13*, 1–12. [CrossRef]

26. Huang, J.; Hu, P.; Wu, K.; Zeng, M. Optimal time-jerk trajectory planning for industrial robots. *Mech. Mach. Theory* **2018**, *121*, 530–544. [CrossRef]

27. Galloway, K.C.; Becker, K.P.; Phillips, B.; Kirby, J.; Licht, S.; Tchernov, D.; Wood, R.J.; Gruber, D.F. Soft Robotic Grippers for Biological Sampling on Deep Reefs. *Soft Robot.* **2016**, *3*, 23–33. [CrossRef]

28. Pettersson, A.; Davis, S.; Gray, J.O.; Dodd, T.J.; Ohlsson, T. Design of a magnetorheological robot gripper for handling of delicate food products with varying shapes. *J. Food Eng.* **2010**, *98*, 332–338. [CrossRef]

29. Rubio, F.; Llopis-Albert, C.; Valero, F.; Suñer, J.L. Industrial robot efficient trajectory generation without collision through the evolution of the optimal trajectory. *Robot. Auton. Syst.* **2016**, *86*, 106–112. [CrossRef]

30. Liu, H.; Lai, X.; Wu, W. Time-optimal and jerk-continuous trajectory planning for robot manipulators with kinematic constraints. *Robot. Comput. Manuf.* **2013**, *29*, 309–317. [CrossRef]

31. Pan, J.; Chitta, S.; Manocha, D. FCL: A general purpose library for collision and proximity queries. In Proceedings of the 2012 IEEE International Conference on Robotics and Automation, Saint Paul, MN, USA, 24–27 May 2012; pp. 3859–3866.

32. KUKA. System Software KUKA System Software 8.3 Operating and Programming Instructions for System Integrators. 2015. Available online: http://www.wtech.com.tw/public/download/manual/kuka/krc4/KUKA%20KSS-8.3-Programming-Manual-for-SI.pdf (accessed on 14 February 2022).

Article

Using Simulation to Evaluate a Tube Perception Algorithm for Bin Picking

Gonçalo Leão [1,2,*], Carlos M. Costa [1,2], Armando Sousa [1,2], Luís Paulo Reis [1,3] and Germano Veiga [1,2]

1   Faculty of Engineering, University of Porto (FEUP), 4200-465 Porto, Portugal;
     carlos.m.costa@inesctec.pt (C.M.C.); asousa@fe.up.pt (A.S.); lpreis@fe.up.pt (L.P.R.);
     germano.veiga@inesctec.pt (G.V.)
2   Institute for Systems and Computer Engineering, Technology and Science (INESC TEC),
     4200-465 Porto, Portugal
3   Artificial Intelligence and Computer Science Laboratory (LIACC), University of Porto,
     4200-465 Porto, Portugal
*   Correspondence: goncalo.leao@fe.up.pt

Abstract: Bin picking is a challenging problem that involves using a robotic manipulator to remove, one-by-one, a set of objects randomly stacked in a container. In order to provide ground truth data for evaluating heuristic or machine learning perception systems, this paper proposes using simulation to create bin picking environments in which a procedural generation method builds entangled tubes that can have curvatures throughout their length. The output of the simulation is an annotated point cloud, generated by a virtual 3D depth camera, in which the tubes are assigned with unique colors. A general metric based on micro-recall is proposed to compare the accuracy of point cloud annotations with the ground truth. The synthetic data is representative of a high quality 3D scanner, given that the performance of a tube modeling system when given 640 simulated point clouds was similar to the results achieved with real sensor data. Therefore, simulation is a promising technique for the automated evaluation of solutions for bin picking tasks.

Keywords: bin picking; industrial robots; modeling; pose estimation; robot vision; simulation

Citation: Leão, G.; Costa, C.M.; Sousa, A.; Reis, L.P.; Veiga, G. Using Simulation to Evaluate a Tube Perception Algorithm for Bin Picking. *Robotics* **2022**, *11*, 46. https://doi.org/10.3390/robotics11020046

Academic Editor: Carlo Alberto Avizzano

Received: 13 February 2022
Accepted: 31 March 2022
Published: 5 April 2022

Publisher's Note: MDPI stays neutral with regard to jurisdictional claims in published maps and institutional affiliations.

1. Introduction

Bin picking consists of using a robotic manipulator to remove, one-by-one, a set of objects that are randomly stacked in a container. It is a complex problem faced by many manufacturing and production systems. One of the main challenges of bin picking is handling objects that are occluded or entangled. When items are entangled, the robot is prone to picking multiple units at once rather than a single one as expected, which can cause disruptions in assembly lines, namely by having objects fall outside the working area. Therefore, bin picking systems dealing with this sort of objects must be able to detect entanglement issues, so that they can operate reliably. This detection can be performed using depth sensors.

In order to accurately evaluate the performance of a robot perception algorithm, it is necessary to use a ground truth system, which consists of meta-data associated with the measurements acquired by a sensor. This meta-data allows for the comparison of the output of a perception algorithm with information that is known to be correct. For the particular case of the perception of the contents of a bin, the ground truth data consists of a point cloud, where each point is associated with the identifier of the object it belongs to. The meta-data can also include a geometric representation of each object.

This paper proposes using simulation to generate ground truth data for bin picking scenarios. Many state-of-the-art research publications on robotics conduct real-world experiments in order to evaluate the efficiency of their solutions. Using a simulator is an attractive alternative since a large quantity of experiments can be conducted in a much shorter amount of time and without the need for expensive robot hardware or manual annotation. Annotated data generated via simulation can be used as input and validation for various methods based on Artificial Intelligence (AI) for bin picking, namely those based on heuristics or machine learning.

This paper focuses on the case of bin picking of entangled tubes that can have multiple curvatures throughout their length. As an alternative to using fixed Computer-aided design (CAD) models of the items, a procedural generation algorithm is presented to randomly create curved tubes, which fall inside a virtual bin. This enables the creation of scenarios where there is intra-class variation of the items to be picked and allows for a first approximation of flexible objects. The Gazebo simulator, coupled with the Open Dynamics Engine (ODE), is responsible for ensuring the tubes' movements and spatial configurations are realistic. A virtual 3D scanner is responsible for acquiring point clouds of the scene.

On previous research, a perception algorithm that estimates the pose and shape of a set of curved tubes was evaluated manually by human annotators [1]. In order to present a practical application of the generator, synthetic point cloud data was used to evaluate this solution in an automated manner. A performance metric is proposed to compare how accurate the perception algorithm is, by comparing its output with the ground truth. A rigorous evaluation of the perception algorithm using real-life sensors would require manually annotating a large amount of point clouds, which would be highly impracticable and thus justifies the need for simulation.

The efficiency of the entangled tubes dataset generator was measured by creating two datasets, each one having 320 point clouds and different types of tubes. In addition, in order to assess the realism of the generated point clouds, the performance of the tube perception algorithm was compared with the results of previous work [1].

The rest of the paper is organized as follows. Section 2 presents related work on synthetic 3D data generation and simulation for bin picking. Section 3 presents the proposed system to generate synthetic point clouds of entangled tubes. Section 4 provides a summary of the tube modeling algorithm that was developed on previous work, while Section 5 presents the proposed method to evaluate its performance. Section 6 provides experimental results of the tube generator and the evaluation of the tube perception algorithm. Finally, Section 7 summarizes the contributions of this work, the main results and several lines of future research.

2. Related Work

2.1. Synthetic 3D Data Generation

The need for the generation of synthetic 3D data is not restricted to the field of robotics. Other areas include autonomous driving [2,3] and machine learning [3]. There are two main approaches to generating 3D synthetic data: using probabilistic models and sensing a point cloud from a simulated environment. The work presented in this paper falls into the second category.

Constructing 3D data with the aid of probabilistic models allows for more efficient memory representation (than using point clouds). Niemeyer et al. build 3D models using RGB images and implicit differentiation [4], while Yang et al. learn a probabilistic distribution for shapes which is then used to learn a distribution for the point cloud associated with the shape [5].

When generating point clouds using simulation, a virtual sensor is commonly used to capture 3D information of the scene. The simulation setting can be a video game [2,6], such as Grand Theft Auto V, or an open-source sandbox, such as Blender [7] or Unity [3]. On the one hand, acquiring 3D point clouds using video games has the strength of a large amount of content being readily available [6]. In order to validate the synthetic ground truth data, Yue et al. compare the data acquired by the sensor with an in-game camera, both placed in the same 3D position [2], whereas Richter et al. analyze the game's communication with the graphics hardware [6]. On the other hand, using an open-source simulator typically has the advantage of being relatively simple to parameterize the settings for the data collection (which allows for a more flexible generation of datasets). Wang et al. [7] use the descriptions returned by a ray casting algorithm in Blender to extract the points' labels.

2.2. Simulation for Bin Picking

There already exists some published work on dataset generation for bin picking using simulation. Schyja et al. [8] propose a framework for generating bins filled with objects using the DirectControl 3 simulation system that lacks the simulation of a 3D sensor. Kleeberger et al. [9] use the V-Rep simulator alongside the Bullet physics engine and change the pose of their bin between the generation of successive test cases. Matsumura et al. [10] perform simulation using the PhysX physics engine to generate datasets with annotations of whether a picking attempt was successful or not. This work presents some novelties with respect to the existing literature:

- All of these works resort to predefined 3D models of the items to be added to the bin, unlike this work where the shape of the items can be procedurally generated based on some parameters defined by the user.
- None of these works discuss in detail the time required for dataset generation with respect to the number of items and their shape.

The utility of simulation is not restricted to dataset generation. When facing the many challenges of bin picking, simulation techniques are often employed to solve one or more of its sub-tasks, including grasp planning [11,12], where grasp configurations are generated and evaluated, and motion planning [13,14], that determines how the robotic manipulator should move.

3. Entangled Tube Dataset Creation

The generation of labeled point clouds is performed using a plugin for Gazebo [15], a 3D simulator commonly used by the robotics community and maintained by Open Robotics. The virtual world created in Gazebo contains a 3D scanner that is positioned looking downwards towards a bin. This sensor is responsible for acquiring point clouds of the scene, using methods that are built-in in Gazebo.

A simulated tube is composed of a linked list of cylinders which are connected by their endpoints. Using multiple cylinders allows for the emulation of curved tubes. The cylinders have a constant radius. To ensure a tube's surface is smooth, a sphere with the same radius as the cylinders is created for each endpoint that connects two cylinders.

In order to create a point cloud for the dataset, the simulated world is initialized with an empty bin. Algorithm 1 is then called in order to create each tube. This algorithm calls two auxiliary methods: Algorithm 2 is used to randomly determine the amount and length of each cylinder of a tube, while Algorithm 3 randomly defines the exact position of each cylinder endpoint. Figure 1 depicts an example of the main steps of the procedural generation of a tube. As it can be seen in the input for Algorithm 1, the generator is highly flexible due to a high amount of parameters that can be adjusted for different bin picking scenarios. Moreover, the dimensions of the bin and the pose of the virtual sensor are customizable.

Each tube is spawned at a certain height above the bin and at a position that enables it to fall inside the bin. After spawning all of the tubes, the plugin waits for all of them to stop moving by checking that the linear and angular velocity of their center of mass is below a certain threshold. This increases the stability of the generation. Making the tubes drop into the bin rather than simply spawning them without gravity provides much more realism for the tubes' spatial configurations. Moreover, spawning all the tubes at once above the bin replicates the strategy employed by manufacturers to fill the bins, which increases the cases of entanglement and makes the test cases more similar to reality. The simulator detects and corrects possible overlaps between tubes.

Algorithm 1 Tube random generation algorithm.

Input:
    n—Number of tubes to generate
    r—Tube radius
    l—Tube length
    cl_{min}—Min cylinder length
    cn_{min}, cn_{max}—Min/max number of cylinders
    $\theta_{min}, \theta_{max}$—Min/max angle between two consecutive cylinders
Output:
    tubes—Set of generated tubes
1: $tubes \leftarrow \varnothing$
2: **for** $m \in \{0, \dots, n-1\}$ **do**
3:     $tube \leftarrow \varnothing$
4:     $cn \leftarrow RandomInteger(cn_{min}, cn_{max})$
5:     $lengths \leftarrow GenerateLengths(cn, l, cl_{min})$
6:     $endpoints \leftarrow GenerateEndpoints(lengths, \theta_{min}, \theta_{max})$
7:     **for** $i \in \{0, \dots, cn-1\}$ **do**
8:        $tube \leftarrow tube \cup CreateCylinderWithRadiusAndEndpoints(r, endpoints[i], endpoints[i+1])$
9:        **if** $i > 0$ **then**
10:           $tube \leftarrow tube \cup CreateSphereWithRadiusAndCenter(r, endpoints[i])$
11:     $tubes \leftarrow tubes \cup tube$
12: **return** $tubes$

Algorithm 2 Cylinder lengths generation algorithm.

Input:
    cn—Number of cylinders
    l—Tube length
    cl_{min}—Min cylinder length
Output:
    lengths—Sequence of cylinder lengths
1: $lengths \leftarrow []$
2: $remainingLength \leftarrow l$
3: **for** $i \in \{0, \dots, cn-2\}$ **do**
4:     $cl \leftarrow RandomInteger(cl_{min}, remainingLength - (cn-1-i) * cl_{min})$
5:     $lengths.Push(cl)$
6:     $remainingLength \leftarrow remainingLength - cl$
7: $lengths.Push(remainingLength)$
8: **return** $lengths$

Algorithm 3 Cylinder endpoints generation algorithm.

Input:
    lengths—Sequence of cylinder lengths
    $\theta_{min}, \theta_{max}$—Min/max angle between two consecutive cylinders
Output:
    endpoints—Sequence of cylinder endpoints
1: $previousEndpoint \leftarrow (0, 0, 0)$
2: $endpoints \leftarrow [previousEndpoint]$
3: **for** $i \in \{0, \dots, Length(lengths) - 1\}$ **do**
4:     $\phi \leftarrow RandomFloat(0, 2\pi)$
5:     **if** $i = 0$ **then**
6:        $\theta \leftarrow RandomFloat(0, \pi)$
7:        $unitDirection \leftarrow (sin(\theta) * cos(\phi), sin(\theta) * sin(\phi), cos(\theta))$
8:     **else**
9:        $\theta \leftarrow RandomFloat(\theta_{min}, \theta_{max})$
10:       $unitDirection \leftarrow (sin(\theta) * cos(\phi), sin(\theta) * sin(\phi), cos(\theta))$
11:       $unitDirection \leftarrow RotateVectorAroundAxisWithAngle(unitDirection,$
    $CrossProduct((0, 0, 1), previousDirection), DotProduct((0, 0, 1), previousDirection))$
12:     $nextEndpoint \leftarrow previousEndpoint + lengths[i] * unitDirection$
13:     $previousDirection \leftarrow nextEndpoint - previousEndpoint$
14:     $previousEndpoint \leftarrow nextEndpoint$
15:     $endpoints.Push(nextEndpoint)$
16: $endpoints \leftarrow ApplyRandomTranslation(endpoints)$
17: **return** $endpoints$

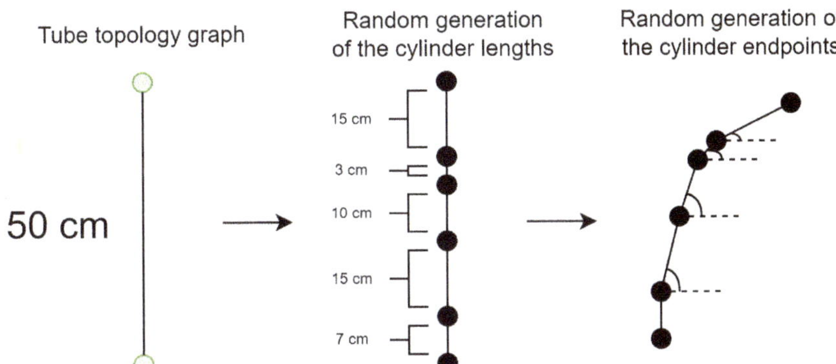

Figure 1. An example of the main steps of the procedural generation of a tube.

A unique color is assigned to each tube that is created. This color is used as a ground truth system to indicate to which tube each point of the cloud belongs. Shadows and light sources are disabled (with the exception of an ambient light) to ensure that only (n + 1) different colors are present in a scene with n tubes: one for each tube and one for the bin.

4. Tube Modeling Algorithm

The tube modeling algorithm that was evaluated was proposed and implemented by Leão et al. [1]. It receives as input a point cloud of the bin and its contents. The algorithm has two outputs:

- A data structure that describes the shape and spatial configuration of the tubes (a set of linked lists of cylinders).
- A subset of the input point cloud where each point is labeled with a color that is unique to each tube that was constructed. The points of the input point cloud that are not present in this output correspond to those for which it is not clear to which tube they belong.

For the purposes of performance evaluation, the latter output was used.

The algorithm was implemented using Point Cloud Library (PCL) [16], a modular C++ library with a vast array of helpful functions for point cloud processing.

The solution performs four main steps: filtering, segmentation, cylinder fitting and tube joining. Figure 2 presents an example of each step of the algorithm. Figure 3 contains the color image of the tubes corresponding to the (uncolored) input point cloud (which is not used by the modeling algorithm).

The filtering step removes the points from the point cloud which do not belong to any of the tubes, but rather to the bin's walls and bottom. In addition, the cloud may be downsampled in order to reduce the processing time of the following steps, and the surface normals may be estimated.

The segmentation step clusters the point cloud into disjoint piece-wise smooth regions, which correspond to visible portions of a tube. This step uses a region-growing algorithm based on the surface normals.

The cylinder fitting step associates a set of cylinders to each cluster obtained from the previous set. Cylinder fitting is performed using a Random Sample Consensus (RANSAC) algorithm.

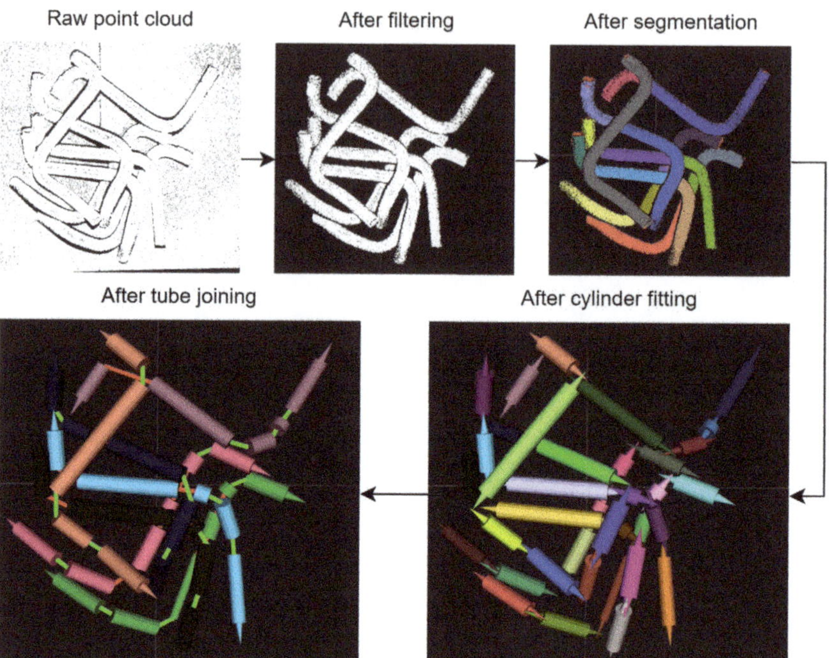

Figure 2. An example of the result of each step of the tube modeling solution.

Figure 3. Corresponding color image of the picking scene for the tube modeling example of Figure 2.

The tube joining step combines the cylinders to form into increasingly longer, more complete tubes. This is performed using a greedy method that considers all pairs of endpoints of distinct cylinders, and orders these pairs using a cost function that considers their Euclidean and angular distance. The pairs are then processed in increasing order of cost. Each *virtual tube* created by this step has a unique color assigned to it, the *label color*.

This modeling solution is described in greater detail in its original publication from 2019 [1].

In order to prepare the labeled point cloud, for each point that was left unlabeled after the tube joining step, it is determined if they are sufficiently close to a cylinder or joint of one of the tubes. If they are close to at least one of the objects, then the point is labeled with the color of the tube that has the closest cylinder or joint to it.

5. Evaluation

The evaluation of the modeling algorithm's accuracy is done by comparing two point clouds returned by this solution: the *filtered cloud*, which is the result of the filtering step, and the *labeled cloud*, which only contains points that were fitted into one of the cylinders. The points of the former cloud have the same color as the input cloud (i.e., the *ground truth colors*), while those from the latter have the same color as the object/tube they belong to (i.e., the *label colors*).

The labeled cloud is a subset of the filtered cloud, since some points may not be used by any of the calls to RANSAC to form a cylinder and may be too far away from all cylinders and joints. These points are labeled with a *null color* $c_\varnothing$, which is different from those of the virtual tubes. Pairs of corresponding points between the two clouds (i.e., points with the same position) have different colors. Thus, the first step needed to compare both clouds is to establish a mapping between the ground truth and label colors. This corresponds to the well-known maximum-cost unbalanced assignment problem in a weighted bipartite graph, where an edge connecting a ground truth color C to a label color c has a weight equal to the number of corresponding points with these colors in their respective clouds. The assignment problem is solved using the Hungarian method [17].

For illustrative purposes, Figure 4 depicts a simplified example of a point cloud with 72 points obtained from scanning a scene with two tubes. The left-hand image contains the point cloud labeled with two ground truth colors, one for each tube. The right-hand image contains one possible labeling of the same point cloud by the tube modeling algorithm, where four colors were defined. Three of these colors are associated with virtual tubes, which means that the tube modeling algorithm (incorrectly) determined that the scene contains three tubes, while the fourth color is the null color. For increased clarity, in both images, distinct shapes are used to represent the points labeled with each color. Figure 5 shows the corresponding bipartite graph for the example, where the edges whose weight is outlined by rectangles represent the optimal assignment: the ground truth colors C_1 and C_2 were associated with the label colors c_1 and c_2, respectively.

Let M be the inferred matching function between the two color spaces and C the number of ground truth colors. For the i-th ground truth color C_i, the corresponding label color is $c_i = M(C_i)$. For all $i \in [1, C]$, the matched label color c_i is never equal to the null color $c_\varnothing$. If the modeling solution creates too many tubes, some of the label colors may not be matched with any of the ground truth colors. Conversely, if too few virtual tubes are built, some of the ground truth colors may not be matched.

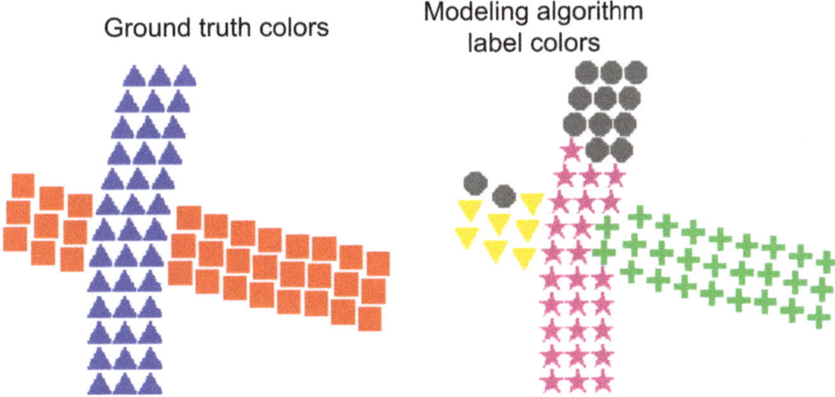

Figure 4. Example of a small point cloud with ground truth and label colors.

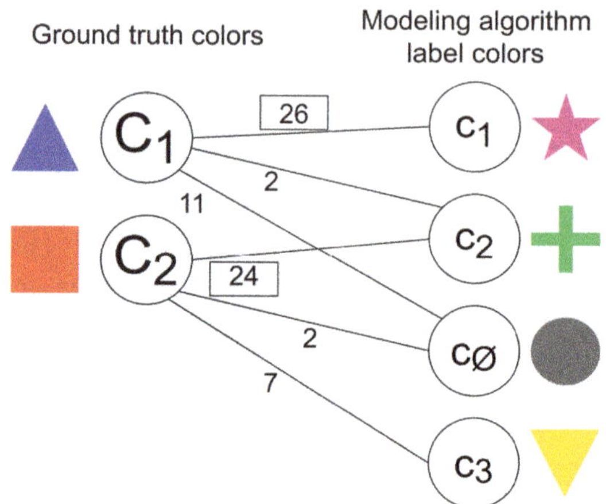

Figure 5. Bipartite graph of Figure 4, which represents the matching between the two sets of colors.

The value used to compute the tube modeling algorithm's performance is presented in Equation (1). It corresponds to the micro-recall metric $recall_\mu$ used for multi-class classification [18]. This metric was used for evaluation since it provides a more intuitive value for the ratio of points that were correctly labeled with respect to other classic alternatives used in multi-class classification.

$$recall_\mu = \frac{\text{number of correctly labeled points}}{\text{total number of points}} = \frac{\sum_{i=1}^{C} tp_i}{\sum_{i=1}^{C}(tp_i + fn_i)} \tag{1}$$

In Equation (1), tp_i denotes the amount of points whose colors in the ground truth and labeled cloud are respectively C_i and $c_i = M(C_i)$, and fn_i are the points with the color C_i and a label color distinct from c_i. In other words, the micro-recall metric divides the number of correctly labeled points by the total amount of points in the filtered cloud. The labeled cloud is compared with the filtered cloud and not with the one used as input for the modeling algorithm to avoid penalizing the solution unfairly for filtering points it does not require for its computations (during the filtering step). In the example from Figure 5, the value for the micro-recall is equal to $\frac{26 + 24}{26 + 2 + 11 + 24 + 2 + 7} = \frac{50}{72} \approx 0.694$.

It should be noted that the performance metric proposed in this section is generic and can be applied to the result of any algorithm that labels the elements of a point cloud according to which item they belong to, by comparing it with the ground truth.

6. Results

Two sets of experiments were conducted. The first set aims to assess the efficiency of the simulator on creating datasets. The second set aims to apply the created datasets to evaluate in a more rigorous manner the tube perception solution described in Section 4.

All of the results were obtained with an Intel Core i7-8750H processor (2.20 GHz), a 32 GB DDR4 RAM (2667 MHz), and a NVidia GeForce GTX 1050 Ti GPU.

6.1. Dataset Generation Efficiency

Figure 6 depicts two examples of simulated scenes in Gazebo with bins filled with 7 randomly generated tubes. As shown in the figure, two datasets were created, each one with a set of tubes with different properties: sets A and B. For each dataset, 20 point clouds (i.e., test cases) were created for different amounts of tubes, ranging from 1 to 15 tubes. In addition, for each dataset, 20 test cases of clouds with 20 tubes were created, to further test the generator's capacities. As a result, each dataset contains 320 test cases. All of the generated datasets were made publicly available: the *Entangled Tubes Bin Picking* dataset is available at https://github.com/GoncaloLeao/Entangled-Tubes-Bin-Picking-Dataset (accessed on 13 January 2022).

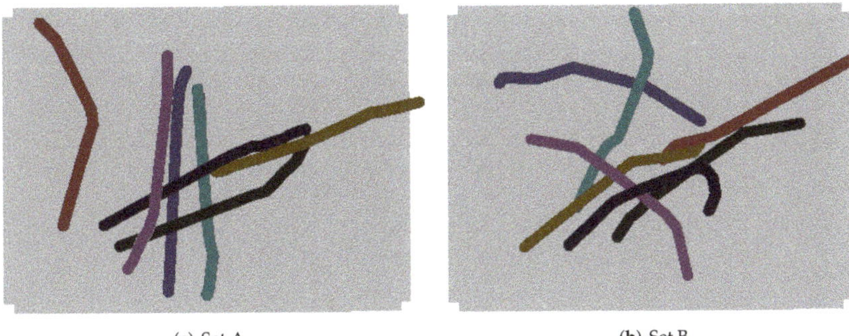

(a) Set A   (b) Set B

Figure 6. Example of a set of 7 tubes for each synthetic dataset.

The tubes of sets A and B have similar properties to the real-life tubes made of Polyvinyl Chloride (PVC) used on the experiments of previous research [1], which are depicted in Figure 7. Point clouds of these tubes captured by real-life sensors on previous work are also publicly available (within the *Entangled Tubes Bin Picking* dataset). These tubes have a diameter of 2.5 cm and a length of 50 cm. They do not contain any bifurcations and are bent with arbitrary angles. The difference between sets A and B lies on the minimum and maximum angle between consecutive cylinders (parameters θ_{min} and θ_{max} of Algorithm 3). For set A, the angles range from $0°$ to $30°$, and for set B, they range from $0°$ to $45°$. Thus, the tubes of set B have more curvatures throughout their length than those of set A.

Figure 7. PVC tubes used on previous research [1] and geometrically similar to those of sets A and B.

For the generation of all the datasets, the dimensions of the bin were 55 cm (length) $\times$ 37 cm (width) $\times$ 20 cm (height). The bin was slightly larger than the one used on previous research to better accommodate larger amounts of tubes. To allow the virtual 3D scanner to perceive the whole working area, it was positioned looking downwards towards the bin, with a vertical distance of 2 m relative to the bin's bottom. The model of the virtual sensor was set to mimic the properties of a Zivid One Plus L scanner, which was used on previous work [1]. It produces depth images with dimensions 1920 $\times$ 1200.

Gazebo's dynamics engine was set to run three times faster than real-time. This factor was set empirically to balance simulation stability and the time needed for the tubes in the bin to stop moving.

As expected, according to Table 1, the time needed to generate a point cloud increases with the amount of tubes in the bin, since more time is needed to allow the tubes to stabilize after being dropped in the bin. On average, set B requires slightly less time than set A to generate a test case. The authors theorize this is due to the tubes of set B occupying, on average, a smaller bounding box than those of set A, due to having larger curvatures. This may reduce the amount of collisions between the tubes, and thus decrease the time needed for the items to stabilize within the bin.

Table 1. Average time needed to generate a test case for each dataset with respect to the number of tubes.

Number of Tubes	Set A	Set B
1	3.13 s	2.68 s
2	6.32 s	5.84 s
3	8.86 s	7.39 s
4	8.77 s	7.57 s
5	8.25 s	8.90 s
6	11.41 s	9.21 s
7	10.39 s	9.81 s
8	10.58 s	10.75 s
9	10.82 s	10.95 s
10	14.20 s	13.73 s
11	14.95 s	13.45 s
12	16.54 s	15.48 s
13	14.53 s	15.85 s
14	16.16 s	16.81 s
15	16.73 s	16.78 s
20	28.10 s	25.02 s

Increasing the simulation speed tends to decrease the tube stabilization time, but using overly large values causes the simulation to become unstable and raises the likelihood of a tube clipping through the bin and falling out of bounds, which invalidates a generation attempt. Thus, fine-tuning the simulation speed may increase substantially the efficiency of this dataset generator.

6.2. Tube Modeling Algorithm Performance

Sets A and B were used to evaluate the performance/accuracy of the tube modeling solution. The parameters used to test the modeling algorithm are identical to the ones used on previous research [1], with the exception of the filtering step. Since the exact position and dimensions of the bin are known, a crop box filter was used to remove the points belonging to the bin, leaving only the points that belong to the tubes. A second filter is then applied to randomly downsample the point cloud by a fixed ratio, in order to reduce the computational effort of the solution. For each of the 20 point clouds of sets A and B with each value for the number of tubes, the tube modeling solution was executed 10 times with a distinct ratio for the downsampling filter between 0.1 and 1.0 (where the cloud is not downsampled), with a step of 0.1.

As depicted in Figure 8, the performance of the perception algorithm is reduced linearly as the number of tubes increases, since the number of cases of occlusions and entanglement increases, making it harder to accurately construct the tubes. This effect was also observed on previous research. In addition, the solution consistently performed better on set A than on set B. This is most likely due to the shape of the tubes on set B being more complex.

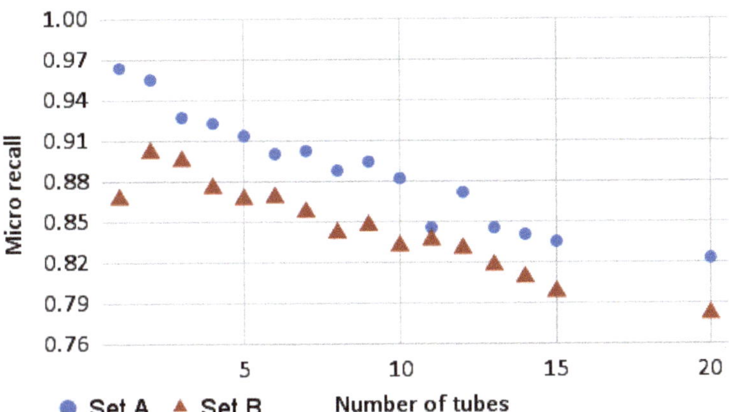

Figure 8. Average micro-recall for sets A and B with respect to the number of tubes using a down-sampling ratio of 0.3.

It is interesting to note that the performance of set B for the cases with a single tube has a significantly lower performance than the cases with 2 to 4 tubes, with an approximate value of 0.867. The authors hypothesize this is due to issues related to the segmentation step of the solution that were observed during the experiments, where the region-growing algorithm was occasionally not able to associate many of the points to a cluster. This prevented the fitting step from creating a sufficient amount of cylinders to cover the whole tube. Since the micro-recall metric gives equal weight to all the points, on clouds with a single tube, the occurrence of this issue is more prevalent on the performance metric, than on larger point clouds.

The issue with the segmentation step problem can be considerably mitigated by fine-tuning the parameters of the region-growing algorithm. A calibration procedure could be developed to adjust the perception algorithm's parameters according to the geometrical properties of the tubes.

This particular case of the performance of the solution being considerably lower on cases with a single tube was also observed on previous research, where the performance metrics were estimated manually by human annotators:

- According to Table 2 of the previous publication by Leão et al. [1], the ratio of "correctly labeled tubes" (obtained by dividing the accuracy metric "Number of correctly labeled tubes" by the actual number of tubes) increased by 25% from 1 to 2 tubes (the largest relative increase from two consecutive tube amounts) and decreased by 12.5% from 1 to 10 tubes, reaching the lowest accuracy value of those between 1 and 10 tubes.
- In a similar trend, the micro-recall for Set B with a downsampling ratio of 0.3 increased by approximately 3.9% from 1 to 2 tubes (also the largest relative increase from two consecutive tube amounts) and decreased by approximately 4.1% from 1 to 10 tubes, also reaching the lowest accuracy value of those between 1 and 10 tubes.

This helps to establish that the micro-recall metric allows for a reliable evaluation of the modeling algorithm, or, at the very least, it contributes to show that it has a similar bias to the performance metrics of previous work. Moreover, the results suggest that using synthetic datasets allows one to reach similar conclusions as those obtained with real sensors.

According to Figure 9, as expected, for both sets, as the downsampling ratio increases (i.e., as the number of points used in the segmentation step increases), the performance of the solution also increases, since the algorithm has more data to work with. Rather than the performance increasing linearly with the ratio, it is particularly sensitive to lower values for the downsampling ratio. For ratios above 0.6 for set A and 0.7 for set B, further increasing the ratio does not lead to significant increases on the micro-recall.

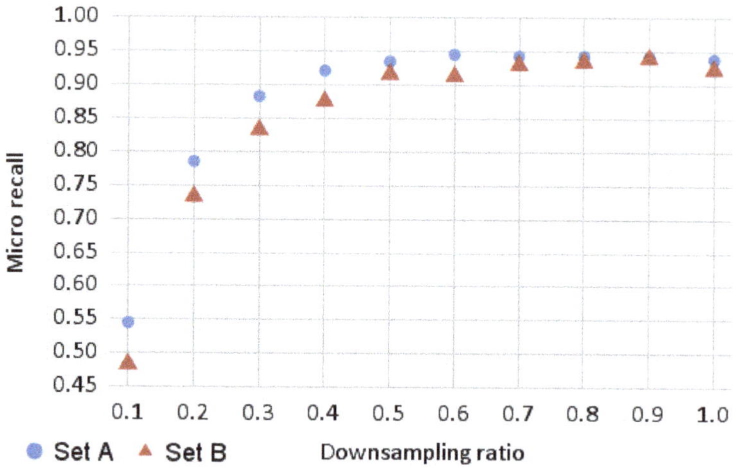

Figure 9. Average micro-recall for sets A and B with respect to the downsampling ratio for bins with 10 tubes.

As shown in Table 2, the execution time of the modeling algorithm increases with both the downsampling ratio and the number of tubes, since increasing any of these parameters leads to the point cloud provided to the segmentation step being larger. This behavior is similar for set B. The increase of the execution time with the number of tubes was also observed on previous work, thus reinforcing the similarities in the results between using simulated and real sensors:

- According to Table 1 of the previous publication by Leão et al. [1], the execution time increased by approximately 38.6% from 1 to 5 tubes and by 32.0% from 5 to 10 tubes.
- In a similar trend, according to Table 2, the execution time for Set A with a downsampling ratio of 0.1 increased by approximately 68.4% from 1 to 5 tubes and by 50.0% from 5 to 10 tubes.

By combining the observations from Figure 9 and Table 2, it can be concluded that there is a trade-off between execution time and modeling accuracy that is regulated by the size of the input point cloud. Thus, it is important to fine-tune the downsampling ratio with respect to the application domain.

Table 2. Average execution time of the tube modeling algorithm for set A with respect to the downsampling ratio and number of tubes.

Ratio \ Number of Tubes	1	5	10	15	20
0.1	0.19 s	0.32 s	0.48 s	0.60 s	0.78 s
0.2	0.19 s	0.41 s	0.69 s	0.90 s	1.23 s
0.3	0.21 s	0.49 s	0.85 s	1.19 s	1.69 s
0.4	0.23 s	0.55 s	1.02 s	1.43 s	2.06 s
0.5	0.24 s	0.61 s	1.12 s	1.69 s	2.37 s
0.6	0.26 s	0.68 s	1.25 s	1.90 s	2.50 s
0.7	0.27 s	0.73 s	1.40 s	2.21 s	2.81 s
0.8	0.29 s	0.82 s	1.56 s	2.43 s	3.11 s
0.9	0.31 s	0.88 s	1.54 s	2.71 s	3.38 s
1	0.32 s	0.91 s	1.65 s	2.84 s	3.66 s

7. Conclusions

The main contribution of this work is a flexible solution that generates labeled point clouds of entangled tubes procedurally. This allows for variations on the shape of each tube in the point cloud, thus producing rich ground truth data. Experimental results showed that it takes less than 15 seconds to generate a bin with 10 tubes. Therefore, in a single hour, a high-quality dataset of 240 point clouds can be built, which makes this generator suitable for input to a wide variety of robotic and AI-based systems, namely those that use heuristics or machine learning.

The second main contribution is a general metric based on micro-recall to evaluate the accuracy of an algorithm that labels each point of a point cloud according to which item it belongs to, such as the tube perception solution developed in previous research [1]. This metric constitutes a significant improvement over the one proposed by previous work, since it is objective and can be performed automatically. Experiments with synthetic datasets and the micro-recall metric produced similar results to those found in previous work, where the data was acquired by physical sensors and the performance was evaluated manually with human annotators. This suggests that the generated point clouds are realistic and that the micro-recall metric is reliable.

This work opens several lines for future research. Firstly, the procedural generator could be extended to generate objects with other shapes, such as tubes with bifurcations, spherical items or objects with holes in them. The evaluation metric proposed in this work is already generic enough to cope with these shapes. Secondly, the dataset generator could be used to further improve the tube modeling solution by providing input to a parameter calibration procedure that takes into account the geometric properties of the items. Finally, as alternatives or complements to the tube modeling solution, supervised learning systems (namely deep learning systems) could be developed to create a model of the objects inside the bin using the test cases provided by the generator for training. These alternative solutions can be capable of modeling a wider variety of objects.

Author Contributions: Conceptualization, G.L., C.M.C., A.S., L.P.R. and G.V.; methodology, G.L., C.M.C., A.S., L.P.R. and G.V.; software, G.L.; validation, G.L.; formal analysis, G.L.; investigation, G.L.; resources, G.L., C.M.C., A.S., L.P.R. and G.V.; data curation, G.L.; writing—original draft preparation, G.L.; writing—review and editing, G.L., C.M.C., A.S., L.P.R. and G.V.; visualization, G.L.; supervision, C.M.C., A.S., L.P.R. and G.V.; project administration, G.L.; funding acquisition, A.S. and G.V. All authors have read and agreed to the published version of the manuscript.

Funding: This work was funded by National Funds through the Portuguese funding agency Fundação para a Ciência e a Tecnologia (FCT), within the PhD studentship 2020.06923.BD. This work was also financed by the ESF—European Social Fund through the Norte Portugal Regional Operational Programme—NORTE 2020 under the Portugal 2020 Partnership Agreement within project RHAQ, with reference NORTE-06-3559-FSE-000116.

Institutional Review Board Statement: Not applicable.

Informed Consent Statement: Not applicable.

Data Availability Statement: The *Entangled Tubes Bin Picking* dataset is available at https://github.com/GoncaloLeao/Entangled-Tubes-Bin-Picking-Dataset (accessed on 13 January 2022).

References

1. Leão, G.; Costa, C.M.; Sousa, A.; Veiga, G. Perception of Entangled Tubes for Automated Bin Picking. In Proceedings of the Iberian Robotics Conference (ROBOT), Porto, Portugal, 20–22 November 2019; Volume 1092, pp. 619–631. [CrossRef]
2. Yue, X.; Wu, B.; Seshia, S.A.; Keutzer, K.; Sangiovanni-Vincentelli, A.L. A LiDAR Point Cloud Generator: From a Virtual World to Autonomous Driving. In Proceedings of the International Conference on Multimedia Retrieval (ICMR), Yokohama, Japan, 11–14 June 2018; Volume 18, pp. 458–464. [CrossRef]
3. Ros, G.; Sellart, L.; Materzynska, J.; Vazquez, D.; Lopez, A.M. The SYNTHIA Dataset: A Large Collection of Synthetic Images for Semantic Segmentation of Urban Scenes. In Proceedings of the IEEE Conference on Computer Vision and Pattern Recognition (CVPR), Las Vegas, NV, USA, 27–30 June 2016; pp. 3234–3243. [CrossRef]
4. Niemeyer, M.; Mescheder, L.; Oechsle, M.; Geiger, A. Differentiable Volumetric Rendering: Learning Implicit 3D Representations without 3D Supervision. In Proceedings of the IEEE Conference on Computer Vision and Pattern Recognition (CVPR), Seattle, WA, USA, 16–18 June 2020; pp. 3501–3512. [CrossRef]
5. Yang, G.; Huang, X.; Hao, Z.; Liu, M.Y.; Belongie, S.; Hariharan, B. Pointflow: 3D point cloud generation with continuous normalizing flows. In Proceedings of the IEEE International Conference on Computer Vision (ICCV), Seoul, Korea, 27 October–2 November 2019; pp. 4541–4550. [CrossRef]
6. Richter, S.R.; Vineet, V.; Roth, S.; Koltun, V. Playing for Data: Ground Truth from Computer Games. In Proceedings of the European Conference on Computer Vision (ECCV), Amsterdam, The Netherlands, 11–14 October 2016; Volume 9906, pp. 102–118. [CrossRef]
7. Wang, F.; Zhuang, Y.; Gu, H.; Hu, H. Automatic Generation of Synthetic LiDAR Point Clouds for 3-D Data Analysis. *IEEE Trans. Instrum. Meas.* **2019**, *68*, 2671–2673. [CrossRef]
8. Schyja, A.; Kuhlenkötter, B. Realistic simulation of Industrial Bin-Picking Systems. In Proceedings of the International Conference on Automation, Robotics and Applications (ICARA), Queenstown, New Zealand, 17–19 February 2015; pp. 137–142. [CrossRef]
9. Kleeberger, K.; Landgraf, C.; Huber, M.F. Large-scale 6D Object Pose Estimation Dataset for Industrial Bin-Picking. In Proceedings of the IEEE/RSJ International Conference on Intelligent Robots and Systems (IROS), Macau, China, 3–8 November 2019; pp. 2573–2578. [CrossRef]
10. Matsumura, R.; Harada, K.; Domae, Y.; Wan, W. Learning based industrial bin-picking trained with approximate physics simulator. In Proceedings of the International Conference on Intelligent Autonomous Systems (IAS), Baden-Baden, Germany, 11–15 June 2018; Volume 867, pp. 786–798. [CrossRef]
11. de Souza, J.P.C.; Costa, C.M.; Rocha, L.F.; Arrais, R.; Moreira, A.P.; Pires, E.J.; Boaventura-Cunha, J. Reconfigurable Grasp Planning Pipeline with Grasp Synthesis and Selection Applied to Picking Operations in Aerospace Factories. *Robot. Comput.-Integr. Manuf.* **2021**, *67*, 102032. [CrossRef]
12. Miller, A.T.; Allen, P.K. Graspit: A versatile simulator for robotic grasping. *IEEE Robot. Autom. Mag.* **2004**, *11*, 110–122. [CrossRef]
13. Leão, G.; Costa, C.M.; Sousa, A.; Veiga, G. Detecting and Solving Tube Entanglement in Bin Picking Operations. *Appl. Sci.* **2020**, *10*, 2264. [CrossRef]
14. Iversen, T.F.; Ellekilde, L.P. Benchmarking motion planning algorithms for bin-picking applications. *Ind. Robot.* **2017**, *44*, 189–197. [CrossRef]
15. Koenig, N.; Howard, A. Design and use paradigms for gazebo, an open-source multi-robot simulator. In Proceedings of the IEEE/RSJ International Conference on Intelligent Robots and Systems (IROS), Sendai, Japan, 28 September–2 October 2004; Volume 3, pp. 2149–2154. [CrossRef]
16. Rusu, R.B.; Cousins, S. 3D is here: Point Cloud Library (PCL). In Proceedings of the IEEE International Conference on Robotics and Automation (ICRA), Shanghai, China, 9–13 May 2011; pp. 1–4. [CrossRef]
17. Ramshaw, L.; Tarjan, R.E. *On Minimum-Cost Assignments in Unbalanced Bipartite Graphs*; Technical Report; HP Laboratories: Palo Alto, CA, USA, 2012.
18. Sokolova, M.; Lapalme, G. A systematic analysis of performance measures for classification tasks. *Inf. Process. Manag.* **2009**, *45*, 427–437. [CrossRef]

Article

A New Hyperloop Transportation System: Design and Practical Integration

Mohammad Bhuiya, Md Mohiminul Aziz, Fariha Mursheda, Ryan Lum, Navjeet Brar and Mohamed Youssef *

Power Electronics and Drives Applications Lab (PEDAL), Ontario Tech University,
Oshawa, ONT L1G 0C5, Canada; Mohd.bhuiya@ontariotechu.net (M.B.); md.aziz@uoit.net (M.M.A.);
Fariha.mursheda@uoit.net (F.M.); ryan.lum@uoit.net (R.L.); navjeet.brar@uoit.net (N.B.)
* Correspondence: Mohamed.youssef@ontariotechu.ca; Tel.: +1-905-721-8668 (ext. 5473)

Abstract: This paper introduces a new Hyperloop transportation system's design and implementation. The main contribution of this paper is the design and integration of propulsion components for a linear motion system, with battery storage. The proposed Hyperloop design provides a high-speed transportation means for passengers and freights by utilizing linear synchronous motors. In this study, a three-phase inverter was designed and simulated using PSIM. A prototype of this design was built and integrated with a linear synchronous motor. The operation of full system integration satisfies a proof-of-concept design. A study of the inverter system in conjunction with a linear synchronous motor for a ridged Hyperloop system is made. The prototype of this system achieves propulsion for the bidirectional movements. Battery state of charge simulation results are given in a typical motoring and braking scenario.

Keywords: three-phase inverter; linear synchronous motor (LSM); magnetic levitation; magnetic propulsion; permanent magnet motors

Citation: Bhuiya, M.; Aziz, M.M.; Mursheda, F.; Lum, R.; Brar, N.; Youssef, M. A New Hyperloop Transportation System: Design and Practical Integration. *Robotics* **2022**, *1*, 23. https://doi.org/10.3390/robotics11010023

Academic Editors: António Paulo Moreira, Félix Vilariño and Pedro Neto

Received: 14 December 2021
Accepted: 29 January 2022
Published: 8 February 2022

Publisher's Note: MDPI stays neutral with regard to jurisdictional claims in published maps and institutional affiliations.

1. Introduction

Today, transportation is a major tool for a growing economy, ranging from daily commutes to large scale freight transportation. However, with the ever-growing increase in population and thus the demand for high-speed transportation raises issues regarding air pollution and climate change. According to data for global emissions in 2020, it was found that 24% of the global greenhouse gas emission is due to fuel dependent transportation [1]. Energy security and climate change are major challenges that need to be addressed for the future generation. In 2016, the Paris Agreement was officially instated for the parties in the agreement to address the climate change problem. This agreement is only a stepping stone to combat this issue. This paper studies the design of a conceptual, novel, and electric mode of transportation known as the Hyperloop. Companies such as Virgin Hyperloop One, Transpod Hyperloop, and HyperloopTT are investing in the research and development of the proposed Hyperloop technology. The Hyperloop system offers a promising alternative to conventional train transportation, presenting a safer, faster, more reliable, and environmentally friendly method of transportation in contrast to conventional transportation. Hyperloop transport is a method of passenger and freight transport that uses vactrain design, incorporating low pressure and air resistance within a tube. Various studies such as [2,3] propose the concepts of utilizing linear motors to facilitate the propulsion and levitation force. The study in [3] presents a Hyperloop system based on electromagnetic propulsion of vehicles (similar technology in magnetic levitated trains) in vacuum tubes to reduce air pressure. Levitation is achieved by generating a repulsive force from the tube levitation system onto the vehicle. Additionally, propulsion is achieved through a traction force generated by moving magnetic field created by linear multi-pole motors [3]. The propulsion force is generated through a spatially moving magnetic field. This is achieved by a three-phase inverter powering and controlling the linear motor [3].

The levitation of a Hyperloop can be achieved either by electromagnetic suspension (EMS) or electrodynamic suspension (EDS) [4]. Due to the complexity of EMS [5], the design discussed in this paper utilizes the concepts of EDS [6]. However, in practicality there are many other factors, such as cost, that come into play. Thus, the use of EDS may not be recommended for a full-scale system. There are two types of linear motors that are commonly studied regarding magnetic levitation trains: the linear induction motor (LIM) such as the one employed in [3], and the linear synchronous motor (LSM). Historically, the LIM is a popular choice for magnetic levitation train. It has attractive features such as cost and low complexity. However, LSM has a high-power factor and provides higher efficiency [7–9]. The magnetic levitation system uses sets of superconducting magnets to create repulsion and attractive forces allowing for levitation and propulsion in conjunction with a linear synchronous motor [10]. The pod (vehicle) in this system travels along a guideway of magnets. Stability and speed control is facilitated by sophisticated control algorithms such as field-oriented control [5,10–12]. The foundation of this system is the use of electromagnetism concepts as applied to thrust generation. The magnetic fields are produced by permanent and superconducting magnets and electromagnets in a static or dynamic mode. Superconducting electromagnets require an electric current that is significantly less than conventional electromagnets [13]. The repulsive and attraction forces allow for levitation and propulsion without contact. A higher lift force to magnetic ratio is ideal for efficient energy consumption. Forces created depend on the placement of polarities of the magnets. Lastly, the control system for the LSM drive follows the field-oriented control approach. This popular control approach has been studied in various papers such as [5,14]. The study in [14] develops electromagnetic thrust and levitation force via field-oriented control. Various other papers, such as the one reported in [15], study the control of AC motors using the d-q axes model. The dynamic model of a permanent magnet linear synchronous motor (PMLSM) is reported in [16]. Hyperloop topics have been reported in some literature works such as [17,18]. The work in [17] reports the achievability of a Hyperloop system. They present the technical issues in building such an infrastructure. They estimate that a Hyperloop pod traveling at 1200 km/h for an estimated weight of 26,000 kg would require 689 kW of power. Additionally, they calculated that the cost for building the Hyperloop infrastructure in Poland of 3000–30,000 km long tracks would cost over $50 billion. In [18], the authors discuss the electric power requirements for a full-scale Hyperloop. Currently, there is limited research on the topic of integration of Hyperloop propulsion and levitation system via linear synchronous motors. This paper is uniquely structured to give the design of the inverter, linear motor, and overall integration of the system.

2. LSM Model and Controller Design

2.1. Modeling of the Linear Synchronous Motor

The core structure of the prototype is the linear synchronous motor (LSM) which makes up the pod and the track. A diagram for an LSM is shown in Figure 1, which is an unrolled form of a traditional rotary motor which consists of a stator and a rotor. The symbol 'τ' represents the pole pitch. It is important to understand the rotary synchronous machine and its properties to correctly form mathematical equations describing the motor characteristics. In terms of rotary machines, characteristics such as torque, angular velocity, and number of pole pairs can be defined. The linear machine can be defined with similar characteristics; however, in respect to the linear machine, the terms are defined as thrust, linear speed, and pole pitch. The stator is powered with three-phase AC current to produce alternating magnetic fields. The rotor is a line of permanent magnets [19]. The referred study uses an LSM that has an ironless core motor ideal for linear motion. The rotor is the moving part (also known as the mover or the pod) lined with permanent magnets and the stator is fixed on the track made up of coils (between the slots shown in Figure 1) receiving the three-phase AC currents [20]. When the stator produces magnetic fields, there are alternating attraction and repulsion forces created between the track and the pod. This allows the pod to propel back and forth on the track without any additional assistance.

To model the linear synchronous motor, the mathematical model studied in [5,10–12,21] was incorporated. For generic analysis, permanent magnet synchronous motor equations were used as an equivalent to the linear synchronous motor. Additionally, this was done because PSIM does not have a linear synchronous motor model and remodeling an LSM was out of scope of this research. However, it is important to note the differences between the linear synchronous motor and its rotary counterpart. The linear motor has different mechanical properties such as thrust force and linear velocity as opposed to torque and angular velocity. In terms of performance, linear motor has non idealities such as the end effect. This is caused by the unrolled, opened structure of the motor. The end effect in linear motors influences the propulsion force and motor efficiency [22–24]. However, the end effect was neglected in this study. The justification for this is because the influence of end effect is not as significant for long primary linear motors moving at high speed [24]. This further simplifies the linear synchronous motor model. The permanent magnet linear synchronous motor (PMSM) voltage equation in 'abc' form is given in (1).

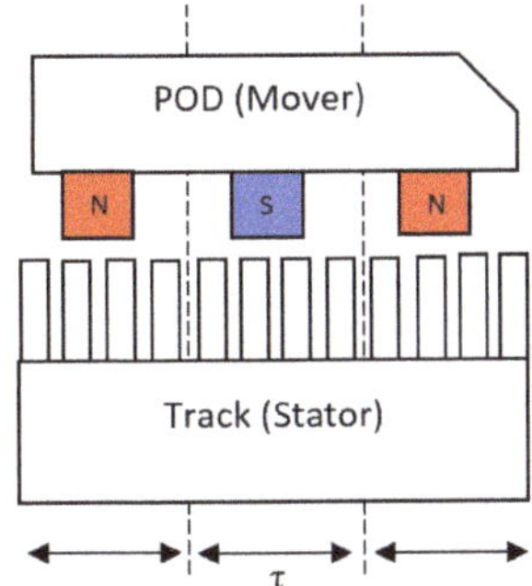

Figure 1. Block diagram for LSM, adapted from [25].

$$[u_{abc}] = [R_{abc}][i_{abc}] + \frac{d[L_{abc}]}{d\theta_r}\omega_r[i_{abc}] + [L_{abc}]\frac{d[i_{abc}]}{dt} + \frac{d[\psi_{Mabc}]}{d\theta_r}\omega_r \tag{1}$$

where L_{abc} is the stator a, b, and c phase self-mutual inductance matrix. This is shown in (2).

$$L_{abc} = \begin{bmatrix} L_{aa}(\theta_r) & M_{ab}(\theta_r) & M_{ac}(\theta_r) \\ M_{ba}(\theta_r) & L_{bb}(\theta_r) & M_{bc}(\theta_r) \\ M_{ca}(\theta_r) & M_{cb}(\theta_r) & L_{cc}(\theta_r) \end{bmatrix} \tag{2}$$

Additionally, Ψ_{Mabc} is the stator a, b, and c phase flux linkages due to the permanent magnets and damper windings on the rotor. This is shown in (3).

$$\Psi_{Mabc} = \begin{bmatrix} \Psi_{Ma}(\theta_r) \\ \Psi_{Mab}(\theta_r - 2\pi) \\ \Psi_{Mc}\left(\theta_r + \frac{2\pi}{3}\right) \end{bmatrix} \tag{3}$$

As seen in (1), the voltage is dependent on the phase self-inductance, which respects the rotor position of the PMSM. The solution of the system can be simplified by using the reference frame theory. This is achieved by taking one set of variables and transforming it to another [11]. To transform the a, b, and c stator variables into rotor dqo axis, the transformation in (4) is used.

$$T_{qdo}^R = \frac{2}{3}\begin{pmatrix} cos(\theta_r) & cos\left(\theta_r - \frac{2\pi}{3}\right) & cos\left(\theta_r + \frac{2\pi}{3}\right) \\ sin(\theta_r) & sin\left(\theta_r - \frac{2\pi}{3}\right) & sin\left(\theta_r + \frac{2\pi}{3}\right) \\ \frac{1}{2} & \frac{1}{2} & \frac{1}{2} \end{pmatrix} \tag{4}$$

The new, transformed dqo voltages are given in (5)–(7).

$$u_{sq} = R_s i_{sq} + \frac{d\Psi_{sq}}{dt} + \omega_r \Psi_{sd} \tag{5}$$

$$u_{sd} = R_s i_{sd} + \frac{d\Psi_{sq}}{dt} - \omega_r \Psi_{sd} \tag{6}$$

$$u_{so} = R_s i_{so} + \frac{d\Psi_{so}}{dt} \tag{7}$$

Using the equivalent circuits, the electromagnetic torque can be derived. However, to simplify the equation, some assumptions can be made. For surface mounted rotors, the reluctance torque of the motor is zero. This is because the stator q and d axis inductances are equal. Assuming the PMSM does not have any damper windings and only built of surface mounted magnets, then only the excitation torque will be present [11]. Thus, the final derived equation for the electromagnetic torque is given in (8). As it can be seen, the electromagnetic torque is dependent on the q – axis component of the stator current.

$$Te = \frac{3}{2}p\{\,\psi_{sd}i_{sq}\,\} \tag{8}$$

2.2. Permanent Magnet Configuration

For the pod, one important consideration is the permanent magnets' arrangement. This paper studies the use of the Halbach arrangement. The Halbach Array is an arrangement of permanent magnets in such a way that the magnetic field is stronger on one side than the other. This phenomenon can be realized by rotating each magnet 90 degrees from the previous orientation so that the polarities of the magnets do not repeat. This causes each magnet to strengthen the magnetic field on one side while a single magnet or multiple magnets facing the same orientation would have uniform magnetic field strength around the cluster of magnets. Employing the Halbach Array theory, the magnetic force of a linear motor can be increased without increasing the size of the permanent magnets. The Halbach Array is optimal for linear motors as only one side of the stator is used for levitation and linear motion. The prototype was built in a way that the permanent magnets are placed near the top and underside of the primary element (rotor) using a 'C"-shaped frame. The magnetic fields of the top and bottom magnet sets were aimed towards each other so that the rotor could achieve higher speed, acceleration, travel accuracy, positioning accuracy, and cycle times through combined attractive and repulsive forces.

2.3. Field Oriented Control

The design of FOC is based on the mathematical model of PMSM mentioned previously. The controller design topology is a closed loop system that includes proportional integral (PI) regulators. More specifically, it includes a speed regulator and two current controllers [5,11]. The block diagram for the control scheme is seen in Figure 2. The block diagram also gives a general overview of the circuit design. The PMSM is controlled and powered via the three-phase inverter. The inverter's source voltage is an external dc voltage source, either from a battery or from the utility rectified to a dc source. At the output of the load there is a speed sensor. The remaining blocks are part of the controller. The process of FOC is straightforward. As it is a closed loop design, it relies on various sensor data to appropriately reach steady state. The steps are as follows. First, set a reference or desired motor speed. Next, obtain three-phase current (i_a, i_b, and i_c) readings to convert it to qd0 form. Additionally, for the transformation, the rotor position of the PMSM is required. This can be obtained by first reading the speed sensor value to obtain the angular velocity.

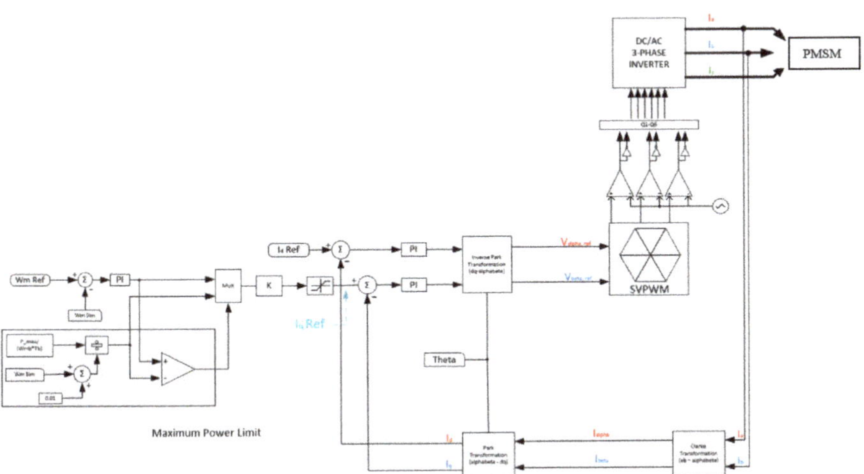

Figure 2. Block diagram for FOC control scheme.

Next integrate the velocity to get position (rotor position θ_r). These four inputs provide the quadrature, direct, and zero currents (i_q, i_d, i_0). These are the measured values. The reference Id current is set to zero. This allows the control of q-axis current (torque). The reference iq is obtained by first calculating the error signal of the speed (measure the difference) then place the error signal into a PI regulator (speed regulator). For the FOC of the PMSM design, the output of the regulator is the iq reference. The error signal of the reference iq and the measured iq is calculated. This is then placed into another regulator (current regulator). A similar process for the Id loop is done. Finally, the resultant signals are placed into an inverse transformation block (dq0 to abc). These signals are now the reference voltage for the PWM signals controlling the switching sequence of the inverter. Additionally, space vector pulse width modulation (SVPWM) technique was implemented for this study. It works by obtaining an optimal switching sequence for the inverter switches. This can be obtained by taking the signals into the alpha beta frame of reference [26]. The inverter output voltages are shown in (9).

$$\begin{bmatrix} Va \\ Vb \\ Vc \end{bmatrix} = \frac{Vdc}{3} \begin{pmatrix} 2 & -1 & -1 \\ -1 & 2 & -1 \\ -1 & -1 & 2 \end{pmatrix} \begin{bmatrix} C_1 \\ C_2 \\ C_3 \end{bmatrix} \tag{9}$$

$$V_s = V\alpha + jV\beta = \sqrt{\frac{2}{3}} \left(V_a + V_b e^{\frac{j2\pi}{3}} + V_c e^{\frac{j4\pi}{3}} \right) \tag{10}$$

The final signals after the PI regulator go into the SVPWM block. This facilitates an optimal switching sequence for the PMSM control. The parameters of the regulator are calculated by using smart control in PSIM and analyzing the system performance.

2.4. LSM Formulas to Generate Specifications

The power can be expressed in terms of the thrust force and the synchronous speed given in Equation (11). The velocity for a linear synchronous motor is given by the frequency and pole pitch of the motor. The thrust can be calculated by determining the mass and acceleration of the linear motor.

$$P_{thrust} = F_{x,thrust} \cdot v_s = ma \cdot 2f\tau \tag{11}$$

Additionally, the thrust density (N/m^2) is described by Equation (12). Where 'p' is the pole pairs, and L_i is the armature stack length.

$$f_{x,thrust} = \frac{F_{x,thrust}}{2p\tau L_i} = ma \cdot 2f\tau \tag{12}$$

The efficiency and the power factor of the system are given respectively by the following Equations (13) and (14).

$$\eta = \frac{F_x v_s}{F_x v_s + 3I^2 R} \tag{13}$$

$$PF = \frac{F_x v_s + 3I^2 R}{3VI} \tag{14}$$

The motor parameters used for simulation are provided in Table 1.

Table 1. Motor parameters.

Symbol	Quantity	Description
p	8	Poles
R_s	0.065	Stator Resistance
L_d	0.001916	d-axis inductance
L_q	0.005	q-axis inductance
P_{max}	50,000	Max motor power
n_{max}	5000	Maximum rpm

2.5. Modeling of Battery and Bidirectional Converter

The DC voltage source is provided by a lithium-ion battery. Accumulative Parameter values can be calculated using (15)–(17). The parameters for a singular battery are given in Table 2. The rated value can be looked up from the manufacture data sheets. K_s and K_p are voltage and capacity derating factors which are set to 1.

$$E_{rated\_total} = N_S \cdot K_S \cdot E_{rated} \tag{15}$$

$$Q_{rated\_total} = N_p \cdot K_p \cdot Q_{rated} \tag{16}$$

$$R_{battery\_total} = \frac{N_s}{N_p} \cdot R_{battery} \tag{17}$$

Table 2. Battery Parameters.

Parameters	Value	Units
N_s No. of cells in series	70	cells
N_p No. of cells in parallel	60	cells
E_{rated} Rated voltage	3.6	V
E_{cut} Discharge cut-off voltage	2.5	V
Q_{rated} Rated capacity	3.35	Ah

The parameters for the battery were calculated based on a prototype design for a 130-kW inverter system, with 750 VDC to power the linear motor. A typical bidirectional DC/DC converter is implemented to observe the generating and motoring modes and its effect on the modeled battery. The converter operates in the boost mode when the PMSM is in the motoring mode of operation (i.e., battery discharging). In regenerative mode of operation, the converter is the buck mode (i.e., battery charging). The block diagram for the control setup is given in Figure 3. The output to this controller determines the mode of operation of the buck-boost DC-DC converter.

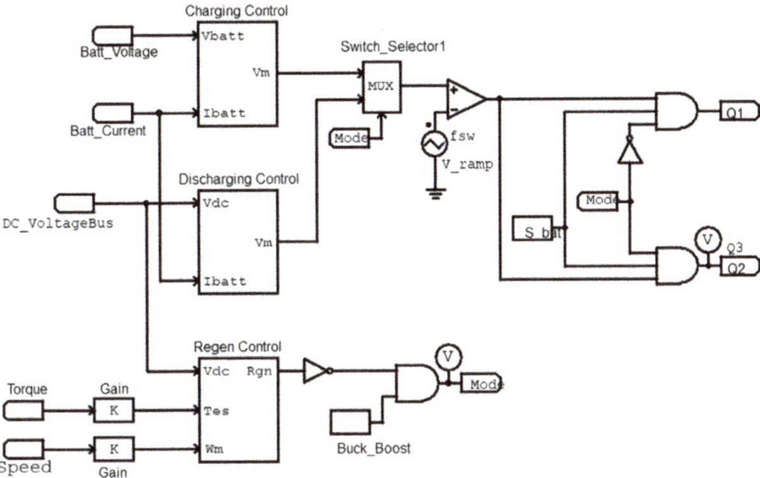

Figure 3. Bidirectional DC-DC Converter Control System.

3. Simulation Results

The simulation setup is provided in Figure 4. The power stage includes a lithium-ion battery pack, a bidirectional dc-dc converter, a three-phase inverter, and the motor. Since the PSIM motor library was limited to only rotary motors, a permanent magnet linear synchronous motor (PMSM) was used. Thus, the angular velocity (as opposed to a linear velocity for an LSM) provided by the PMSM block was used to sense the speed for the controller. In terms of the control stage, typical FOC algorithm was used. However, to implement the regeneration system, a torque estimate was required to sense if the motor was in regenerative breaking or motoring mode. Thus, the additional control algorithm provided by PSIM was used.

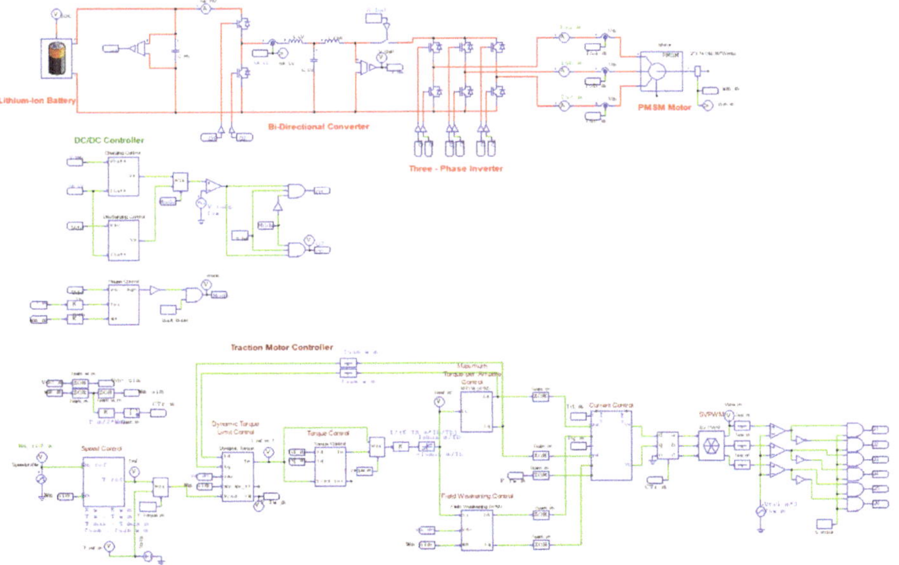

Figure 4. Circuit setup for FOC of PMSM.

The controller will then provide the required PWM signals to produce a three-phase current waveform with minimum harmonics and to acquire a wave that looks as sinusoidal as possible. The following figures are simulation results of the power inverter. Figure 5 shows the simulation results of the AC currents in motoring mode. The speed response is given in Figure 6.

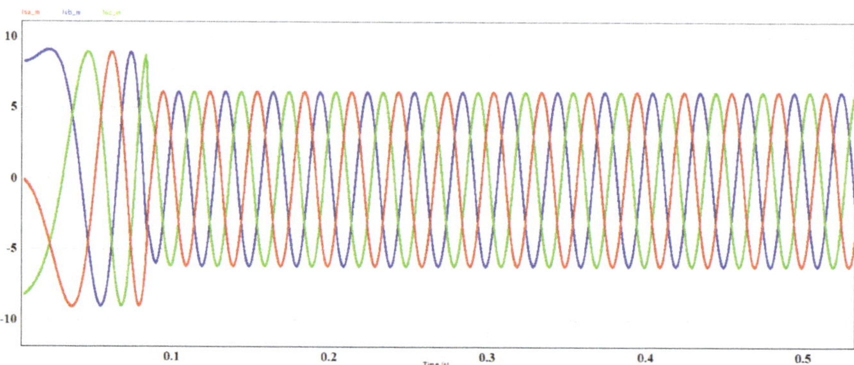

Figure 5. Three-phase AC line currents (amperage vs. time).

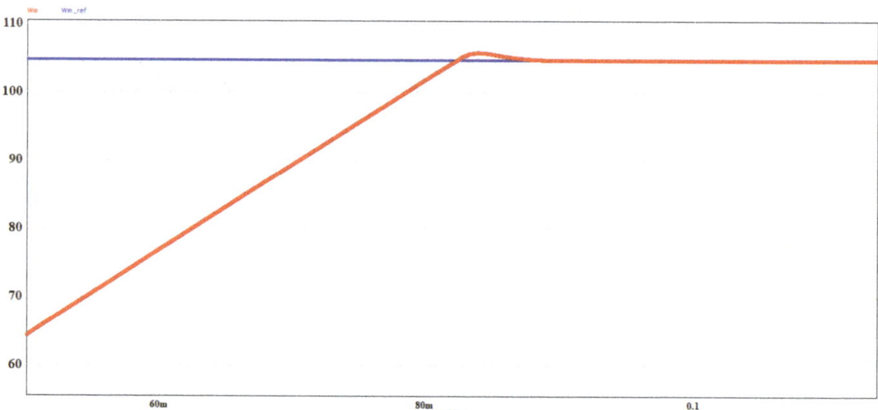

Figure 6. Speed response (rads/s vs. time).

Figure 7 highlights the directionality of the current in relation to the battery. This direction is dependent on if the pod is in fact driving, regenerative braking, or idling. For example, beginning at 0.5 s when the vehicle enters regenerative braking, we can see the current flowing into the battery and therefore returning energy to the system.

In Figure 8 the waveform represents the SOC during the various drive states of the simulation. The figure first shows a decreasing SOC which indicates that the vehicle is drawing current from the battery to drive. An increase in SOC then occurs as regenerative braking is employed returning potential energy to the battery module. Lastly, we see that a period of minimal drain upon the SOC in this time frame is representative of the LSM idling until driving is resumed.

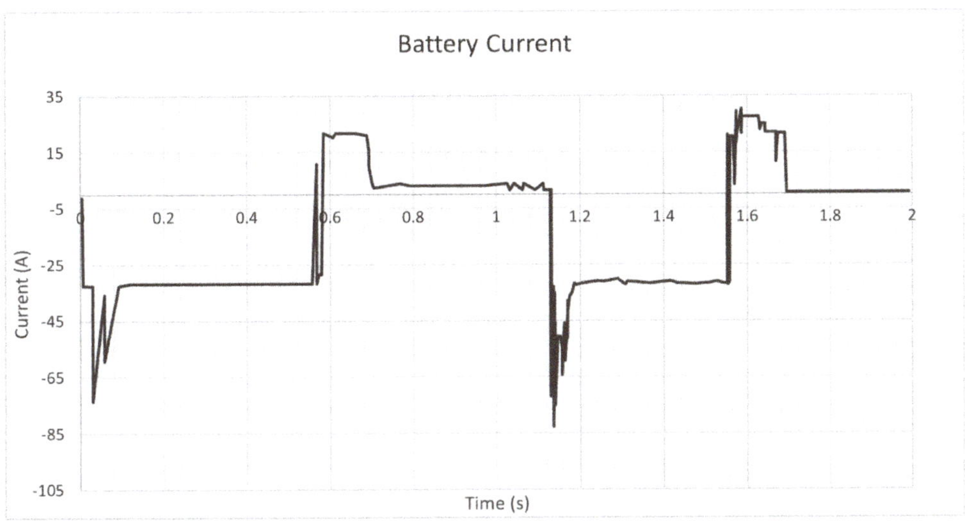

Figure 7. Battery current for regenerative breaking (amperage vs. time).

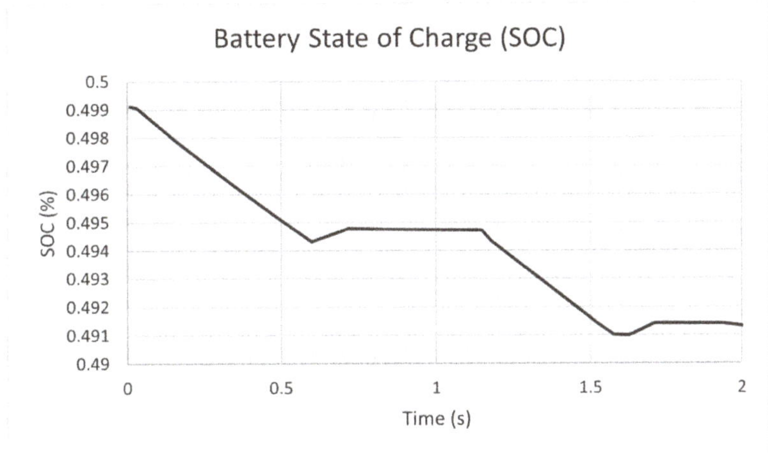

Figure 8. SOC of the battery.

Finite element analysis of the pod is given in Figure 9 and shows magnetic field safe values for humans as per IEEE C95.1.

Next, the simulation is run so that the ideal distance can be determined. Once the single coil is set up, the simulation is expanded to four coils as shown in Figure 10. The effect of the adjacent coils can be observed and recorded accordingly. Figure 11 illustrates the magnetic field strength of the active coils. These levels adhere to the IEC-60118 used in North America and Europe.

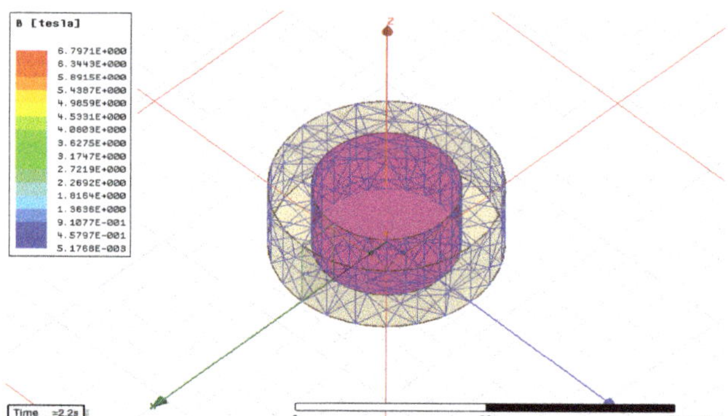

Figure 9. Single Coil Setup in ANSYS Maxwell.

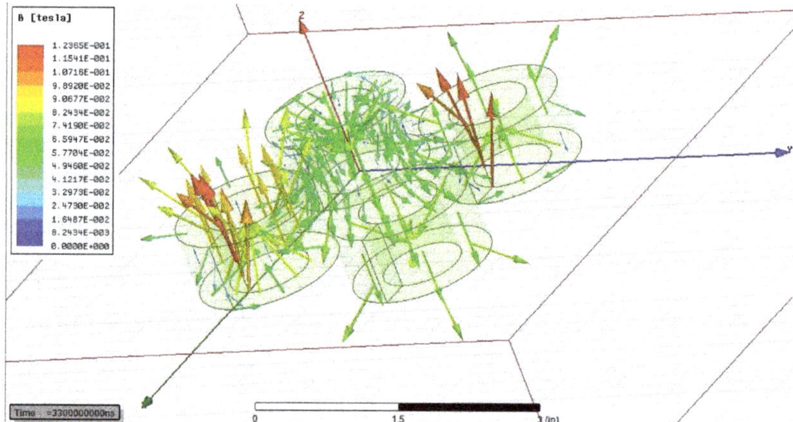

Figure 10. Four coil simulations.

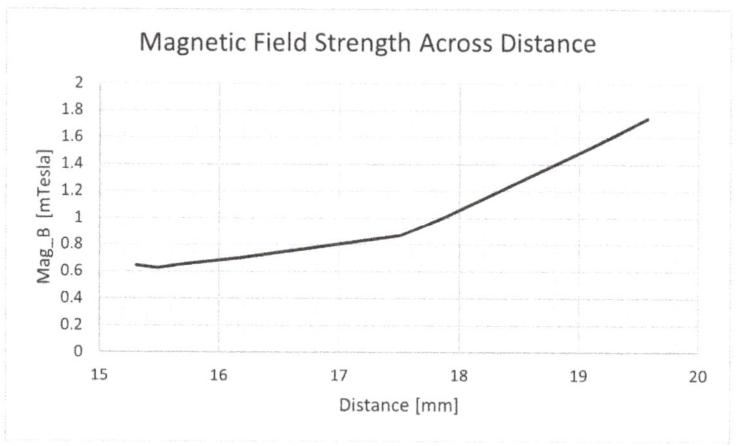

Figure 11. Magnetic field strength across Z distance.

4. Experimental Analysis

Autodesk Eagle was used to design the printed circuit board (PCB). The hardware prototype of the inverter is circuit driven by the MIC4609 driver. A Texas Instrument microcontroller (TMS320F2808) was used to generate the PWM signals based on the proposed FOC. This prototype design was done to provide a proof of concept. The controller and the PCB circuit have similar components to the original proposed design. The prototype for the power inverter is shown in Figure 12. It consists of 6 surface mount IGBTs (RGT50NS65DGT) connected in a typical inverter circuit orientation. Using the data sheet bootstrap capacitors and diodes were chosen as recommended.

Figure 12. Three-phase inverter prototype.

The waveforms generated by the output of the inverter are given in Figures 13 and 14. Figure 13 shows the preliminary inverter output voltage. In Figure 14 the purple waveform is the voltage line feeding the load (linear motor). The green sinusoidal waveform below it is the AC current waveform of a single phase. The results are what is expected for the circuit output. The interface for controlling the linear motor was created using C++. It allows the control of position, velocity, and acceleration. It also provides real-time feedback of the motor's status.

The developed inverter circuit was used to power and control the linear motor (provided by Bosch Rexroth). The prototype of the integration is shown in Figures 15 and 16. Communication with the synchronous linear motor was achieved through Ethernet Protocols. The MCP015A-L040 drive was used to facilitate prototype integration.

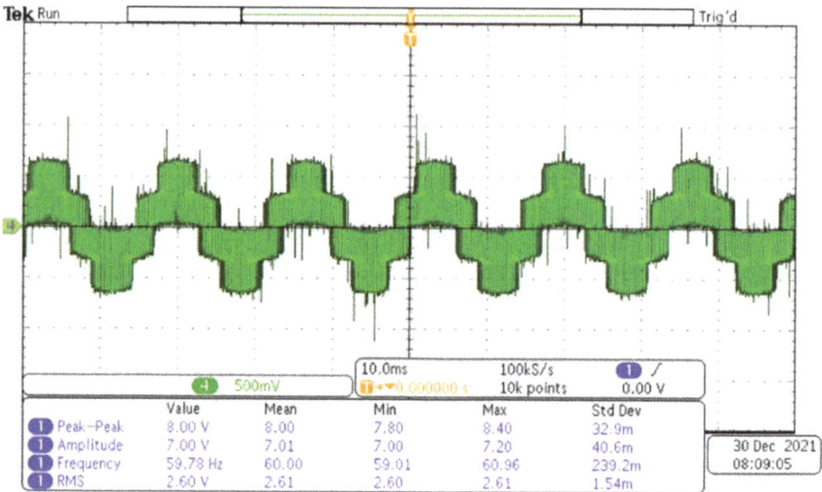

Figure 13. Line to line voltage waveform (Vab); before the dv/dt filter of the motor.

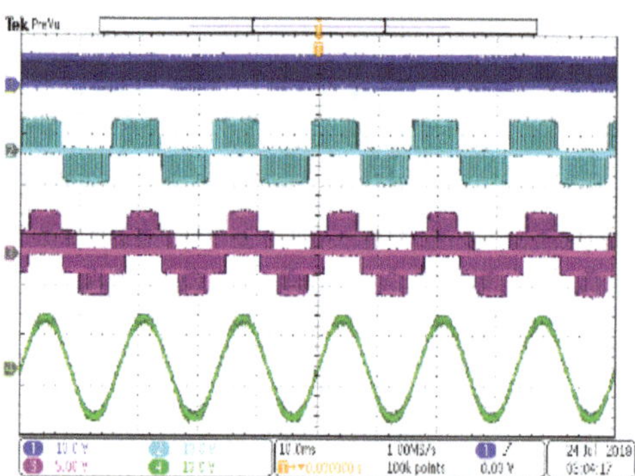

Figure 14. Prototype inverter circuit waveforms.

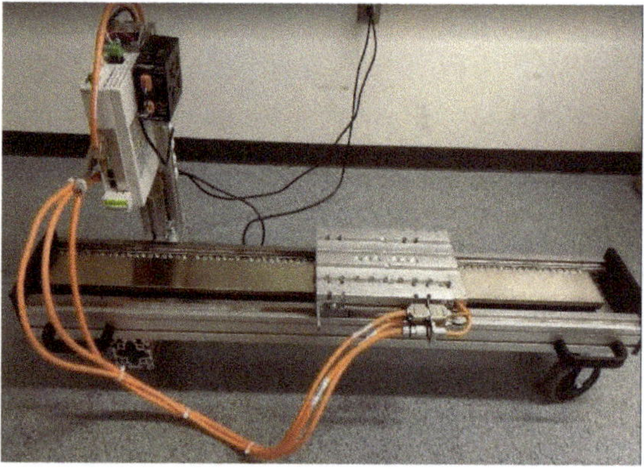

Figure 15. Prototype linear motor.

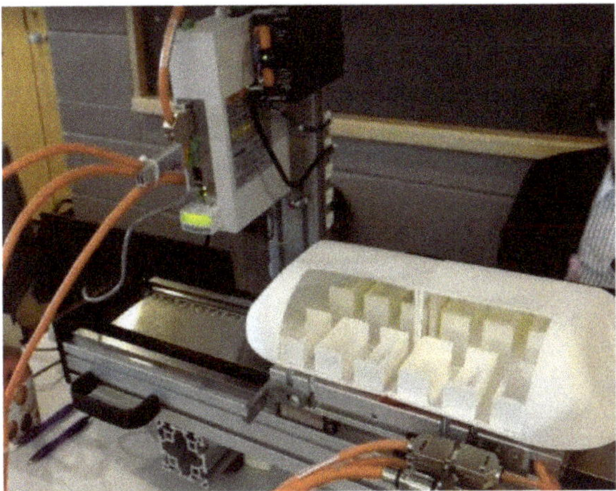

Figure 16. Prototype linear motor with test pod.

5. Conclusions

A successful component design and system integration of the Hyperloop pod's functionality is presented using PSIM simulation, C++ software interface, and a prototype hardware from "off-the-shelf" components. Simulations of field-oriented control for an equivalent rated rotary synchronous motor was built in PSIM. With intensive hardware and software integration methods, the goal to provide linear bidirectional movement for the prototype was achieved. The integrated system includes the linear synchronous motor, three-phase power inverter, and a battery pack. This was implemented on a small-scale prototype model. The three-phase inverter was designed using PSIM's platform and tested by generating case study simulations that validated its functionality before integration. Additionally, the battery model was also implemented in PSIM and sized accordingly. Measured results from the inverter prototype are in good agreement with the simulation results. The operation of the full system integration proves that the design of the inverter and linear synchronous motor achieve a linear motion, with bidirectional movement for the pod prototype, in motoring and braking modes. The presented design provides a good starting point for a linear synchronous motor-based Hyperloop. However, the scale it was designed for is currently not capable of a full-sized passenger transportation. The nature of the system comes with high infrastructure build time and costs. Thus, various design trade-offs were needed to be made in terms of cost, size, power capabilities, etc. Nonetheless, future work for this study includes improving the design in various aspects such as fine-tuning and verifying the system in a hardware in the loop environment, designing it for a larger test track, and improving the prototype design capabilities to include additional data such as speed and acceleration. Hardware in the loop technology will be used in the hopes to reduce cost and prototyping time. A longer test track would provide a better understanding of the performance and efficiency. Additionally, this test track would provide a better study for the dynamic characteristics. This would require an improved prototype design with additional sensors to provide speed and acceleration values. Lastly, as this is the first iteration of the prototype, the design choices and trade-offs were very minimal. The prototype was designed for a proof of concept. The next iteration will be designed and better engineered to achieve an optimal design. Future work will also detail mechanical modeling and passenger ride comfort calculations. To summarize, the scope of this study is to provide some insight on the propulsion system design methodology and initial prototype results for the novel Hyperloop transportation system.

Author Contributions: Conceptualization, M.M.A. and M.Y.; methodology, M.B.; software, M.M.A. and F.M.; validation, N.B.; formal analysis, F.M.; investigation, M.B.; resources, M.Y.; data curation, R.L.; writing—original draft preparation, R.L.; writing—review and editing, M.B.; visualization, F.M.; supervision, M.Y.; project administration, M.Y.; funding acquisition, M.Y. All authors have read and agreed to the published version of the manuscript.

Funding: This work was supported by the Natural Science and Engineering Research Council of Canada in conjunction with Transport Canada (Fund number 210850). This was an invited paper and the APC discount was provided by MDPI. We thank MDPI for this opportunity.

Institutional Review Board Statement: This study did not require ethical approval. Not applicable.

Conflicts of Interest: The authors declare no conflict of interest.

Nomenclature

LIM	Linear Induction Motor
LSM	Linear Synchronous Motor
PMSM	Permanent Magnet Synchronous Motor
EMS	Electromagnetic Suspension
EDS	Electrodynamic Suspension
POD	Referring to Hyperloop Capsule/Vehicle
FOC	Field Oriented Control
SVPWM	Space Vector Pulse Width Modulation
u_{abc}	Stator a, b, c phase to neutral voltages
R_{abc}	Stator a, b, c phase resistances
i_{abc}	Stator a, b, c phase currents
L_{abc}	Matrix of stator phase self and mutual inductances
L_{aa}, L_{bb}, L_{cc}	Stator a, b, c phase self-inductances
$M_{ab}, M_{ac}, M_{ba}, M_{bc}, M_{ca}, M_{cb}$	Mutual inductances of stator a, b, c phases
Ψ_{Mabc}	Stator a, b, c phase flux linkages from PMSM
T_e	Electromagnetic Torque
SOC	State of Charge
E_{rated}	Rated voltage of battery cell
Q_{rated}	Rated capacity of battery cell
$R_{battery}$	Internal resistance of the battery cell
N_p	Number of cells in parallel of battery pack
N_p	Number of cells in parallel of battery pack

References

1. IEA. Transport: Improving the sustainability of passenger and freight transport, International Energy Agency. Available online: https://www.iea.org/topics/transport (accessed on 22 September 2019).
2. Ji, W.Y.; Jeong, G.; Park, C.B.; Jo, I.H.; Lee, H.W. A Study of Non-Symmetric Double-Sided Linear Induction Motor for Hyperloop All-In-One System (Propulsion, Levitation, and Guidance). *IEEE Trans. Magn.* **2018**, *11*. [CrossRef]
3. Janzen, R. TransPod Ultra-High-Speed Tube Transportation: Dynamics of Vehicles and Infrastructure. *Procedia. Eng.* **2017**, *199*, 8–17. [CrossRef]
4. Hasirci, U.; Balikci, A.; Zabar, Z.; Birenbaum, L. Experimental performance investigation of a novel magnetic levitation system. *IEEE Trans. Plasma. Sci.* **2013**, *5*, 1174–1181. [CrossRef]
5. Sayeed, J.M.; Abdelrahman, A.; Youssef, M.Z. Hyperloop Transportation System: Control, and Drive System Design. In Proceedings of the IEEE Energy Convers. Congr. Expo. (ECCE 2018), Portland, OR, USA, 23–27 September 2018; pp. 2767–2773. [CrossRef]
6. Zhang, Z.; She, L.; Zhang, L.; Shang, C.; Chang, W. Structural optimal design of a permanent-electro magnetic suspension magnet for middle-low-speed maglev trains. *IET Electr. Syst. Transp.* **2011**, *2*, 61–68. [CrossRef]
7. Lim, J.; Jeong, J.H.; Kim, C.H.; Ha, C.W.; Park, D.Y. Analysis and Experimental Evaluation of Normal Force of Linear Induction Motor for Maglev Vehicle. *IEEE Trans. Magn.* **2017**, *11*. [CrossRef]
8. Jeong, J.H.; Lim, J.; Ha, C.W.; Kim, C.H.; Choi, J.Y. Thrust and efficiency analysis of linear induction motors for semi-high-speed Maglev trains using 2D finite element models. In Proceedings of the 2016 IEEE Conference on Electromagnetic Field Computation (CEFC), Miami, FL, USA, 13–16 November 2016. [CrossRef]
9. Wang, H.; Li, J.; Qu, R.; Lai, J.; Huang, H.; Liu, H. Study on High Efficiency Permanent Magnet Linear Synchronous Motor for Maglev. *IEEE Trans. Appl. Supercond.* **2018**, *3*. [CrossRef]

0. Boldea, I. *Linear Electric Machines, Drives, and Maglevs Handbook*, 1st ed.; CRC Press: Boca Raton, FL, USA, 2017.
1. Yesilbag, E.; Ergene, L.T. Field oriented control of permanent magnet synchronous motors used in washers. In Proceedings of the 2014 16th International Power Electronics and Motion Control Conference and Exposition, Antalya, Turkey, 21–24 September 2014; pp. 1259–1264. [CrossRef]
2. U.S.E.P. Agency. Global Greenhouse Gas Emissions Data. Available online: https://www.epa.gov/ghgemissions/global-greenhouse-gas-emissions-data (accessed on 21 March 2020).
3. Voltes-Dorta, A.; Becker, E. The potential short-term impact of a Hyperloop service between San Francisco and Los Angeles on airport competition in California. *Transp. Policy* **2018**, *71*, 45–56. [CrossRef]
4. You, C.; Zhang, R.; Wang, X.; Du, Y.; Ge, Q. Vector control of maglev PMLSM based on minimum loss SVPWM method. In Proceedings of the 2016 19th International Conference on Electrical Machines and Systems (ICEMS), Chiba, Japan, 13–16 November 2016.
5. Sadat, A.R.; Shadabi, H.; Sabahi, M.; Sharifian, M.B.B. Tracking of X-Y direction positions with using permanent magnet linear synchronous motors. In Proceedings of the 2014 22nd Iranian Conference on Electrical Engineering (ICEE), Tehran, Iran, 20–22 May 2014; pp. 527–532. [CrossRef]
6. Wang, K.; Ge, Q.; Shi, L.; Li, Y.; Zhang, Z. Development of ironless Halbach permanent magnet linear synchronous motor for traction of a novel maglev vehicle. In Proceedings of the 2017 11th International Symposium on Linear Drives for Industry Applications (LDIA), Osaka, Japan, 6–8 September 2017. [CrossRef]
7. Kowal, B.; Ranosz, R.; Klodawski, M.; Jachimowski, R.; Piechna, J. Demand for passenger capsules for Hyperloop High-Speed Transportation System -case study from Poland. *IEEE Trans. Transp. Electrif.* **2021**. [CrossRef]
8. Tbaileh, A.; Elizondo, M.; Kintner-Meyer, M.; Vyakaranam, B.; Agrawal, U.; Dwyer, M.; Samaan, N. Modeling and Impact of Hyperloop Technology on the Electricity Grid. *IEEE Trans. Power Syst.* **2021**, *5*, 3938–3947. [CrossRef]
9. Pan, S. Development of Permanent Magnet Tubular Linearmotor and Position Feedback Device Based on Hallsensor. Univ Wollongong Thesis Collect. 1954–2016. Available online: https://ro.uow.edu.au/theses/4852 (accessed on 14 December 2021).
10. MATLAB, "PMLSM". Available online: https://www.mathworks.com/help/physmod/sps/ref/pmlsm.html (accessed on 14 December 2021).
11. Ong, C.-M. *Dynamic Simulations of Electric Machines*; Prentice Hall: Hoboken, NJ, USA, 1998.
12. Giangrande, P.; Cupertino, F.; Pellegrino, G. Modelling of linear motor end-effects for saliency based sensorless control. In Proceedings of the 2010 IEEE Energy Conversion Congress and Exposition, Atlanta, GA, USA, 12 September 2010; pp. 3261–3268. [CrossRef]
13. Platen, M.; Henneberger, G. Examination of leakage and end effects in a linear synchronous motor for vertical transportation by means of finite element computation. *IEEE Trans. Magn.* **2001**, *37*, 3640–3643. [CrossRef]
14. Lu, J.; Ma, W. Research on end effect of linear induction machine for high-speed industrial transportation. *IEEE Trans. Plasma Sci.* **2011**, *39*, 116–120. [CrossRef]
15. Gieras, J.F.; Piech, Z.J.B.; Tomczuk, B. *Linear Synchronous Motors: Transportation and Automation Systems*; CRC Press: Boca Raton, FL, USA, 2011.
16. Abassi, M.; Khlaief, A.; Saadaoui, O.; Chaari, A.; Boussak, M. Performance analysis of FOC and DTC for PMSM drives using SVPWM technique. In Proceedings of the 16th Int. Conf. Sci. Tech. Autom. Control Comput. Eng. STA 2015, Monastir, Tunisia, 21–23 December 2015.

MDPI

St. Alban-Anlage 66

4052 Basel

Switzerland

Tel. +41 61 683 77 34

Fax +41 61 302 89 18

www.mdpi.com

Robotics Editorial Office

E-mail: robotics@mdpi.com

www.mdpi.com/journal/robotics